丁衡高 院士，惯性技术和精密机械专家，是中国战略导弹惯性技术奠基人之一，中国惯性技术学科发展的主要推动者，国家微米纳米技术倡导人。1931年2月3日出生，江苏省南京市人。1952年毕业于南京大学机械系，1961年获苏联列宁格勒精密机械光学学院技术科学副博士学位。历任有关研究室、研究所技术负责人，国防科工委科技部副部长，国防科工委主任。现任中国惯性技术学会名誉理事长、中国微米纳米技术学会名誉理事长，1994年遴选为中国工程院首批院士。自1961年起，长期从事制导武器的陀螺仪、加速度计、惯性平台系统等的研制工作。突破了气浮轴承等惯性器件的关键技术，成功地应用于多种战略导弹、运载工具及多种测试设备，获全国科学大会奖及国防科技重大成果一等奖。负责潜地固体战略导弹的液浮惯导系统的研制与生产，获国家科技进步奖特等奖。曾获何梁何利科学与技术进步奖及国家863计划特殊贡献奖。1988年被授予中将军衔，1994年晋升为上将军衔。中国共产党中央委员会第十二届中央候补委员，第十三届、十四届中央委员，全国政协第九届常委。

▲ 2005年4月5日，中国微米纳米技术学会成立大会合影
（前排左十二为丁衡高院士）

▲ 2000年4月，参加在日本广岛举办的6th World Micromachine Summit会议
（前排左四为丁衡高院士）

◀1994年，丁衡高院士（前排中）到清华大学精仪系实验室指导工作

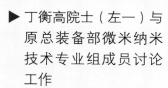

▶丁衡高院士（左一）与原总装备部微米纳米技术专业组成员讨论工作

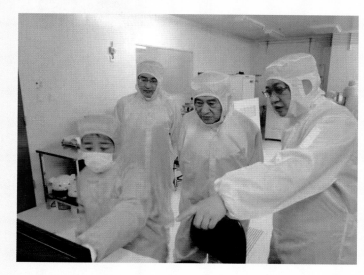

◀2010年12月，丁衡高院士（右二）在中电13所指导科研工作

▶ 2011年5月,丁衡高院士(右三)在北京微电子技术研究所指导科研工作

◀ 2012年5月,丁衡高院士(左二)在上海交通大学指导科研工作

▶ 2018年3月,丁衡高院士(左)在淄博高新区MEMS研究院指导科研工作

丁衡高院士
微米纳米技术文集

中国宇航出版社

·北京·

版权所有　侵权必究

图书在版编目（CIP）数据

丁衡高院士微米纳米技术文集 / 丁衡高著. -- 北京：中国宇航出版社，2024.7

ISBN 978-7-5159-2377-2

Ⅰ.①丁⋯　Ⅱ.①丁⋯　Ⅲ.纳米材料－加工－文集　Ⅳ.①TB383-53

中国国家版本馆 CIP 数据核字(2024)第 084112 号

责任编辑	朱琳琳	封面设计	王晓武	

出版发行　**中国宇航出版社**

社　址	北京市阜成路 8 号　邮　编　100830	版　次	2024 年 7 月第 1 版
	(010)68768548		2024 年 7 月第 1 次印刷
网　址	www.caphbook.com	规　格	710×1000
经　销	新华书店	开　本	1/16
发行部	(010)68767386　(010)68371900	印　张	34　彩　插　4 面
	(010)68767382　(010)88100613（传真）	字　数	510 千字
零售店	读者服务部	书　号	ISBN 978-7-5159-2377-2
	(010)68371105	定　价	98.00 元
承　印	北京中科印刷有限公司		

本书如有印装质量问题，可与发行部联系调换

序　言

1994年，丁衡高院士首次提出"微米纳米技术"概念，全力倡导开展这项事关国家安全、国民经济和社会发展的基础性、前瞻性、战略性高技术研究，至今已过去30年。30年来，我国微米纳米技术快速发展，在基础研究、技术开发、产业发展等方面取得了令人瞩目的成就，部分领域已走在国际前列，呈现出良好的发展态势。

饮水思源，丁衡高院士作为这一新兴技术研究的倡导者、开拓者和参与者，为30年来我国微米纳米技术的进步和发展做出了不可磨灭的开创性贡献。

丁衡高院士创造性地提出了"微米纳米技术"概念。20世纪90年代初，微机电系统技术在国际上开始崭露头角，我国MEMS技术研究则刚零星起步。另一方面，国际上纳米技术始见端倪，以原子力仪为代表的新技术为人类观察和操纵原子层面提供了有力手段，我国纳米技术方面的研究还大多集中在纳米材料和纳米表征方面，而纳米器件则鲜有研究。在此关键时刻，丁衡高院士以其战略科学家的敏锐思维和远见，创造性地提出了"微米纳米技术"概念。他认为，人类社会在改造自然方面，已从宏观层次进入微米层次，进而发展到纳米层次，由此产生的微米纳米技术必将迎来快速发展时期，我国应及早开展这方面的研究工作，并强调，微米技术与纳米技术应统筹融合发展。他指出，纳米技术是用单个原子或分子来构造特定功能产品的科学技术，它将开发物质的信息资源与结构潜力，使单位体积的物质储存和处理信息能力实现飞跃，从而导致人类认识世界、改造世界的能力发生重大突破；纳米技术将成为微米技术拓展认识事物的窗口，而微米技术可以作为工具来访问和分析纳尺度世界，集成应用在微纳系统中。他进一步解释说，微米纳米技术这个提法，既考虑到以微机电系统为代表的微米技术已取得进展的现实性，又着眼于刚刚起步的纳米技术的前瞻性、基础性，这样就能科学地将"Top down"和"Bottom up"两类技术发展路径贯穿在我国微米纳米技术发展战略中。微米纳米技术这一概念的提出，在当时（20世纪90年代初）不仅很快统一了国内学术界的认识，而且促进了本学科的发展。这一概念也得到了国际同行的广泛认同。

丁衡高院士倾力推动将我国微米纳米技术上升到国家战略高技术层面。为了更好地推动国内微米纳米技术这门新兴学科的发展和及早应用到军事技术中，丁衡高院士于1994年11月2日，给中央领导同志写信并呈送所著《面向21世纪的军民两用技术——微米纳米技术》，他的关于在国家层面尽早开展微米纳米技术及相关学科研究工作的建议，得到中央的高度重视与充分肯定，国务院主要领导和分管科技工作的副总理批示支持开展研究、跟踪和应用。次年，为促进微米纳米技术领域的学术交流和技术协作，他在军口成立了国内首个微米纳米技术专家咨询组，指导建立国内首个微米纳米技术国防科技重点实验室，并倡导成立了中国微米纳米技术学会，我国微米纳米技术研究藉此进入蓬勃发展的新阶段。

2000年1月26日，他与清华大学时任精密仪器系主任周兆英教授一起再次致书中央，力陈加速开展我国微米纳米技术，尤其是MEMS技术研究的重要性和紧迫性。在中央领导的重视和支持下，国家科技部于当年成立了国家863计划微机电系统技术发展战略研究专家组，年近七旬的丁衡高院士受邀挂帅担任组长。在他主持下，专家组提出了"发展高技术、促进产业化""在世界高技术领域占有一席之地"的我国MEMS发展战略指导思想和"需求牵引、技术推动、重点突破"的发展方针。专家组同志用一年的时间圆满完成了发展战略研究工作，制定了我国MEMS研究发展战略纲要，提出了切合实际的发展战略与技术途径，特别强调了以惯性、信息和生物三大技术方向为主攻目标，加强相关基础能力建设，包括 MEMS 加工、封装测试等，还提出建立以微系统应用为牵引、以批生产的厂家为主体、以市场为导向、产学研相结合的微米纳米技术创新体系。这些工作极大地推动了我国微米纳米技术的研究、开发和产业化进程。

丁衡高院士提出以高性能微惯性器件为重点的我国 MEMS 发展战略。微米纳米技术是由相关专业、学科与先进工程技术相结合，基础研究与应用探索研究紧密联系的新兴尖端科学技术；它不仅涵盖信息、材料、生物、医疗等领域，还涉及如制造学、显微学、物理学、化学等学科；其研究内容包括材料结构与性能研究、功能开发应用研究等，十分广泛。究竟选择什么研究方向和目标作为我们的首选、突破口和重点，是我国微米纳米技术工作者急需回答的问题。对此，在 20 世纪 90 年代初，丁衡高院士预见到微米纳米技术对惯性技术将带来颠覆性影响，提出"九五""十五"期间要把高端微惯性器件研制作为发展我国军用微米纳米技术的重点和突破口，围绕微型惯性测量组合（MIMU）等相关技术开展攻关；"十一五"期间，他又指出要在高端微惯性器件走向应用的同时，对射频微机电系统（RF MEMS）技术及微能源技术要予以充分重视。他指出，基础研究中的概念、建模分析，尤其是关键共性的技术实现手段，包括加工工艺、测试与封装等研究是 MEMS 技术研究开发的重点。他还对 MEMS 的应用途径和长远发展问题进行了深入探索，强调：研制 MEMS 器件需要标准工艺；研制不同 MEMS 器件需要开发特殊工艺；专用集成电路（ASIC）和 MEMS 集成很重要，要使 MEMS 器件具备自检测、自标定和自补偿功能；圆片级封装是 MEMS 封装的发展趋势；MEMS 只有做到微纳结合才有生命力等。正是由于他的高瞻远瞩，我国微米纳米技术研究尤其是 MEMS 技术领域从一开始就战略清晰、方向正确、目标明确，不仅建立和完善了良好的研究环境与加工平台，还为后续稳定快速发展奠定了基础。

近年来，丁衡高院士与惯性技术界有关同志回顾了我国惯性

MEMS 的发展历程与成就，分析了国内外惯性 MEMS 技术的快速发展状况及成功典型应用情况，并以国内企业将 6 轴集成 MIMU 应用于智能辅助驾驶，在车载定位这个细分赛道走在国际前列为例，总结了发展惯性 MEMS 的若干经验：把高性能惯性 MEMS 作为发展军用微米纳米技术重点的战略是正确的；惯性 MEMS 的颠覆性主要体现在创新应用上，不仅大大推动了传统装备的制导化和智能化水平，还催生了大量新装备和新产品；高性能惯性 MEMS 的大批量推广应用是依靠持续的技术突破和敢为天下先的创新精神。他还为惯性 MEMS 的发展指明了方向，即集成化、智能化、高效率。

丁衡高院士悉心指导我国微系统技术及微纳器件研究和应用。他始终置身于科研实践一线，指导、参与许多具体项目、课题的设计、研究和实验。例如，在微陀螺仪、微加速度计、MIMU、微惯性/卫星组合导航系统等关键技术攻关，以及微纳结构设计、高精度体硅深刻蚀、ASIC 电路、误差自补偿、深组合导航算法等核心技术突破中均对相应的研发团队给予了具体指导，成效显著。

指导研制的 MIMU 系列化产品，体积、功耗相比传统器件均降低 1 个数量级以上，在航空炸弹、巡航导弹、远程轰炸机、反坦克弹等型号任务中获得成功应用；指导研制的抗高过载 MEMS 加速度计已在侵彻弹药中成功列装，打破了国外禁运，使我国智能引信技术达到世界先进水平。

他最早提出开展基于 MEMS 的微纳卫星技术研究。在他指导下，清华大学突破了基于 MEMS 的空间姿态测量新原理、新方法，研制了具有自主知识产权的系列微型化姿态传感器，精度达到国际先进水平；清华大学、浙江大学分别于 2004 年 4 月和 2010 年 9 月成功研制国内首颗纳卫星和皮卫星，它们分别是当时世界上成功在轨运行的最小的纳卫星和皮卫星。

指导研制的 MEMS 微型光谱仪，已形成战场急救快速检测、深海石油井下在线检测等原创性核心装备，填补了国内外空白；指导研发和标定的世界首台套流体壁面剪应力测试仪，解决了大型客机气动和潜艇水动减阻优化设计等重大工程问题。

丁衡高院士一贯强调微纳融合是 MEMS 创新发展的关键，鼓励青年学者开展前沿基础研究，并就具体科研问题进行深入探讨。2008年，他指导的博士后在国际上首次发现了纳电极阵极化带效应。对此，丁衡高院士科学地提出：微纳电极阵产生的极化带电场，是特定条件下的微纳结构效应；从热力学的观点看，结构效应在这里显现为系统负熵，是输入能量中可利用的有效能量；构建结构效应与负熵之间的理论关系，通过负熵的理论从热力学的角度思考高效微纳系统设计问题意义重大。在本文集《微纳电极阵等离子体微系统》一文中的"四点启示"，对微系统的研究与发展具有普遍指导意义。

丁衡高院士高度重视我国微米纳米技术创新人才培养。他受聘作为清华大学兼职教授、博士生导师二十年整（从1987年至2006年），培养了多名博士研究生和博士后。清华大学评价他"充分运用个人多年的研究成果和实践经验，对学校科研方向的选择、技术攻关的指导和研究生的培养等方面都发挥了很大作用，表现出他具有坚实的学科专业知识、丰富的工程实践经验和高水平的研究能力"。他还是东南大学（1997年）、南京航空航天大学（2004年）、西北工业大学（2006年）、重庆大学（2014年）等学校的名誉教授，上海交通大学微纳科学技术研究院名誉院长（2002年至2010年），传感技术联合国家重点实验室学术委员会主任（2003年至2008年），新型微纳器件与系统技术国防实验室学术委员会主任（2008年至2015年）。他重视发挥青年科技工作者的作用，鼓励创新，扎实开展原创性基础研究，提出在研究计划中专门设立青年微米纳米研究基金，以年轻博士为支持主体，开展研究工作。他几乎深入我国每个微米纳米技术主要研究机构进行调查研究，发掘人才。在他的鼓励与指导下，一大批年轻的科研工作者锐意创新，脱颖而出，多数已成为我国微米纳米技术研究和产业发展的中坚力量及国家级科技领军人才。

值此开展我国微米纳米技术研究30周年之际，为了充分展示丁衡高院士在我国微米纳米技术发展历程中所做的努力和贡献，传承和发扬他一贯的科学精神、创新思维和严谨作风，中国微米纳米技术学会组织出版《丁衡高院士微米纳米技术文集》。本书整体上按照发展战略研究、科学技术综述、科学技术研究这三类题材，收集了丁衡高院士已公开发表的关于微米纳米技术领域的讲话致辞、学术报告和科技论文共 36 篇，以飨读者。

中国微米纳米技术学会

2024 年 6 月

目 录

发展战略研究

面向 21 世纪的军民两用技术——微米/纳米技术/（1995 年 6 月）
.... 3
微型惯性测量组合/（1996 年 2 月） 17
微机电系统的科学研究与技术开发/（1997 年 9 月） 26
微小卫星应用微小型技术有关问题的思考/（1997 年 11 月） ... 39
微系统与微米/纳米技术及其发展/（2000 年 5 月） 47
MEMS 技术发展战略研究中需考虑的几个主要问题/
（2000 年） 60
关于微米/纳米技术的认识与思考/（2000 年 12 月） 76
微米/纳米技术当前发展动向/（2001 年 10 月） 87
抓住机遇，促进微米纳米技术新发展/（2003 年 2 月） 96
微纳技术进展、趋势与建议/（2006 年 12 月） 104
微米纳米科学技术发展及产业化启示/（2007 年 12 月） 124
MEMS 器件研制与产业化/（2009 年 10 月） 141
在"MEMS 在机械与运载工程领域的应用研究"咨询项目启动
会上的讲话/（2012 年 5 月） 145

三十年不断发展的 MEMS 惯性传感器/（2023 年 10 月） ······ 152

科学技术综述

微型惯性测量组合的关键技术/（1996 年 2 月） ············ 163
微机电系统技术的实际应用——微型仪器/（2000 年 4 月） ··· 173
微惯性仪表技术的研究与发展/（2001 年 12 月） ··········· 183
《微纳米加工技术及其应用》序/（2004 年 12 月） ·········· 193
微型化陀螺研究进展和展望/（2005 年 9 月） ············· 195
《微纳系统》译丛总序/（2012 年 6 月） ················· 209
微纳电极阵等离子体微系统/（2024 年 5 月） ············· 224

科学技术研究

叉指式硅微加速度计的结构设计/（1998 年 11 月） ········· 241
微型光学陀螺仪中声表面波声光移频器的研究/（1999 年 2 月）
································· 253
一种微型隧道效应磁强计的设计/（2000 年 8 月） ·········· 262
Investigation of the System Configuration for Micro Optic
　Gyros/（2005 年） ······························ 269
A MEMS Hybrid Inertial Sensor Based on Convection
　Heat Transfer/（2005 年 6 月） ···················· 282
Micromachined Gas Inertial Sensor Based on Convection
　Heat Transfer/（2006 年 1 月） ···················· 294
静电悬浮微机械加速度计设计/（2007 年 2 月） ··········· 312

目 录

A Study of Cross-axis Effect for Micromachined Thermal Gas Inertial Sensor/（2007 年） ………………………… 324

Sensor Fusion Methodology to Overcome Cross-Axis Problem for Micromachined Thermal Gas Inertial Sensor/（2009 年） ………………………………………………………… 336

Modeling and Experimental Study on Characterization of Micromachined Thermal Gas Inertial Sensors/（2010 年） ………………………………………………………… 355

A Micromachined Integrated Gyroscope and Accelerometer Based on Gas Thermal Expansion/（2013 年） ………… 377

A Micromachined Gas Inertial Sensor Based on Thermal Expansion/（2014 年） ……………………………………… 392

A Temperature Compensation Method for Micromachined Thermal Gas Gyroscope/（2015 年） ………………………… 417

Thermal Characteristics of Stabilization Effects Induced by Nanostructures in Plasma Heat Source Interacting with Ice Blocks/（2022 年） ………………………………… 432

Reconfigurable Plasma Composite Absorber Coupled with Pixelated Frequency Selective Surface Generated by FD-CGAN/（2022 年 12 月） ……………………………… 492

发展战略研究

面向 21 世纪的军民两用技术
——微米/纳米技术

（1995 年 6 月）

摘　要　微米/纳米技术是本世纪出现且发展迅猛的一项高新技术。本文在论述了微型机电系统和专用集成微型仪器的制造工艺和发展动态的基础上，进一步论述了小型、微型和"纳米"卫星的概念和应用前景。本文还对纳米技术的实质及其产生的革命性影响进行了讨论。微米/纳米技术的应用前景充分显示了其鲜明的军民两用性，将深刻地影响国民经济和国防科技的发展，是整个国家科技发展战略应研究的重要课题。

关键词　微米/纳米技术

自微电子技术问世以来，人们不断追求越来越小、越来越完善的微小尺度结构的装置，并对生物、环境控制、医学、航空、航天、精确制导弹药、灵巧武器、先进情报传感器以及数字通信等领域，不断提出微小型化方面的更新更高的要求。按照一种习惯的划分，装置尺度在 1 mm（10^{-3} m）～10 mm 范围的称微小型（mini -）机械；在 1 μm（10^{-6} m）～1 mm 范围的称微型（micro -）机械；在 1 nm（10^{-9} m，或 10 Å）～1 μm 范围的称纳米（nano -）机械。用于制造上述尺度结构的工具有两类：分子处理工具和批量加工工具。化学家和分子生物

本文发表于《中国惯性技术学报》1995 年第 2 期。

学家采用越来越精良的分子处理工具来制造和处理精细的分子结构（如蛋白质）。借助于扫描隧道显微镜①（scanning tunneling microscope，STM）和原子力显微镜②（atomic force microscope，AFM），物理学家把化学和分子生物学的处理方法结合起来，开辟了从专用集成电路（application specific integrated circuits，ASIC），到微型机电系统（microelectron‒mechanical systems，MEMS），到专用集成微型仪器（application specific integrated microinstrument，ASIM），再到纳米技术③（nanotechnology）的开发和批量加工工具的技术。这一系列技术可以概括为微米/纳米技术（micro/nano‒technology）。

当前，微米/纳米技术在国际上已初露头角，它使人类在改造自然方面进入一个新的层次，即从微米层次深入到原子、分子级的纳米层次。正像产业革命、抗菌素、核能以及微电子技术的出现和应用所产生的巨大影响一样，纳米技术将开发物质潜在的信息和结构潜力，使单位体积物质储存和处理信息的能力实现又一次飞跃，在信息、材料、生物、医疗等方面导致人类认识和改造世界能力的重大突破，从而给国民经济和军事能力带来深远的影响。

微米/纳米技术作为本世纪出现的高技术，发展十分迅猛，并由此开创了纳米电子学、纳米材料学、纳米生物学、纳米机械学、纳米制造学、纳米显微学及纳米测量等等新的高技术群。如同当代高新技术

---

① STM，利用探针与表面间的隧道电流直接观察原子、分子。它的分辨率高（0.01 nm），达到原子量级水平。发明人 Binnig 和 Rohre 获得 1986 年的诺贝尔奖。

② AFM，利用探针与表面之间的原子力代替隧道电流，可以直接观察原子、分子，而且对导体和非导体均适用。

③ 纳米技术，即纳米级（0.1～100 nm）的材料、设计、制造、测量和控制技术。

都往往具有军民两用性一样，纳米技术也是面向 21 世纪的一项重要两用技术，有着广阔的军民两用前景。微米/纳米技术民用和军用方面已开发出的一些成果和许多可能的应用是吸引人的。美、日、西欧等国家和地区均投入相当的人力和财力进行开发。目前，微米/纳米技术在军事应用方面的工作主要集中在微型机电系统和专用集成微型仪器，对纳米卫星也进行了论证和探索。这些都很值得我们重视。

1 微型机电系统

在过去 35 年电子革命的过程中，微电子技术的产生和发展使门电路的尺度不断缩小，使得手持蜂窝式移动电话、地球低轨道上重仅 10 kg 的电子新闻广播系统，以及可以与 70 年代大型计算机相匹敌的个人计算机成为现实。这一切以及 90 年代其他电子奇迹的关键，都在于灵活的批量加工工艺，它可以使数以百万计的灵巧的微小型零部件能够同时制造。数兆位的存储器芯片、微处理器以及射频分系统，已取代门电路成为主要的部件。

10 年前，人们在意识到用半导体批量制造技术可以生产许多宏观机械系统的微米尺度的样机后，就在小型机械制造领域开始了一场类似的革命。这就导致了微型机电系统（MEMS）的出现，如微米尺度的压力传感器、加速度传感器、化学传感器和各种阀门等。

微米级制造包括尺寸小于 $1\ \mu m$（10^{-3} mm）材料的制造和使用。微米级制造工艺包括光刻、刻蚀、淀积、外延生长、扩散、离子注入、测试、监测及封装。纳米级结构的尺寸范围为 $10\sim 100$ nm 或更小。纳米级制造包括微米级制造中的一些技术（如离子束光刻），但也包括为了利用材料的本质特性以期获得理想的结果而对材料进行原子量级的修改及排列的技术。1991 年 3 月 22 日，美国国家关键技术委员会向美国总统提交了《美国国家关键技术》报告。其中，第 8 项为"微米级和纳米级制造"[1]。报告指出："微米级和纳米级制造涉及显微量级（微米级制造）和原子量级（纳米级制造）的材料及器件的制造和使用"；"对先进的纳米级技术的研究也可能导致纳米级机械装置及传感器的生产"；"微米级和纳米级技术的发展已使人们能开发出一类新的显微量级尺寸的器件。这些器件能在诸如环境控制、医学等不同的领域工作。它们的低成本及比现有器件高的灵敏度可能使许多领域会有突破"。

美国发展微型机电系统关键性的研究中心基本上都分布在各个大学，如康奈尔大学、斯坦福大学、加州大学伯克利分校、密执安大学及威斯康星大学等。美国国防部高级研究计划局（ARPA）重视并积极赞助 MEMS 的国防应用，现已建立了一条 MEMS 标准工艺线来促进新型装置的迅速开发和小批量生产。

加州大学伯克利分校和美国其他研究所，已在微电子技术的基础上用微米加工方法制作出微小齿轮和微型电机。日本和德国也在积极推进微型机械加工技术。德国还研制出一种称之为 LIGA① 的微结构成形工艺。利用 LIGA 工艺的微型产品大大改善了 MEMS 的现状，使 MEMS 向实用化跨出了一大步。

目前，MEMS 已从实验室探索走向工业应用，并正在迅速发展。已研制成一些引人注目的器件，其中许多几乎是肉眼看不见的，这些新型器件包括回转式电机、线性执行机构、加速度计、谐振器、传动装置、操纵杆之类的工具以及其他各式各样的部件。迄今为止，传感

---

① LIGA 是德文 X 射线光刻、电铸成型和铸塑三个工艺过程的缩写。LIGA 工艺基本上是利用 X 射线光刻技术，在任何数量的材料上构成深层微结构的方法。此项工艺已被广泛用于开发静电电机、硅齿轮和传感器用的膜片。

器已经表明极具商业应用的希望,可用于加速度计、惯性制导系统、化学传感器等[2]。

在80年代,发达国家就注重发展MEMS。据介绍,日本每年投资1.5～2.0亿美元用于发展这一技术。在这方面,美国也不甘落后,ARPA制定了一项将MEMS技术从实验室进入军事应用的三年计划,并拨出2400万美元支持这一计划。据文献[3]提供的信息,美国Draper实验室在1991年研制出一种惯性测量元件(IMU)样机,尺度为$2\ cm \times 2\ cm \times 0.5\ cm$,重5 g,陀螺漂移误差为10(°)/h或更小。Draper实验室当前的努力目标是:1～10(°)/h,并瞄准0.5(°)/h的新目标。专家们深信还可以达到0.1～0.01(°)/h的精度。重约10 g,精度1(°)/h,价格10美元的惯性测量元件将会有巨大的商业市场。

2 专用集成微型仪器

"微米/纳米技术"一词的含义,包括了从亚毫米到亚微米范围内的材料、工艺和装置的综合集成。微米技术专家已成功应用的芯片制造技术——这是近几十年来实现微小型化的主要途径——在"纳米"技术的开发中起着重要的支持作用。

微米/纳米技术的实际应用便是微型工程(microengineering)。微型工程包括具有毫米、微米、纳米尺度结构的传感器和作动器的设计、材料合成、微型机械加工、装配、总成和封装问题。利用这项技术可以把传感器、作动器以及信号和数据处理装置集成在一块普通的基片上。MEMS与微电子技术的综合集成,导致了专用集成微型仪器(ASIM)概念的出现。ASIM的制造,是从半导体工艺即专用集成电路(ASIC)技术和MEMS技术发展而来的。具有亚微米特点的专用集成微型仪器会使亚毫米器件降低研制与试验费用、缩小体积、减轻重量，同时还可以降低对电源和温控的要求，降低对振动的敏感性和通过冗余提高可靠性。

同 ASIC 一样，ASIM 可以用微电子工艺技术的方法批量制造。但是 ASIM 比 ASIC 更为复杂，因为它既有固定的部件又有活动的部件，而且采用诸如生物或化学活化剂之类的特殊材料。确切地说，ASIM 是一种高水平的微型器件。这种器件把几种微机电系统组装在一块硅基片上，或者说就如同一个有许多基片模块的组件，这就是一个高度复杂的分系统。

目前正开发中的 ASIM 的原型已可以探测出局部地区和遥远地区的环境。信息可以通过基片上的通信系统传送到相邻的微型仪器，或传送到中央处理机。

ASIM 将代替现有航天器和各级运载火箭上目前使用的分系统，然后再发展成为独特的空间系统结构。ASIM 将在航天器和航天系统技术方面引起一场革命，出现超小型卫星系统，最终实现"纳米卫星"。

ASIM 现已进入商业市场，如压力传感器、（汽车）气囊控制器。美国国家航空航天局（NASA）已着手实施一项低费用（不足 1 亿美元）的"发现号"微型卫星计划。霍普金斯大学、洛斯·阿拉莫斯国家实验室和过去从事"战略防御计划"（SDI）的一些部门也正积极考虑小型卫星的研制任务。

鉴于许多商业应用的需要，微型机电系统和专用集成微型仪器标准加工工艺线正在发展，它可用以生产可靠、低成本的航天系统用的专用集成微型仪器。

3 小型、微型和"纳米卫星"

专用集成微型仪器技术能应用于小型卫星的一些单个功能部件,导致各种超小型卫星分系统的出现。下一步就是不可避免地在航天系统中使用专用集成微型仪器。预计在 5~10 年内能得到适合于航天应用的 MEMS[4]。

于是,就有了小型、微型和"纳米卫星"的概念。

小型卫星(minisatellite):能用小型运载火箭发射的常规设计的航天器;重量范围约为 10~500 kg。

微型卫星(microsatellite):在所有的系统和分系统中全部体现微型制造技术成果并能执行卫星应有的功能;重量范围约为 0.1~10 kg。

"纳米卫星"(nanosatellite):依靠一种分布式的体系结构完成自身功能,并将尺度减至最小可能的微型卫星。重量范围约小于 0.1 kg。以硅或其他半导体为衬底的专用集成微型仪器,能应用于制导、导航、控制、姿态控制、热控制、推进、能源和通信等航天器系统。美国一些公司已在研究在芯片上制造卫星的方案,并得到了 NASA"小企业创新研究"合同的支持。

"纳米卫星"代表了卫星发展、构型和运行体系结构的一种新模式。当某种设计方案经过飞行试验验证之后,使用单一半导体标准工艺线就可制造出成批的复制品。这就使得低成本地研制和试验新卫星,低成本地制造卫星,以及增加可靠性和灵活性成为可能。相比之下,常规卫星由成百上千个分立部件组装而成,而这些部件又是由许多制造点生产的。"纳米卫星"概念的出现,使分布式航天体系结构成为可能。这种体系结构相对免除了单个航天器失效的影响,从而增加未来航天系统的生存力和灵活性。一种简单的"纳米卫星"可以由外表面带有太阳能电池和天线的、在硅基片上堆砌的专用集成微型仪器组成。"纳米卫星"也代表了卫星利用方式转变的一个范例。"纳米卫星"最好的应用是布设成局部星团和分布式星座。如果在太阳同步轨道施放"纳米卫星",在 18 个等间隔的轨道面上,每个轨道面上等间隔施放 36 颗"纳米卫星",一共 648 颗这样的卫星就可以保证在任何时刻,对地球上任何一点的连续覆盖。"纳米卫星"可以大批量生产,成本效益比一般的卫星好得多,从而保证同时使用几百颗卫星。设想"纳米卫星"可用于低地球轨道通信和地球观测。

把微米/纳米技术引入航天系统,需要材料、微电子制造、三维微型机械加工以及各种特殊的航天器分系统(结构、推进、温度、导航、动力、通信)等不同领域专家的密切合作。

4 纳米技术

计算机工业一直不断地追求超高集成度的芯片。当芯片线宽尺度进一步缩小时,在某个转折点,也许是 150～100 nm 处,也许是在更小的尺寸上,因量子力学效应的增强,将会遇到很大困难。[①] 但是,科学家仍然为克服这种困难,在不断地探索,继续把这门先进的新技术推向前进。

纳米技术的实质在于,多少年来人们在不断做出越来越小的装置,直到趋近分子尺度。达到这个转折点,人们不再能把装置做得更小了,

---

① 由于电子具有波粒二象性,通常,当电子处于尺度大于 0.1 μm 的空间时,可以把电子视为粒子,而当尺度小于 0.1 μm 时,电子作为一种波动所呈现出来的量子效应便不能忽略。原先视电子为粒子的微电子技术的理论基础不再有效,需要研究利用电子的量子效应原理制作器件,称为量子器件,或称为纳米器件。

除非他们从分子开始，即在装配器中对分子进行装配。

根据分子工程（molecular engineering）的概念，人们现在可以按照需要对物质从分子水平上构筑；分子工程有理论，即量子力学；也有分析计算的手段，即基于概率论的多坐标复杂偏微分方程计算机求解方法；还有可视化的研究工具，体现"灵境"①（virtual reality）的仿真。数字电子技术的基本特征是以完美的控制和离散方式快速处理信息，从而产生信息革命。信息革命的核心是信息属性劳动资料的创造②，如能处理任何离散形式信息的可编程数字计算机。今天，又出现了纳米技术。纳米技术的核心是装配分子，或者说，按人们的意志直接操纵单个原子、分子或原子团、分子团，制造具有特定功能的产品。持乐观态度的科学家，如斯坦福大学的 K. EricDrexler 曾预测，在 2010 年到 2020 年间可能实现一个原子存储一位计算机信息。据《新科学家》杂志报道，日本日立公司 1993 年 12 月份宣布，已制成在室温下工作的单电子存储芯片，而且是一种非丢失性存储器。和现有的存储芯片相比，同样存储 1 bit 信息，新存储器的功耗只是前者的百万分之一，面积为前者的万分之一。

纳米技术革命的基本特征是以完美的控制和离散方式（原子和分子）快速排布原子或分子的结构，从而产生物质处理技术的革命。纳米技术革命本质上是更深层次的信息革命。采用分子器件制作的全新

---

① 通过传感器和操纵装置看到一个计算机模拟景象，并可在其中随意采取行动的技术。灵境技术不仅可以用来探索外层空间，而且还可以用来探索分子内部的微观世界。

② 马克思在《资本论》第一卷中说过，在劳动资料中，机械性的劳动资料（其总和可称为生产的骨骼系统和肌肉系统），能显示一个社会生产时代的具有决定意义的特征。受马克思这一论点的启发，我们可以把信息属性的劳动资料的创造，看作是信息革命时代社会生产的重要特征。

的"纳米计算机",其数字逻辑图像可以建立在比90年代计算机小得多的尺度基础上,而且速度更快,效率更高。如果说,90年代计算机芯片的大小犹如一幅巨大的风景画,那么纳米计算机就像画中的单个建筑物。

5 微米/纳米技术具有鲜明的军民两用性和前沿性

微米/纳米技术作为面向21世纪重要的军民两用技术,将深刻影响国民经济和国防科学技术的发展。美、英、日等国高度重视发展微米/纳米技术。美国国家关键技术计划把"微米级和纳米级制造"列为"在经济繁荣和国防安全两方面都是至关重要的技术"。美国国家基金会把微米/纳米技术列为优先支持的项目。1993年美国宇航公司组织了20余位航天、电子等领域的专家,对微米/纳米技术在航空航天领域的应用进行了广泛的讨论,并发表了《革命性低成本航天系统"纳米卫星"的概念》文集。

前面已经介绍的ARPA制订的MEMS发展计划,一直在采用与制造微电子器件相同的工艺和材料,充分发挥小型化、多元件和集成微电子技术的优势,设计和制造新型机电装置。ARPA正在以下方面推进MEMS的发展与应用:个人导航所需的小型惯性测量装置,大容量数据存储器件,小型分析仪器,非侵入式医疗传感器,光纤网络开关,以及环境与安全监视用的分布式自动传感器等。ARPA支持开发了一种运动探测部件,它是具有一定灵敏度和稳定性的个人惯性定位装置所必需的。扩充现有的GPS,以MEMS为基础的惯性跟踪器可以提供个人定位信息。ARPA的另一个MEMS项目已演示了一种以MEMS为基础制造的加速度计,能承受火炮发射炮弹所产生的近10^5 g的加速度。有了这种加速度计,就可以为目前的非制导弹药提供一种经济的制导系统,从而使这种武器减少了采购和后勤维护费用。ARPA还支持对国内商业和学术研究用户建立定期的、共享的MEMS制造服务。这种服务可以让几百个研究和工业用户节省费用并快速制造MEMS装置。

MEMS主要的民用领域是:医学、电子工业和航空航天。例如,将来可制造出静电驱动的微型电机,用来控制计算机及通信系统。在环境、医学及机构应用中,微型传感器可用来测量各种化学物质的流量、压力及浓度,美国现已研制出可进入人体直肠的MEMS。MEMS技术的发展激励着人们寻求MEMS在军事技术中的应用,包括现实的和富于幻想的军事应用。在美国国防部、空军和陆军的赞助下,RAND公司在1993年完成了《微型机电系统的军事应用》研究报告[2],探索了微型机电系统的潜在军事应用。报告设想的微型机电系统的军事应用是:

有害化学战剂报警传感器 在特定的微机电系统上加一块计算机芯片(售价20美元),就可以构成袖珍式质谱仪,在化学战环境中用来检测气体。目前使用的质谱仪,一台的价值为17 000美元,重68 kg以上。

敌我识别 目前的敌我识别系统采用反射带、有源信标或应答器,这些设备很容易被侦听或截获。而微型机电系统则散布于飞机蒙皮上,或车辆的外表面,能以较低的功率自动对询问信号做出回答。

灵巧蒙皮 利用微机电系统可以做成具有可程序控制乃至特性动态可调的材料。例如,可以制作灵巧的潜艇蒙皮,它能立刻确定当时的速度并且命令中央计算机进行精密操纵调整,从而将噪声减至最小。

分布式战场传感器网络 用无人驾驶飞机将微机电系统探测器散布在战场的广阔地域,同时确定每个探测器的位置并进行询问,把各个探测器给出的编码数据储存起来,将结果传送给作战指挥官。应用微型计算机技术使探测器和灵巧武器掌握敌方目标的方位和特征,并能接收远处指挥官的指令,这种微机电探测器网络可以对敌坦克和步兵构成威胁。

微机器人电子失能系统 微机器人电子失能系统相当灵活,能以相当精确的方式布设,它能"感觉"敌方电子系统的位置,进而渗进该系统,使之丧失功能。

昆虫平台 将微机器人电子失能系统预先植入昆虫的神经系统,控制它们飞向敌方目标搜索情报,也可以利用这些昆虫使目标丧失功能或杀伤士兵。

50年代末,诺贝尔物理学奖获得者 R. Feynman 曾指出:"如果有一天可以按人们的意志安排一个个原子,那将会产生怎样的奇迹?"今天,科学家已实现了对单个原子的操纵,不仅有可能创造先进的数字计算机,而且利用纳米材料的性能可以实现具有特殊功能的仪器。虽然制造具有特定功能的产品尚待时日,但开发应用前景十分诱人,国际性的竞争已经展开。

日本认识到纳米技术在商业、军事和医学方面的长远潜力,已建成第一个分子装配器。这个国家不邀请外国科研人员参与这项实际工作。欧洲有关纳米技术的一项计划已在法国的一个实验室开始起步。为了保持欧洲的竞争能力,这项计划同样是保密的。

日本的第二代分子装配器已开始产出少量的分子装置。商业上可用的产品样品已有传感器、分子电子装置和科学仪器。日本已有一项长期计划的蓝图,目的是使未来羽毛丰满的分子制造能以相当低的价格用普通材料做出任何产品。

分子制造所具有的明显的军事潜力,引起了军事兴趣,因而刺激了其研究与发展计划。美国开发微米/纳米技术的经费中有一半左右来自国防部系统。

战略家们在他们的头脑里、杂志中和计算机上，针对纳米技术对国家安全的影响进行作战模拟（wargame）。他们关注分子制造及其潜在产品的军事影响，认为有许多理由来推动纳米技术的军事应用研究计划。

微米/纳米技术是一项新技术，是面向21世纪的重要的军民两用技术，它的出现无疑将深刻影响国民经济和国防科学技术的未来发展，所以是整个国家科技发展战略应研究的重要课题。

参 考 文 献

[1] National Critical Technologies Panel. Report of the National Critical Technologies Panel. Washington：The Superintendent of Documents，U. S. Government Printing Office，1991.

[2] Brendley K. W，Steeb R. Military Applications of Microelectromechanical Systems. RAND，1993.

[3] Elwell J. Progress on Micromechanical Inertial Instruments.（handout）. CSDL－P－3179（Aug. 1991），555 Technology Square，Cambridge，MA.

[4] Janson S. Helvajian H，Robinson E. Y. The Concept of "Nanosatellite" for Revolutionary Low－Cost Space Systems. 44th CONGRESS OF THE INTERNATIONAL ASTRONAUTICS FED－ERATION，October 16－22，1993/Graz，Austria，IAF93－U. 5. 573. International Astronautics Federation，1993.

Dual – Purpose Technology for Military and Commercial Applications Aiming at the 21th Century: The Micro/Nano – Technology

DING Heng – gao

(The Science and Technology Industry Committee of National Defense)

Abstract The micro/nano – technology is a new hi – tech which arises and is forging rapidly in this century. Based on the description of the fabrication techniques for Micro Electromechanical Systems (MEMS) and Application Specific Integrated Microinstrument (ASIM) and developments in the world, this paper discusses the concepts of minisatellite, microsatellite and nanosatellite and their promising prospects. Topics presented also include an explanation of the essence of the nanotechnology and its profound impact. The promising applications of micro/nano – technology indicates that this technology has obvious amphibiousness for military and commerce. It will deeply affect the developments of national economy and the science and technology industry of national defense, and is one of the most important programs that should be studied in the national science and technology development strategy.

Keywords Micro/Nano – technology

微型惯性测量组合

（1996年2月）

摘　要　本文论述了惯性测量器件和组合的发展简史、机遇，选择微型惯性测量组合牵引我国微米/纳米技术发展的依据，以及采纳的技术路线。

关键词　微型惯性测量组合

0　引言

我在《面向21世纪的军民两用技术——微米/纳米技术》一文中指出："10年前，人们在意识到用半导体批量制造技术可以生产许多宏观机械系统的微米尺度的样机后，就在小型机械制造领域开始了一场类似的革命。这就导致了微型机电系统（MEMS）的出现，如微米尺度的压力传感器、加速度传感器、化学传感器和各种阀门等。"[1]

今后几年内，可能会出现芯片尺寸的陀螺、加速度表、惯性测量组合和其他传感器取代一些机电系统的情况。鉴于微型惯性测量组合（Micro Inertial Measurement Unit，MIMU）在推动国民经济和军事技术发展方面的重要性，本文将就MIMU发展的简要历史，当前面临的机遇，以及我们选择MIMU作为微米/纳米技术发展需求牵引的依据和采纳的技术路线，作专门的论述。

本文发表于《仪器仪表学报》1996年第17卷第1期。

1 惯性测量术语的演变

恩格斯说过:"一门科学提出的每一种新见解,都包含着这门科学的术语的革命。"我认为这一见解能很好地帮助我们理解从上世纪中叶的机械陀螺,到本世纪 50 年代的振动陀螺,再到今天的 MIMU 的发展。

1852 年,法国物理学家傅科提出了"陀螺"这个术语,做出了最早的机械陀螺仪。1911 年,美国人斯佩里组织了一个专门公司开始陀螺仪的工业生产。从 30 年代开始,陀螺仪已广泛用于航空。继普通滚珠轴承支承的机械陀螺之后,出现了许多新型机械陀螺,如气浮陀螺、液浮陀螺、静电悬浮陀螺、磁悬浮陀螺等,都是为了减小轴承摩擦力矩的需要。所有这些机械陀螺,都有一个高速旋转的质量,只不过支承方式不同而已。为了减小漂移速度,转子又必须保持足够的动量矩,所以这一类陀螺仪表,都不易微小型化。

为了制造微小型化的陀螺仪,就需要利用新的物理现象,建立新的陀螺概念。第一个新陀螺概念是"振动陀螺",它来自一个仿生学的启发:长脚蚊和某些双翅昆虫在飞行过程中靠一对振动臂获得飞行稳定性。后来又出现了音叉振动陀螺的实验装置。

几乎就在这同时,振动陀螺的研究工作出现了另外一条不同的途径。人们发现利用压电效应可以圆满地完成振动激发、频率控制和信号拾取的任务,于是出现了振动梁角速率传感器。

1913 年,萨格耐克(Sagnac)利用一个由环状回路形成的光学干涉仪来研究转动对光传播的影响。他的独特实验开始了又一种新陀螺的孕育过程:利用光学干涉仪测量转动角速度。1963 年,美国首次做出激光陀螺实验装置。

今天，振动陀螺、加速度表和激光陀螺技术正在发生重大变化。例如，德雷珀实验室（Charles Stark Draper Laboratory）推出的微型有常平架的振动陀螺、音叉陀螺、力再平衡加速度表、石英谐振加速度表；卫星通信公司（SatCon Technology Corporation）的静电悬浮微机械陀螺；通用电气-费伦蒂（GEC–Ferranti）公司的压电振动速率陀螺；诺思罗普（Northrop）公司的固态激光陀螺、微光学陀螺和硅微机械摆式加速度表等等[2]。这些微小型仪表的出现，迅速扩大了惯性测量组合在军事和民用领域的应用。促成这一重大变化的推动力，是微观尺度加工技术的演进与革命。

2 发展的机遇

微观尺度生产领域制造技术（微米/纳米技术）的演进与革命，为包括微机械惯性测量器件在内的 MEMS 的发展提供了历史性的机遇。

著名物理学家 Richard Feynman 是 1965 年诺贝尔奖获得者。1959年 12 月 29 日，在美国物理学会年会的一次谈话中，他预见到制造技术从宏观到微观将沿着 Top–Down 的途径发展。这就是：用大机器做出小机器，用这种小机器又能做出更小的机器。这种发展对世界经济和科技发展产生巨大影响，人们在微观尺度领域造就了一代代的批量加工工具。

微观尺度生产领域制造技术革命性的发展，体现在三个方面[3]：

1) 扫描隧道显微镜（STM），可以实现原子分辨率；

2) 电子束、离子束、X 射线束等束制造技术，可以使器件的特征线宽做到 $100\sim250\text{Å}$；

3) 分子工程与常规平面工艺的密切结合。

这样，除微电子电路外，人们不仅努力制造微光学、微机械、微化学装置之类的微装置，而且更为重要的是人们还一直在朝向获得原子的容许极限而努力。

把微米尺度器件的制造推进到纳米容许极限，即将微米制造推进到微米/纳米制造，是在人类运用粒子束（光子、电子、离子）光刻所具备的能力范围之内[3]。美国半导体行业协会预测：1995年集成电路的制造工艺精度能够达到 0.35 μm，1998 年 0.2 μm，2001 年 0.18 μm，2004 年 0.13 μm，2007 年 0.1 μm，2010 年将达到 0.07 μm。

继大规模集成电路之后，微观尺度生产领域制造技术下一个逻辑发展步骤，将是 MEMS 的发展与应用。大规模集成电路的成就，究其实质，只不过是对电子的运动进行通断控制。MEMS 将为电子系统提供通向外部世界所需的窗口，使它们可以感受并控制运动、光、声、热及其他物质力量[4]。

微观尺度生产领域制造技术发展的另一个推动力来自分子制造，即纳米技术。这是一条不同于 Top-Down 的 Bottom-Up 途径，即从最小的构造模块——一个分子开始，进行物质的构造。这是一种用分子作为部件，用分子装配器装配，并用计算机控制的、人能把握与应用的分子尺度工具的加工系统。只要给出适当的原材料，它就能装配出所需的任何物质的材料。这种在微观尺度领域里从分子尺度向上建造的技术，给工程技术带来新的革命内容，如同农业机械和能源机械给工程技术带来过的革命一样。在美国《时代》周刊1995年7月17日一期题为《未来的技术已经来临》一文中，纳米技术被列为改变未来的十项技术中的一项。

微观尺度生产领域制造技术的发展，促进了信息处理方式的革命。第一次革命是把计算机从冷却剂的束缚下解放出来。这一场革命使计算机出现在我们的办公桌上，从而带来很大的社会冲击。今天正在发生的第二场革命，将使现在的计算机从桌面上再小型化下来，即从桌面上消失掉，嵌入所有需要信息处理的机电装置中，并把信息处理和装置的控制推进到更接近装置要用的地方去。

对应于集成电路的是"微电子"技术，对应于微机械系统的是"微机械"技术。这两门技术相结合，形成了一代全新的微机电系统。

这一发展的起点，始于 60 年代早期用各向异性蚀刻技术制造出来的第一个微压力传感器。自那以后，微电子制造技术直接影响着微机械的进展。70 年代，一些工业公司致力于微加工技术的商品化，因而出现了第一代微机械产品。得克萨斯（Texas）仪器公司就是在这时成功地将计算机微打印组件推向了市场。70 年代末，可控的各向异性凹槽蚀刻技术得到进一步发展，再结合 80 年代对超大规模集成电路工艺方法的开发，以及首先在德国出现的高深宽比的微机械加工工艺，就为硅微机械的创新提供了历史的机遇。于是，在微机械的研制与生产中，1987 年出现了与"集成电路标准工艺线"相对应的概念："硅微机械标准工艺线"。

近几年来，工业界和学术界对微机械传感器、作动器的研究异常活跃。为了测量湿度、温度、流速、粘滞度、压力、加速度、化学反应和许多其他物理参数，已制造了各种各样的微传感器。

大规模集成电路批量加工技术与硅表面微机械加工集成方面最近的进展，使得目前已有可能把微传感器和微作动器的支持电路（即放大、补偿、变换、多路传输和接口功能）都集成在芯片上。

3 MIMU 的发展

陀螺仪测量运动物体的姿态或转动的角速度，加速度表测量速度的变化；陀螺的功能是保持对加速度对准的方向进行跟踪，从而能在惯性坐标系中分辨出指示的加速度；对加速度进行两次积分，就可测出物体的位置。由 3 个正交陀螺、3 个正交加速度表和 1 个坐标转换计算机，就可综合集成为一种惯性测量组合（IMU），它可以提供运动物体的姿态、位置和速度的信息。

最近几年来，利用光学技术和/或振动质量的微型惯性仪表已变得切实可行。它没有连续转动的部件，在寿命、可靠性、成本、体积和重量方面优于常规惯性仪表，适用于较恶劣的环境。

据美国航空航天学会（AIAA）的一篇研究报告[5]，用单晶硅片化学刻蚀方法，在一块 4 in 的硅片上可以批生产多达 4 000 个独立的微型惯性仪表。

惯性测量组合制造技术的进步，可用诺思罗普公司的三种产品来说明。1974 年"鱼叉"导弹系统 IMU 的质量为 3.2 kg，1980 年"不死鸟"导弹系统 IMU 的质量只有 1.8 kg，而 1985 年"先进中程空空导弹"系统 IMU 的质量进一步降到 1.4 kg。这一趋势还将继续下去，预计当其采用的微光学陀螺技术成熟时，有可能降到 70 g[6]。

德雷珀实验室正在研制的 MIMU 预计最终尺寸为 2 cm×2 cm×0.5 cm，重约 5 g[5]。其中陀螺的漂移率约为 10 (°)/h 或更好一些，适用于短时间的导航。下一步的目标是漂移率为 1 (°)/h。

MIMU 在军事应用和民用方面都有广泛的应用前景和明确的应用目标。据 ARPA 分析，2000 年的 MEMS 市场发展预测是：压力传感器将占总需求的 25%，光学开关为 21%，惯性传感器为 20%，液体调控为 20%，数据存储为 5%，其他为 9%。微型惯性仪表可能的应用包括：汽车（定位测量仪表、防滑刹车系统、安全气囊展开、自动调整），小型卫星和航天器（发射重量是一个重要的费用因素），精度更高的炮弹，摄录一体机，通用航空，医疗电子设备，以及玩具。现已有一些微加速度表，如美国模拟设备公司的 ADXL50 加速度表，可用于汽车防撞和节油。采用这种加速度表不仅可提高汽车的安全性，还可节省 10% 的汽油，仅此一项国防部系统每年就可节约几十亿美元的汽油费用。美国和日本的一些汽车制造公司已决定在 1995 年制造的汽车的安全气囊中采用这样的微加速度表。诺瓦传感器公司这类可用于汽车的微加速度表的年产量为 600 万件。

DARPA 在 1995 财年的计划中，积极支持的一项 MEMS 计划是研制具有一定灵敏度和稳定性的个人惯性定位装置，并利用 GPS 提高其性能。DARPA 积极支持的另一项计划，已演示了一种基于 MEMS 的加速度表，能承受火炮发射炮弹时产生的接近 100 000 g 的重力加速度。这一技术将用于改造非精确制导弹药，降低采购和后勤费用。洛克威尔（Rockwell）公司与德雷珀实验室联合，在 MIMU 研制方面处于领先地位。据专家们预测，2000 年以前将有中等精度的 MIMU 产品进入市场。研制出纽扣大小的补充 GPS 的微型惯性导航系统，已不再是非常遥远的事情，估计其成本可能只有 5 美元。MIMU 与 GPS 接收机配合使用时，能提供廉价的战术导航，与红外成像技术和光学雷达相结合，将创造明天的灵巧武器。

4 我们的发展设想

微米/纳米技术是一项重要的军民两用技术。我国微米/纳米技术的研究开发工作尚处在起步阶段。研究内容主要集中在 MEMS 方面，并取得了一些初步成果。应根据国际发展趋势和我国的需求，选择我们经过努力可以较快见成果的军民两用的 MEMS 系统。

我国在惯性技术的发展上有比较好的技术基础，有一支高水平的队伍，并已提出了"九五"期间开展中等精度微型惯性仪表研究的初步计划。因此，选择 MIMU 作为"九五"微米/纳米技术预研计划的需求牵引是合适的。

"九五"期间以 MIMU 中的微型惯性仪表为主攻方向，围绕其中的关键技术，以及材料、特殊工艺和微型惯性仪表设计等方面开展工作，力争突破微型惯性仪表的关键技术。采取综合集成、伞形辐射的方式组织关键技术攻关。有的单位按 Top – Down 技术路线开展下行研究，少数单位沿 Bottom – Up 技术路线开展上行研究，并找出与 Top – Down 技术路线的结合点。

在 MEMS 加工方面有两条技术路线：一条是微电子标准工艺线的创新应用；另一条是 LIGA 技术路线。我们在微电子加工技术方面的基础较好一些，设备稍先进，增加必要的先进设备就可以开展 MIMU 的研究试验工作。选择微电子集成电路标准工艺线已形成优势的单位，建立微米/纳米技术加工中心（即微机械标准工艺线），实行开放，为 MIMU 加工服务。跟踪研究 LIGA 及其他高深宽比微机械加工工艺，同时培养这方面的人才。加强微米/纳米技术发展战略研究和微机理的理论研究，以利于创新。对目前看得比较准的加工中心建设、MIMU 研究、LIGA 跟踪等，尽快创造条件。其中加工中心和 LIGA 技术跟踪所需的技术改造，可列入国家重点实验室专项。

MIMU 是微传感器、微执行器和微控制电路的综合集成，是微机械、微光学与具有惯性测量功能的大规模集成电路芯片的综合集成。发展 MIMU 是面向综合集成的挑战。我们必须通过综合集成努力实现我们的目标。

参 考 文 献

［1］ 丁衡高. 面向 21 世纪的军民两用技术——微米/纳米技术//微米/纳米技术文集［M］. 北京：国防工业出版社，1994 年 11 月.

［2］ Proceedings of the Workshop on Microtechnologies and Applications to Space Systems，Held at the Jet Propulsion Laboratory on May 27 and 28，1992，JPL Publication 93 – 8. NASA – CR – 195688.

［3］ Ivor Brodie and Julius J Muray. The Physics of Micro/Nano – Fabrication. 1992 Plenum Press，New York.

［4］ Kaigham J Gabriel. Engineering Microscopic Machines. Scientific American，September 1995，150 – 153.

［5］ John Elwell. Progress on Micromechanical Inertial Instruments. AIAA – 91 – 2765 – CP，1482 – 1485.

［6］ Anthony Lawrence. The Micro Inertial Measurement Unit（MIMU）Using Advanced Inertial Sensors. Fourteenth Biennial Guidance Test Symposium，MSD – TR – 89 – 21 Volume 1，AD – A216 925MF，97 – 107.

微机电系统的科学研究与技术开发

(1997年9月)

摘 要 着重论述了当前国际上微机电系统的相关基础理论研究、微机电系统制造工艺技术的发展、微机电系统发展动态与趋势，最后强调了探索适合我国国情的微机电系统发展途径的重要性。

关键词 微机电系统；微机械；微惯性测量组合

1996年6月5日，作者曾在中国工程院的学术会议上，作了题为"微机电系统和微惯性测量组合"的报告。本文是在该报告稿基础上重新整理而成的。

1 微机电系统的相关基础理论研究

微电机系统（MEMS）可以分成多个独立的功能单元，输入的物理或化学信号由传感器转换为电信号，经过信号处理后，通过执行器与外界作用。微惯性测量组合（MIMU）由微陀螺、微加速度计及相关的专用集成电路组成，是一种重要的MEMS，主要用于运载平台和弹药的导航与制导。

MEMS开辟了一个新的技术领域。MEMS研究不仅涉及元件和系统的设计、材料、制造、测试、控制、集成、能源以及与外界的联

本文发表于《清华大学学报（自然科学版）》1997年第37卷第9期。

接等许多方面，还涉及微电子学、微机械学、微动力学、微流体学、微热力学、微摩擦学、微光学、材料学、物理学、化学、生物学等基础理论。

在尺寸微小的范围内，许多物理现象与宏观世界的情况有很大差别，可能产生若干新的效应，有必要逐步揭示各种效应的物理实质，对微观机理展开深入研究。力的尺寸效应和表面效应说明，宏观领域内作用微小的力和现象，在微观领域则有可能起重要作用。在微小尺寸领域，与特征尺寸 L 的高次方成比例的惯性力、电磁力（L^2）等的作用相对减小，而与尺寸的低次方成比例的粘性力、弹性力（L^3）、表面张力（L^1）、静电力（L^0）等的作用相对增大。这是 MEMS 常用静电力致动的理由。随着尺寸的减小，表面积（L^2）与体积（L^3）之比相对增大，因而热传导、化学反应等加速，表面间的摩擦阻力显著增大。材料性能和摩擦现象也受到制作工艺的影响。当然，MEMS 的理论研究不止于此，还有如微流体力学以及 MEMS 建模等问题。

MEMS 的系统集成包括微传感器或执行器与控制处理电路的集成、集成基座、系统分割、能源供给、接口与通信等等。微系统的测量涉及材料的机械性能、微构件和微系统参数与性能测试。微薄膜构件的弹性模量、泊松比、拉伸强度、残余内应力、破坏韧性、疲劳强度等与块材的不尽相同，需在测量的基础上建立微构件材料的数据库和微系统的数学力学模型，还应重视微缺陷的测量。目前的微系统控制都很简单，而微机器人是微系统研究的一个新目标，它将能进入人体和狭小空间进行工作，需要多变量的协调控制。

2　MEMS 制造工艺技术的发展

MEMS 制造技术首先是在微电子平面加工工艺基础上发展起来

的，后来又先后有了深反应离子刻蚀（DRIE）、LIGA 技术和准分子激光等多种工艺创新。在应用时这些工艺相互补充，各有所长。

2.1 微电子平面加工工艺

目前面市的一代 MEMS 具有简单、易于大规模生产的特点，因而价格便宜。这一优势来源于 MEMS 的制造采用与微电子器件和电路相同的材料（硅）和工艺（平面工艺），因而容易与控制电路集成。采用硅微电子平面工艺制造 MEMS 的一个典型的例子，是美国 Analog Devices 公司设计的一种加速度计，它基于电容的变化来敏感与测量加速度，电容板的刻蚀深度为 2 μm。该公司大量生产的这种硅微加速度计，把微传感器（机械部分）和集成电路（电信号源放大器、信号处理和自校正电路等等）集成在硅片上 3 mm×3 mm 的范围内。在一块 10 cm 的硅片上可同时制造成百个微加速度计。

用常规的微电子平面加工工艺制造的 MEMS、微电路和微结构的典型厚度在 2～10 μm 之间。当然，硅不是胜任各种需求的最好的材料。对一些特定应用而言，如微型行星齿轮，采用的结构材料为镍铁合金（NiFe）。

2.2 硅熔合键合与深反应离子刻蚀相结合的工艺

Analog Devices 公司的硅加速计由于电容板的厚度为 2 μm，使敏感加速度的精度受到限制。Stanford 大学和 Lucas NovaSensor 公司合作，正在利用硅熔合键合与深反应离子刻蚀结合起来的新加工工艺，制造一种精度更高的加速度计，电容板的宽度为 6 μm，间隔为 4 μm，厚度为 60 μm，深度比为 15∶1。这种新工艺把"表面"微机械加工与传统的"体"微机械加工的优点结合起来，即把一般集成电路制造工艺的设计灵活性和兼容性与体工艺的坚固性和三维成型能力结合起来。

对于微机械结构加工来说，这种结合的工艺包括两个主要工艺步骤，第一步用浅刻蚀和硅熔合键合形成嵌入的内腔，第二步进行深入腔体的深反应离子刻蚀。在两个步骤之间，用一般的集成电路工艺（如CMOS 工艺）制造信号处理电路。深反应离子刻蚀（DRIE）采用氯基或氟基等离子体可以刻蚀出近乎垂直壁面的深层结构。目前已能刻蚀出 200 μm 的深度。

2.3 LIGA 技术

LIGA 技术是 X 射线深层光刻、微电铸和微塑铸三个工艺的组合，是德国 Karlsruhe 研究中心微结构技术研究所开发的。

LIGA 工艺利用 X 光在聚合物中雕刻出深图案，然后对这些图案进行电镀，产生微型三维金属结构或金属模具。在这些模具中注入各种材料（包括多种聚合物），可以进行微机械装置的大批量生产。LIGA 工艺可制造任何线宽、形状的微结构，结构高度可高达 1 mm，线宽尺寸可小到 0.2 μm，微结构深宽比可达 50～500，表面粗糙度为 30 nm 的产品。工艺适用的材料包括聚合物、金属、合金、陶瓷等。目前用 LIGA 工艺研制的产品有：微涡轮、微泵、微加速度计、医用微型镊子、微光纤开关、微光谱仪等。

LIGA 工艺要求能量很高的同步辐射 X 光源，因此比一般刻蚀方法昂贵、费时。日本先进制造技术开发协会在 1992 年建立了 LIGA 技术委员会，成员包括 7 家私人公司、3 家国立实验室、1 所大学和 1 个微机械中心。这个委员会对 LIGA 技术的国际发展情况进行了一次持续 3 年的调查研究，并在研究的最后阶段就日本工程师、科学家和大学教授对 LIGA 的态度进行了问卷调查。这次调查的结论是：LIGA 技术的市场前景仍不明朗。尽管如此，美、日及西欧一些国家和地区仍把 LIGA 技术作为一项有希望的工艺给予一定重视。

当前，与 LIGA 技术相比较，硅微加工技术发展更成熟、更经济，更容易实现与微电子电路的综合集成。因而在现阶段，MEMS 工艺技术的发展，仍然具有以硅基技术为重点的特征。

2.4　准分子激光加工工艺

准分子激光器输出处在电磁波谱的紫外区域。几乎所有物质都吸收紫外光，因而准分子激光与材料之间存在有效的耦合效应。许多聚合物都吸收氯化氙准分子激光 308 nm 和氟化氙准分子激光 351 nm 波长的辐射；陶瓷和金属吸收氟化氪准分子激光 248 nm 波长的辐射；而玻璃、石英和钻石吸收氟化氩准分子激光 193 nm 波长的辐射。由于强的紫外光吸收，准分子激光辐射的整个能量就集中在材料的一个薄层内，高能量密度超过某个阈值时，就会发生冷烧蚀现象。控制激光脉冲数量即可获得所需的烧蚀（刻蚀）深度。

准分子激光器的短波长辐射使非常小的尺度的精确成像成为可能，而这正是微加工所要求的。更短的波长可以获取更小的结构尺度。准分子激光深度紫外光光刻的高端，用于制造分辨率接近 0.2 μm 的集成电路。

从 1975 年发明准分子激光器以来，科学家和工程师就致力于研究准分子激光在材料加工中的应用。早期的准分子激光可靠性不够，费用也不具竞争力。随着准分子激光技术的不断改进，现在情况已有很大改观。按照德国哥丁根 Lambda Physik 公司的技术，10 亿个激光脉冲的使用费用已从 1988 年的 15 万美元降至 1994 年的 2.4 万美元，而在可靠性方面，已达到几千小时的平均故障间隔时间。德国美茵兹微技术研究所（IMM）最近发展了使用准分子激光烧蚀与 LIGA 技术结合的新加工工艺。准分子激光用于加工电镀膜片初始三维结构的第一道工序。

3 MEMS 的发展动态与趋势

3.1 微型惯性器件及其组合的发展

1977 年，美国 Stanford 大学、加州大学伯克利分校和 Draper 实验室开始在硅片上采用微电子加工工艺生产微硅加速度计和微硅音叉陀螺。1978 年，美国 Northrop 公司开始探索在硅片上制造光波导的技术。1991 年研制出微型光学陀螺（MOG）样机，精度达到 10（°）/h。1979 年，美、日等国采用多功能集成光学芯片研制干涉型光纤陀螺。其基本工作原理是依据光的传播规律和萨格耐克相移效应，其输出相移和输入速度成正比。1989 年，Draper 实验室已研制出双框架式微硅陀螺。其基本工作原理是加速度计附加抖动装置。抖动装置有角振动和线振动两种，因而从本质上讲，微硅陀螺是一种振动式角速度传感器。已制成的加速度计和陀螺量程分别为（$10 \sim 10^5$）g 和 $50 \sim 500$（°）/s，室温下偏置稳定性为 1 mg 和 20（°）/h。1993 年 Draper 实验室将技术转让给 Rockwell 公司，实现了批量生产。Draper 实验室现已研制出由三个微硅陀螺、三个微硅加速度计和附加电子电路构成的 MIMU，每个陀螺和加速度计的边长都小于 1 mm，这 6 个传感器安装在立方体的 3 个正交平面上。包括专用集成电路在内的高密度封装，尺度为 2.0 cm×2.0 cm×0.5 cm，重 5 g，功率小于 1 W。

1994 年，Litton 公司、Honeywell 公司和 Draper 实验室成功地将光波导移相器做成多功能集成光路芯片，广泛用于多种光纤陀螺。

美国国防部高级研究计划局（DARPA）正在开发采用光纤陀螺的 MIMU 与全球定位系统（GPS）的组合系统。GPS 信号用于校正惯性漂移误差，当 GPS 信号被干扰后，惯性系统能自主工作。此项计划称"GPS 制导包"（GPS Guidance Package，GGP），GPS 接收机由 Litton

公司和 Rockwell Collins 分部合作开发，光纤陀螺则由 Texas Instruments 公司和 Honeywell 公司合作开发。今天，光纤陀螺惯导系统，如 Litton LN-210（包含三个光纤陀螺和三个微硅加速度计），正在替代导弹所用的环状激光陀螺系统，并将对军用和民用飞机的环状激光陀螺形成挑战。此项计划第二阶段的工作集中于减小尺度和重量。

1996 年 6 月 16 日至 25 日，美国导航学会第 52 届年会在波士顿举行。会议报告中涉及 MIMU 的内容有：海军 12.7 cm 炮弹制导用的 MEMS/GPS 组合系统（Draper 实验室），精密微机械陀螺的发展途径（俄罗斯应用力学研究所），用于先进战术导弹的 MEMS（Hughes 导弹系统公司），数字石英 IMU 的标定（Rockwell 公司），精密微机械惯性传感器（俄罗斯应用力学研究所），硅加速度计（Draper 实验室）和用于未来战术导弹或微型卫星的隧穿型微惯性传感器（Draper 实验室）。

3.2　DARPA 和 NASA 的 MEMS 应用计划

DARPA 电子技术办公室副主任、负责 MEMS 计划的加布里埃尔说，装有 MEMS 的精确弹药在两年内投入作战应用。今后 3 至 5 年内趋于成熟的几项技术包括：武器安全、解除保险和引爆；平台稳定；位置探测；生物医学装置和显示设备。小型分析仪器和人员/运载器导航系统 5 至 7 年内可以在市场上买到。敌我识别、有源结构和海量数据存储等应用在 7 年内也将成熟。他还说，完整的导航系统能从人的手掌那么大减小到芯片尺寸，价格也从 1 万美元降到 100 美元。

美国 RAND 公司受 DARPA 委托，1992 年 10 月至 12 月进行了"未来技术驱动军事作战革命"的研讨，93 名技术和军事专家经过认真工作，最后确定，以执行多种战场探测任务的十分小的系统、生物分子电子学、新军事后勤技术、计算机空间安全和防护以及提高单兵作战能力 5 个技术领域作为 DARPA 新研究倡议的候选项目。他们认为前三项是 DARPA 的技术机遇。

美国国防转轨战略的核心是发展两用技术，为此制订了技术再投资计划（TRP），主要目的在于推动政府同工业界的合作。TRP 由 DARPA、商业部国家标准与技术研究院、国家科学基金会、美国国家航空航天局（NASA）、陆军、空军、海军、能源部和运输部等 9 家组成的国防技术转轨委员会管理。该计划从 1993 年开始执行。MEMS 由于其鲜明的两用性及其在军民两用方面的作用，1995 年被列为 TRP 优先发展的 12 个关键技术领域之一，政府对 MEMS TRP–1 投资 500 万美元。该项目的经费由政府与项目参加单位对等投资。

NASA 于 1994 年 9 月制订了面向 21 世纪的"新盛世"计划，经过认真评选，确定开发和验证的 6 个方面的革命性的技术领域是：

1) 自主技术。航天器能自主完成导航控制、数据处理、故障判断及部分的重构和维修工作，以便大大减小对地面测控、通信等支持系统的依赖。

2) 微电子技术。在现有微型化的基础上进一步集成，实现革命性的突破。例如，将航天器所有分系统的全部电子线路和部件集成在一个仅为骰子般大小的体积内。

3) 仪器技术与结构。

4) 微机电系统。主要有效载荷的微型化，力求缩小各个科学仪器的体积（如研制微型集成相机和成像光谱仪等），目前已能将一台小型地震仪做成只有 25 美分硬币大小。

5) 通信系统。大力发展光波频段的通信技术，使通信接收机、发射机、天线、放大器、开关和传输线路等都有革命性的进步。

6) 分系统模块化和多功能系统。实现航天器各分系统的模块化、集成化和多功能。

3.3 微型卫星和汽车工业对 MEMS 的需求牵引

英国萨瑞卫星技术公司自 1981 年以来，设计、建造并发射了 12 颗微型卫星。实践证明，应用 MEMS 后，卫星体积更小，研制过程加快、研制费用更省。NASA 正在着手实施一项不足 1 亿美元的"发现号"微型卫星计划。此外，美国航空航天公司也在构思纳米卫星（或称"芯片级卫星"）的设计与制造。该公司估计本世纪末将研制成功成本为 10 万美元的纳米卫星，整个开发经费约 5 000 万美元。

除争夺军事技术领先地位外，潜在的市场需求也是争夺 MEMS 技术领先地位的一个重要因素。汽车工业对 MEMS 的需求就是一个很好的例子。现代汽车的电子含量以每年 10% 的比例增长，而传感器含量则以每年 20% 的比例增长。近几年汽车采用微加速度计和微压力传感器等 MEMS 器件，在节油和安全方面取得了明显效果。美国系统规划公司估计，1998 年美国法律将规定汽车都必须使用气袋，以保证紧急情况下的人身安全。届时将有 1 500 万辆汽车要装备气囊，加上节油等方面的需要，将形成一个很大的市场需求。受 DARPA 委托，系统规划公司进行的 MEMS 市场预测研究表明，全球 MEMS 市场将呈现迅速上升的趋势，1990 年 MEMS 的全球市场总额为 4.8 亿美元，1995 年为 15 亿美元，2000 年将增到 139 亿美元，而 139 亿美元的 MEMS 市场将带动产值为 1 000 亿美元的新的或改进的系统的市场。

4 探索适合我国国情的 MEMS 发展途径

我国的科技专家对 MEMS 在军民用方面的重要性已有共识。我国 MEMS 研究工作尚处于起步阶段，初步有了一支研究队伍，并取得一些成果。MEMS 要跟上世界发展步伐，跻身 21 世纪先进技术之列，需要做很多努力。需有关部门密切配合，选准发展途径，突破关键技术。

鉴于 MEMS 在推动国民经济和军事技术发展方面的重要性，国防科工委已把 MEMS 列入"九五"国防预研计划。按照"有限目标，突出重点，军民两用"的原则，"九五"期间把 MIMU 中的微惯性器件作为主要方向，围绕其中的关键技术，在材料、特殊工艺和微惯性器件设计等方面开展工作。采用目标牵引、重点突破、综合集成、伞形辐射的方式组织关键技术攻关。现阶段主要从微电子微细加工技术的创新应用入手，开发 MIMU 的惯性器件。同时注意加强微机理的理论研究，以利于创新和突破。

德国威斯巴登的 Arthur D. Little 国际管理咨询公司进行的研究表明，德国微电子工业发展迟缓的一个重要原因是组织分散，在高等级的化学、光学和成像分析系统，净化设备以及研究开发设施和制造公司之间，没有形成有效合作的网络，妨碍了创新过程，结果延缓了德国微电子的发展。正是基于这样的教训，开发 LIGA 技术，成了德国政府试图使德国从竞争微电子工业世界领先地位失败的阴影中振兴起来，夺取微机械领域领先地位的一项重大举措。X 射线扫描仪是实现 LIGA 精密深层光刻的基础。德国 JENOPTIK 公司已制造了 5 台用于 LIGA 技术的 X 射线扫描仪，其中 3 台在德国使用，1 台在法国使用，1 台在美国使用。正在制造的第 6 台将设置在我国台湾省。同时为了加快技术转移，Karlsruhe 研究中心负责把 LIGA 技术转移给德国私人工业。为此，它把所有专利转让给 MicroParts 公司，这是一家由钢铁、化学和电力工业公司联合经办的私人公司，销售用 LIGA 工艺制造的部件。MicroParts 公司已获许应用 LIGA 技术制造下一代喷墨打印机的喷嘴，这种新型打印机将具有 96 mm^{-1} 分辨率，喷嘴密度将是目前一代喷墨打印机的 4 倍。

DARPA 正在支持一项实验，即将美国各地的表面微加工机构集中在北卡罗来纳微电子中心（Microelectronics Center of North Carolina，MCNC）的半导体加工设施上进行加工。这项实验是为了在微电子工艺基础上建立一个共享的 MEMS 标准工艺线（foundry），以较少的费用为企业和学术机构定制少量微机械系统的样品。这与 DARPA 在微电子工业方面推行的做法相似。北卡罗来纳微电子中心曾是微电子技术发展进程中的一个重要研究开发基地。DARPA 将它改造为 MEMS foundry，是微机电系统技术发展方面国家计划指导作用重要性的又一个实例。

日本通产省从 1991 年开始执行为期 10 年的"微机械技术计划"。今年是计划执行的第 5 年。这是一项将微机械技术的开发置于日本国家优先地位的计划，10 年期间将投资 2.5 亿美元支持 24 家公司的研究工作，包括汽车零件、电子、医疗装置和机器人等工业部门。计划的目标有两大项：1）制造能在管道中行走的微机器人的作动器和传感器，这种机器人用于诊断与确定故障点；2）研究制作一种能深入人体大脑进行外科手术的携带照相机和其他传感器的智能导管。虽然计划中也包含了日立公司利用半导体微细加工技术制造汽车工业所需的加速度计，和住友电气公司用 LIGA 技术制造管道检测机器人使用的陶瓷麦克风的内容，但计划最显著的特征是强调通过非光刻的传统的机械加工工艺线（如金属与塑料部件的切削、研磨）实现机械的微型化。这是一条用大机器造小机器，用小机器造微机器，用微机器造 MEMS 的技术途径。支持此项计划的日本人相信，未来不只属于硅，硅仅是人们要使用的材料中的一种。

近年来，我国从事 MEMS 研究发展工作的专家分别对美国、日本、德国和法国相应领域的情况与经验进行了实地考察,并组织了交流和探讨。从作者了解的国际上开发 MEMS(日本称微机械,德国称微系统)的情况看,美国科学家侧重于在微电子技术的基础上,从微芯片上取得制造工艺的突破;日本则偏重从机械加工工艺实现微机械的制造;而德国的特色是在 LIGA 工艺的应用上取得了进展。我国科研人员应研究日、美、德三国的特点,结合我国的实际,探索合适的途径,以期取得更大的成就。

参 考 文 献

[1] Hogan H. Invasion of the micromachines. New Scientist,1996,150 (2036):28 - 33.

[2] Stix G. Micron machinations. Scientific American,1992,267 (5):72 - 80.

[3] Philip J. Klass,fiber - optic gyros now challenging laser gyros. Aviation Week & Space Technology,1996,145 (1):62 - 64.

[4] Carving Electrical,Mechanical components out of silicon. Signal,1996,50 (11):27 - 29.

[5] Casani E K,Wilson B. The new millennium program:technology development for the 21st century. AIAA,96 - 0696.

[6] 周兆英,叶雄英,李勇,等. 微型系统和微型制造技术 [J]. 微米纳米科学与技术,1996,2 (1):1 - 11.

Research and Development of Micro Electro Mechanical Systems

Ding Heng – gao

(Tsinghua University)

Abstract This paper presents the research on fundamental theory in the field of MEMS (Micro Electro Mechanical Systems). It also shows relevant fabrication technologies by pointing out the recent trends and developments of MEMS. Special emphasis is finally given to the approach of exploring, studying and developing MEMS in China.

Keywords Micro electro mechanical systems; Micromachines; Micro inertia measurement unit

微小卫星应用微小型技术有关问题的思考

(1997 年 11 月)

摘 要 本文着重论述了微小卫星在军事应用及航天器发展中的重要意义,分析了影响微小卫星性能、体积和重量等的主要因素,最后提出了微小卫星应用微小型技术特别是微机电系统应注意的若干问题。

关键词 微小卫星;MEMS(微机电系统)

1 关于航天器的发展

根据任务的需求,人们总希望随着科学技术的进步,选择那些性能好、研制生产过程快而费用少的航天器,这种选择是根据科学技术可能达到的水平,综合多种因素得出的。"需求牵引,技术推动"是航天事业发展的动力。日益增长的社会需求和科学技术上的新成就都有力地促进了航天器的发展。目前面临的突出矛盾是航天器的技术要求高,费用也十分昂贵。现在航天器每千克入轨后的发射费用高达五万至十万美元甚至更多,这是很惊人的。只有降低费用,提高费效比,才能使航天器有更大的发展。现在国外卫星发展的总趋势,是采用先

本文根据作者 1997 年 11 月 20 日在北京"微小卫星应用微小型技术"学术研讨会上所作特邀发言重新整理而成。

进的材料、工艺和设计思想，利用微小型技术进行优化设计，提高功能密度。我们应结合国情，跟踪研究国外有关卫星技术的发展问题。

传统的设计观念是优先考虑性能，即追求最好的性能，追求最大的有效载荷，发射费用则要到了设计后期才能确定。目前，从经济可承受性考虑，需要更新这种设计观念。在设计的开始阶段就应进行全成本核算，优先考虑性能价格比和风险度，否则就没有竞争力。

当前航天费用高，主要有两方面的原因。一是卫星本身昂贵；二是运载器、发射和测控服务费用高。解决航天费用高的问题，也应从这些方面入手。一方面从卫星本身考虑，提高功能，减轻重量，减小体积。在减少发射费用方面，可以通过一箭多星和多次重复使用的运载器来缓解。另外，自主技术值得重视，即自检测、自诊断和自主运行等，不需要地面站测控，从而降低费用。这一技术途径很重要，在军用和民用上都有重要的价值。据有关专家介绍说，英国萨瑞卫星技术公司（SSTL）研制成的控制系统（控制柜尺寸约为 $1.6\ m\times 0.8\ m\times 0.5\ m$）可以控制 8 颗星。地面站将信息输入到星上，星根据输入的信息进行自主测控，并通过与 GPS 交联，自主地修正轨道。但 GPS 的缺点是它在有些特殊地段不起作用。另外还有获取地磁信息的自主导航系统，但精度较低。从目前来看，这些方法都还不能确定卫星发射的入轨点，而要靠地面站来测定。

小卫星的定义有多种，通常以重量来衡量，例如将重量在 $100\sim 1\ 000\ kg$ 的卫星称为小卫星（Smallsat），$10\sim 100\ kg$ 的称为微小卫星（Microsat）、小于 $10\ kg$ 的称为纳米卫星（Nanosat）等。最近几年，功能密度也成为所谓的衡量指标，但由于技术进步，大卫星的功能密度同样可以提高，因而它并不足以体现小卫星的本质特点。综合国外情况，可将研制小卫星的技术路线分为两类。一类是以英国萨瑞卫星技术公司为代表，它主要是大量应用成熟的先进技术，使用商业化部件、设备，缩短研制周期，减少成本，发展实用型小卫星，具有适应性强、重量轻、功能好、研制周期短和价格低廉等一系列优点；另一类是以美国国家航空航天局"新千载计划"（New Millennium Program）为代表，重点研制发展新技术的技术试验型微小卫星，强调运用实验室的技术，不是现有的成熟技术，而是正在试验阶段的新技术，更强调技术创新，以求跳跃式地提高航天器的功能密度。这类卫星能快速完成单一任务，又能多星组合完成较复杂的航天任务。"新千载计划"的第一个目标是先研制一颗重约 100 kg 的技术验证型小卫星；第二个目标是在下世纪初研制重约 5～10 kg、以科学试验为主要目的、载有先进的微型仪器、可以连续地返回信息流的微小型卫星。掌握这些微小卫星及其试验的先进技术后，就可以向军用和民用的各个方面延伸应用。所以这个计划得到了美国上下各方的支持。

以军事应用为例，现代战争中的信息战和电子战越来越重要，而卫星在其中的作用也日益突出，它是军队平时和战时进行电子侦察、战情广播、通信联络的重要手段，是军队 C^3I 系统的重要组成部分。针对目前在轨的大卫星需要经费多、研制周期长，有人提出未来的军事卫星体系可采用分布式的小卫星星座方式，通过发射多星进行多星组网，具有单星功能有限、灵活、研制周期短、经费少、被破坏的卫星节点易于及时补充等特点，从而引起西方军事强国的关注。

2 影响微小卫星性能、体积和重量的主要因素

第一个是功率产生能力的限制。这里包括两点，其一是太阳能电池的能量与面积成正比。如果太阳能电池的面积是 10 cm×10 cm，能量转换效率为 15%，则它能产生的最大功率为 2 W，估计最终的轨道

平均功率可达到 1 W。如采用砷化钾,能量转换效率可达 17% 或 18%。其二就是蓄电池。能源系统需要蓄电池,短期工作的卫星,有的只需要蓄电池供电,不需太阳能电池。一般情况下,总的能源系统约占到卫星总重量的 30%,这些问题都有待能源系统的科技进步来解决。根据计算,在纳米级卫星上,如果仪器的各项技术水平能提高的话,1 W 的功率是够用的。

第二个是星载动力系统(空间推进系统)对卫星体积和重量的影响。可控卫星的动力系统(含推进器和燃料等)的重量和体积与卫星的大小及工作时间长短有关。据报道,未来十年航天任务中(不包括载人航天任务),发射后的航天器上的动力系统将占发射后总重量的 40%~50%,分析认为推进器在小卫星上的影响比在大卫星上要大。为了满足任务需求和显著减少星载动力系统的重量,多种新型的推进器已研制成功。如推力为 670 N 的双组元推进器,仅重 56 g,长 10 cm;姿控用的冷气喷气推进器,仅重 6 g,特征尺寸为 1 cm,推力为 4.4~44 mN。几种新概念的系统也正在开发之中,如"数字推进器",它是在单晶硅片上设计制造由多个小推进器组成的阵列式推进系统,重量很轻并且结构简单。

第三,对于长期运行的航天器,辐射屏蔽要求是限制航天器质量和尺寸的另一个因素。辐射剂量是高度、轨道倾角和有效屏蔽厚度的函数,在海平面上,没有任何辐射屏蔽的条件下,辐射剂量仅约 0.3 rad/年。但在 700 km 高的太阳同步轨道上,即使使用 3.5 mm 厚的铝或 4.3 mm 厚的硅作为屏蔽,其年辐射剂量也高达 1 000 rad。

第四,限制微小航天器尺寸的其他因素有:有效载荷和通信天线角分辨率,最小热质量和最小弹道系数。随着航天器尺寸的减小,表面积和体积之比、表面积和质量之比增大。故而轨道的衰减率比一般小卫星大,下降得就比较快,由此它生存的寿命比较短。这样,由于卫星在轨周期相对短,如果要求长期使用,就必须不断发射新的卫星予以补充,因此批量生产微小卫星就比较适合。美国 Motorola 公司在"铱"星系统上采用了新的批量加工技术。新的"铱"星生产线,计划一批投 50 到 60 颗星,周期由过去的一年缩短到 22 天,最终目标达到每 5 天装配一颗卫星,这是很大的技术进步。"铱"星因为轨道低,单星寿命不长,需要不断补充,批量生产可解决这个问题。近年来,科学技术的巨大进步和航天事业的发展需求促使人们研究、制造小卫星或微小卫星。在小卫星研制中,微小型技术已发挥了重要作用,在微小卫星的应用技术中,微小型技术特别是微机电系统(MEMS)尤为重要。

3 在微小卫星上应用 MEMS 应注意的问题

一般地说,MEMS 是在半导体平面工艺的基础上产生并发展起来的,但是它又不同于一般半导体工艺。从工艺上说,MEMS 的特点之一是要求在基片上得到高深宽比的构型。即得到深度较大的沟槽、孔等,且能保持垂直度和平滑,并且能尽量减小应力。

关于 MEMS 设计技术,不少国家正在研究之中。如美国的 MIT 开发了 MEMCAD 软件,但还未到成熟应用的阶段。这里结合空间应用,将 MEMS 系统集成到空间系统应考虑的几项重要工作简介如下:

1) 认识空间环境对微结构和系统的影响。这方面的工作包括对材料、结构性质及可能失效的模式的研究。在常规条件下这个问题也要解决,而在空间条件下这个问题就更为突出。

首先遇到的是冲击问题。现在美国 AD 公司研制的用于汽车上的微型加速度计,一般能承受 2 000 g。但在空间条件下可能要受到大得多的冲击。比如说，卫星上都有火工品，有的火工品点火时的冲击达上万 g。还有一个区别是，对于火工品的点火冲击，一般的大尺寸卫星上的机械节点有助于衰减高频率的冲击影响，而微小卫星的结构致密，只有少数节点，火工品与仪表间的距离很短，且不说是冲击，就是震动都能对微结构产生较大的影响。由于冲击或震动，系统的谐振频率可能就变了，从而将改变器件的性能。美国喷气推进实验室（JPL）研究的一种火星探路器，它在正式的着陆舱着陆以前要先释放一个微型着陆舱，在飞行过程中，火工品需点火 80 次，这样频繁的冲击必定给系统带来影响。

其次是材料的疲劳问题。虽然地面的应用也面临着疲劳问题，但空间系统面临各种效应，如冲击、温度等效应组合起来形成的疲劳问题则很突出。特别是空间应用的器件和系统，在有热循环时对寿命影响最大，我们还要考虑长寿命使用问题。材料的疲劳必将影响器件和系统，从而引起它的一些性能的变化。

最后是空间辐射对 MEMS 的影响。除总剂量外，还有瞬态效应和剂量率的影响。辐射对电子器件有影响，对机械器件和各种材料也有影响，在辐射环境中易受辐射损伤，致使性能退化。这些都是 MEMS 在空间应用中面临的重要问题，但是只要我们采取适当的方法，经过可靠的设计和封装，还是可以克服的。

2) 针对空间环境的需要研究新的设计方法和工具。MEMS 设计中的难点，在于力学性能和电性能之间的相互影响，如果不考虑相互影响，就不可能进行正确设计和分析。我们在工作中也遇到同样的问题。但目前还没有见到有效的设计工具。现在一般是通过试验，测试其功能，相应地修改设计来解决，而缺乏切合实际的设计方法。设计问题在一般的环境中存在，但空间环境中的问题更多、更复杂，要求更高。这对我们提出了新的课题。

3）关于封装问题。要解决好试验方法和封装工艺，国外在封装工艺上也遇到过问题，即封装后器件或系统的性能可能改变了。由于MEMS是多芯片器件，往往把它和其他的芯片放在一起封装。通过实践观察有的可行，有的则不行。特别是涉及封装以后带来的变化。有的要单独封装，即要根据特殊的条件和要求来封装，要根据器件的失效模式来考虑封装的工艺。这个问题看似简单，但具体如何封装，在什么条件下封装，封装了以后如何进行有效的监测，这都是要研究和探索的问题。

下面简要介绍JPL在有关MEMS方面的工作，作为我们研制工作的参考。

1）单轴加速度计，重6 g，灵敏度达10^{-7} g；微陀螺的重量是100 mg，精度1~10（°）/h的偏置稳定度。

2）利用隧穿技术的红外传感器，工作时无须冷却。

3）微型气象站，它是一个小型传感器的集合体。传感器中包括：用热电偶测温，从-70到70 ℃，精度可以到0.1 ℃；用激光多普勒测风速，精度可以达到0.1 m/s；还有测量露点的仪器。把这几个器件集合起来就可实现一个微型气象站。

4）微地震测量仪，重量100 g，在4 Hz带宽时灵敏度可以达10^{-9} g/Hz；

5）微推进器，这也是比较尖端的问题，流体通过直径为几微米的孔或管道，完全在界面层里流动。已有的评估流体动力学的手段和方法已不适用，所以要从理论上弄清楚，要研究流体的表面效应，研究微流量中间热传导问题和压力差问题。关于微阀、微泵，则还有如何防止泄漏等问题。

6）在基础研究工作方面，JPL 的做法也值得我们注意。比如他们开展了一些典型零件的试验工作，如悬臂梁、桥和薄膜等零件的各种特性研究，以及不同工艺对微结构的影响问题。又如在应力试验，包括火工品冲击试验中同样采取了典型零件的试验方法等，这些都值得我们参考。

以上主要就航天器的发展，小卫星、微小卫星的特点及空间系统应用 MEMS 可能存在的问题作了简要介绍。高新技术的迅速发展给我们提供了一个机遇，我们要抓住这个机遇。首先是通过跟踪、研究和分析，明确微小卫星的发展方向、技术途径及关键技术，大力协同，制定一个具体的实施计划。

微系统与微米/纳米技术及其发展

(2000年5月)

1　微系统和微米/纳米技术研究

美国物理学家、诺贝尔奖得主，理查德·P. 范曼（Richard. P. Feynman），在1959年12月的一次美国物理学会年会上预言，未来制造技术将沿着Top-Down（由大到小）和Bottom-Up（由小到大）的途径发展，并将由此引发科学、技术、工程和应用的巨大变革。Top-Down意指用大机器做出小机器，再做出更小的机器[1]，亦即可以做出越来越小的机器。Bottom-Up意指从最小的构造模块分子开始，进行物质的构筑，即"按照希望的方式排列原子"。范曼的论述标志着继微电子技术之后，微观世界的又一次革命性变革。

物理学、化学、材料科学和微电子技术的巨大进步，为微制造技术的飞跃提供了前提条件。恩格斯曾说过："一门科学提出的每一种见解，都包含着这门科学的术语的革命"。这一见解能帮助我们理解从传统制造技术到微制造技术，到微米/纳米技术，再到微系统乃至纳米技术的发展。

微系统和微米/纳米技术使人类在改造自然方面进入一个新的层次。开发物质潜在信息和结构潜力，将使单位体积物质储存和处理信

来自全国微米/纳米技术学术会议，会议时间：2000年6月14日。

息的能力实现又一次飞跃，导致人类认识和改造自然能力的重大突破。21世纪微系统将逐步从实验室走向实用化，必将对国民经济和国防建设产生重大影响。

(1) 微系统及其特点与特性

微系统是微米/纳米技术的重要组成部分，是微米/纳米技术当前发展重点之一，欧洲也将此称为微系统（micro-system），日本称为微机械（micro-machine），美国则叫作微机电系统（MEMS）。微系统是由机械、电子、光学及其他一些功能元件，集成在单片或多片芯片上构成的微型智能系统。一般具有电、机械、化学、生物、磁，或其他一些性质，有感知、处理和/或致动功能，且可进行批量生产。微系统可用来改变我们感知和控制物质世界的方法。图1为微系统的模型。

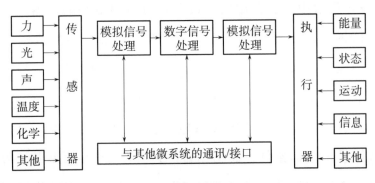

图 1　微系统模型

我认为微系统、微机械和微机电系统三种叫法没有本质差别，实质上都是一种微小型装置，都是采用微小型加工技术制造的。叫法上的差异反映了各国研究范围和技术手段的侧重。欧洲强调微系统技术的系统方面和多学科性；美国微电子水平有优势，自然由微芯片上取得制造工艺突破；日本机械加工水平比较高，则偏重于从机械加工工艺实现微机械制造。不少人认为称作微系统比较贴切，它反映了这种装置的实质，即从系统的角度来认识这门技术。但也有人认为微系统包含的范围太广了。

微制造技术是微系统技术的基础，包括设计、材料、工艺、测试技术，以及微机理研究；微器件是微系统的基本组成单元，包括微传感器、微执行器、微结构、信号处理与控制、接口和能源等，微系统就是建筑在上述基础上的，见图2。

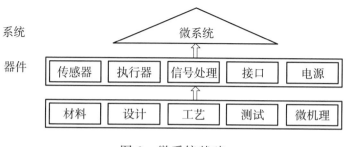

图2 微系统基础

要区分微系统和其应用系统，不要混淆起来。应用系统可以是较大的系统，它应用了微制造技术、微型器件或微系统，如机器人、微型飞行器和微卫星等。

微系统有许多独特的优点，如体积小、重量轻、性能稳定、可批量生产、性能一致性好、惯性小、功耗低、谐振频率高、响应时间短等，能完成大尺寸系统不能完成的任务。从研究开发来说，值得注意的是微系统的多学科综合性、突破性、带动性和创新性等特点。

多学科综合性 微系统是多学科前沿的综合，涉及微电子学、微机械学、微动力学、微流体学、微热力学、微摩擦学、微光学、材料学、物理学、化学、生物学等，还涉及元器件和系统的设计、制造、测试、控制、集成、能源、材料，以及与外界联接等许多方面，研究开发微系统需要诸多学科和技术专家密切配合。

突破性 微系统对传统大尺寸机电系统概念上是一个突破,特征尺寸小到微米和/或纳米量级,大尺寸机电系统的某些物理特性可能会不适用,需要进行微小尺寸机械的基理研究,突破传统理论的限制。首先要突破大尺寸机械的概念,开展微机理研究,包括尺度效应和表面效应、微流体力学效应、力学和热力学效应。还需开展微机械特性和微摩擦学、微光学、微系统表征与计量方法等研究。也应注意机电热、固液气、物理和化学等相结合的交叉学科的研究方法。

带动性 微系统是一项带动科学技术、国民经济和国家安全等方面发展的关键技术。将对21世纪的军事、科学技术、生产和生活方式产生革命性影响,并引发新兴产业和武器装备变革。采用微系统或微技术的新武器装备将改变以往武器装备越来越大的趋势,而出现一批微小型武器,如微型飞机、微型机器人、纳卫星和皮卫星等。未来的高科技战争中,基于微系统的军事技术装备必将起举足轻重的作用。

创新性 微系统本身就是科技创新的产物,其概念、机理、设计、加工等方面都不同于传统的大系统。设计方面,宏观世界的结构知识与法则,有些对微系统会不适用;材料方面,除微电子工业使用的材料外,微传感器和微执行器还需开发其他材料,如陶瓷和聚合物材料等,以及全面认识材料的机械特性,如残余应力、杨氏模量、泊松比、屈服力、极限应力和疲劳强度等;制造方面,在微电子工艺基础上,发展了体硅和表面硅加工工艺,后来又有了深反应离子刻蚀(DRIE)、LIGA、准分子激光和光成形等多种工艺创新。只有跨学科的合作才能产生创新的思路,微系统在国际上还起步不久,正是创新的大好机遇。

(2) 纳米技术与发展

纳米技术是一门多学科交叉的技术,研究在 0.1~100 nm 尺度空间电子、原子和分子的运动规律和特性。纳米技术本质在于以逐个原子的方式在分子层次上进行操作的能力，产生具有根本不同的新分子组织形式的大型结构。纳米技术涉及的材料和系统，由于它们的纳米级尺寸，结构和元件均呈现新的重大改进的物理学、化学和生物学性质以及现象与过程。可能导致诸如材料与制造、纳米电子学、医疗与卫生、环境与能源、化学与制造工业、生物技术与农业、计算与信息技术，以及国家安全等领域新的重大突破。它的效益还涉及其他许多部门应用、科学与教育和全球贸易与竞争等多个方面。

美国政府为寻求未来纳米技术的领先地位，2000年2月公布了《美国国家纳米技术倡议》[2]，并且放在发展科学技术最优先的地位，2000年的经费为2.7亿美元，2001年为4.95亿美元，比2000年增长2.45亿美元。美国总统克林顿2000年1月21日的一次讲话中说，"我的预算包括一项价值5亿美元的新的重要的国家纳米技术倡议……我们的研究目标有些可能要20年甚至更长时间才能达到，这正是这项倡议对于联邦政府具有重要作用的理由。"工业界领导人认为：下个世纪，纳米科学与技术将改变几乎每一种人造物体的特性。材料性能这样重大的改进及制造方式的变化，将引发一场工业革命。

《倡议》提出的2001财年纳米技术五项优先研究领域是：1）纳米科学与工程长期基础研究；2）纳米尺度材料结构单元与系统部件的合成和加工；3）纳米器件概念和系统体系结构研究；4）纳米结构材料与系统在制造、电力系统、能源、环境、国家安全和保健方面的应用；5）培养和造就新一代熟练技术人员。

美国在合成、化学制品和生物技术方面领先；日本纳米器件和凝固纳米结构有优势；欧洲在分散体、涂料和新型测量仪表方面较强。美国IBM公司已经朝制造纳米级发动机迈出了第一步，研制出一种可以在固体表面控制液体流动的旋转分子。桑迪亚实验室用自组装方法研制出一种表面巨大、具有完全规则纳米结构的超薄涂层，孔隙被设计成允许一定尺寸的分子通过。这种涂层可以用作化学传感器，检测分子的灵敏度比普通材料高 500 倍。我国在纳米材料和器件研究方面取得了初步成绩，如纳米硅薄膜、巴基管、多孔阳极氧化铝膜研究和生物芯片等。

2 微系统的国内外成就

我国 80 年代末开始微系统研究，90 年代末已有 40 多个单位的 50 多个研究小组，在新原理微器件、通用微器件、新工艺和测试技术，以及初步应用等方面取得显著进展，初步形成以下几个研究方向：(a) 微型惯性器件和微型惯性测量组合；(b) 微型传感器和微执行器；(c) 微流量器件和系统；(d) 生物传感器、生物芯片和微操作系统；(e) 微机器人和 (f) 硅和非硅加工工艺。

下面简单介绍国内外微机电系统研究和应用的成就及问题。

1) 惯性器件。清华大学、东南大学、复旦大学等正在研制微加速度计、微陀螺仪和微型惯性测量组合。清华大学 1997 年年底研究成功扭摆式微加速度计，1998 年和 1999 年又先后研究成功振动轮式微陀螺仪和梳齿式微加速度计。梳齿式微加速度计精度达 mg 级，可抗 5 000 g 冲击。掌握了 10^{-18} F 电容的测量技术，设计了高动态力平衡回路，具有较高的零偏稳定性。振动轮式微陀螺仪已经能够输出敏感信号。利用引进的微加速度计和微陀螺正在研制微型惯性测量组合样机。

美国 Draper 实验室的微加速度计精度优于亚 mg，微陀螺仪优于 10 (°)/h；微型惯性测量组合为 1~10 (°)/h。现正在研制微型惯性测量组合与 GPS 的组合装置，关键技术之一是 GPS 接收机的微型化。

2) 微传感器和微致动器。我国已研制出压阻式、电容式、力平衡式等多种原理、结构和性能的微压力传感器及压力-温度复合传感器,且可小批量生产。还研制出微型谐振传感器、微型振动传感器、微型触觉传感器、微型真空力敏传感器和隧穿红外探测器等。微驱动器也已研制了静电微马达、电磁微马达、压电微马达、微型电磁悬浮直线电机等。上海交通大学的1mm电磁微马达可连续工作1小时、寿命1年,力矩1.5 μNm,转速2 000 r/min,重12.5 g。

我们去年在瑞士电子和微技术中心看到,他们研制的一种微型激光器阵列(芯片),可以垂直发射激光,效率高达$(29.3\pm1.2)\%$;芯片相机有数字式和模拟式两种,采用CMOS电路。芯片尺寸为50 mm×70 mm,也有小至6.4 mm×4.8 mm的。还有激光成型模具,采用复制法可以大批量生产。

3) 微流量器件和系统。我国已研制出多种结构的热致动微泵、压电致动微泵和小型泵。微泵最大流量为365 μL/min,最大背压118 cm水柱。还研制出多种微型阀及气体、液体微流量传感器等。研制出喷嘴直径为20～50 μm电阻电热式微推进器微喷嘴样机,推力为1.2 mN。美国麻省理工学院正在研制燃气涡轮发动机、涡轮发电机和火箭发动机。他们研制的一种微型涡轮喷气发动机外径为12 mm、长3 mm、压比4、转速2.4×10^6 r/min、功率输出16 W、推力0.125 N、重量1 g、燃油消耗7 g/h,预计2001年进行安装新发动机的无人机飞行试验,如图3所示。

4) 生物传感器、生物芯片和微仪器。我国研制了微电泳芯片、DNA芯片、微光谱分析系统和微型机械光开关等。生物芯片是指在芯片上组装成千上万不同的DNA或蛋白质等生物分子微阵列,实现生物分子信息大规模检测,又称芯片实验室(lab on chip)。生物芯片是

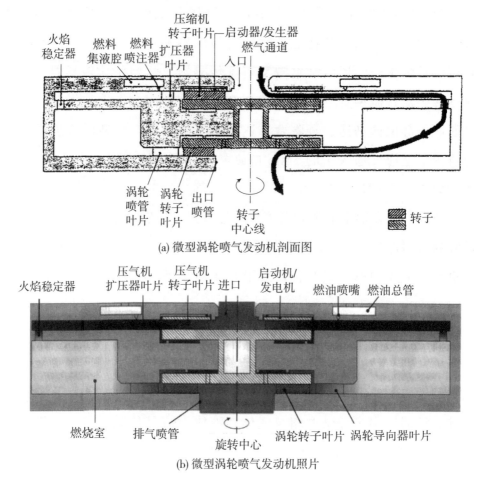

图 3　微型涡轮喷气发动机剖面图和微型涡轮喷气发动机照片

微系统的一个重要发展方向。东南大学 DNA 微阵列芯片的空间分辨率为 30 $\mu m \times$ 30 μm，集成度达 6.5536×10^4 个/cm^2，每步合成正确率为 99.5% 以上。

5) 应用系统。微系统推动了微小型化武器装备的发展。当前，正在开展微型航空器、微机器人、纳卫星和皮卫星等应用系统研究，并以此带动微系统技术和其他微技术的进一步发展。

上海大学研究了两种细小管道内爬行微机器人，一种是细小管道

(φ10 mm）作业微机器人，另一种是电磁力驱动的管道（φ20 mm）微机器人。清华大学、南开大学、广东工业大学等也在从事微机器人研究工作。今年初，日本电子公司宣布研制出一种蚂蚁大小的微机器人。这种微机器人长 5 mm、宽 9 mm、高 6.5 mm，重量仅 0.42 g，可提起比它重一倍的物体，并能以 2 mm/s 的速度移动。可在发电厂的狭窄管道周围爬行、检查以至修理管道。

1999 年 7 月，美国国防高级研究计划局对 4 种微型航空器进行了飞行检验。微型航空器将作为排级装备由战场上的单个士兵操作，用来提供所需的局部侦察或其他传感器信息，还可以用于标识、目标定位和通信，最终也可能用作武器。

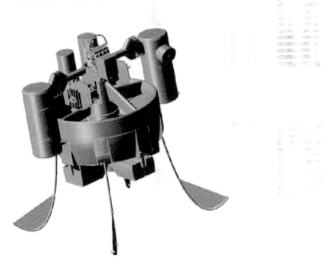

图 4　15 cm 的微型飞机

皮/纳卫星等航空航天领域应用的微系统主要有：微加速度计（图 5）、微陀螺、红外传感器、微磁强计和微气象台站等。今年 1 月 26 日，美国把一对绳系"皮卫星"释放到低地球轨道。每个皮卫星重量不足 230 g，尺寸为 10.2 cm×7.6 cm×2.5 cm。两个皮卫星用 30 m 长的绳子系留在能够进行微波通信的距离内。这次发射的皮卫星的基本任务是验证微机电系统技术、用微机电开关阵列进行无线电频率切换试验,以及进行彼此在轨通信和与固定在地面 50 m 抛物面天线上的第三颗皮卫星进行通信试验。

图 5　纳卫星用的微加速度计

由于微米/纳米技术和微系统技术本身的复杂性,尤其是不同于大尺寸系统的机理问题,使得其发展起来有很多困难,主要是:1) 微机理研究涉及多学科,需要多学科、多技术人才群和各种基础研究设施;2) 加工工艺多样化和设计手段相对落后;3) 产业化进展缓慢,仍以大学和研究所为主。可能是高风险和回报不确定性的原因,企业界尚未参加进来。我国微系统发展除存在类似问题外,还存在投入不足、力量分散、研究队伍后继乏人和缺少全国性计划等问题。

3　对我国进一步发展微系统技术的意见

这里,简要谈谈对我国进一步发展微系统技术的几点意见。

(1) 跟踪世界先进水平，密切关注发展前沿；开展发展战略研究，选准突破口和技术途径

我国微系统技术要赶上世界先进水平，首先要全面了解世界微系统技术发展情况，知道先进水平是什么，发展前沿在哪里。在此基础上紧密结合我国实际，开展发展战略研究，做好技术发展的顶层设计，确定突破口和技术途径。吸取、借鉴国外行之有效的方面，避免走弯路。我们一方面要重点加强前瞻性和创新性基础研究，抓住技术发展的生长点；另一方面要重点加强国民经济和军事装备急需的关键技术研究，促进微系统技术产业化。

(2) 坚持"需求牵引，技术推动"和"有所为，有所不为"的原则

首先，要根据我国国情，坚持"需求牵引，技术推动"和"有所为，有所不为"的原则。以国民经济和国防需要的迫切程度为牵引，从技术层面上研究确定微系统领域的研究开发项目。对国民经济和社会发展有重大影响的微系统要有所为。对打赢未来高技术条件下局部战争急需的微系统要有所为，微系统在武器装备方面的应用大致有两方面，一是采取技术嵌入方式，对现有武器系统进行改进，提高其能力和延长服役时间，例如，开发弹药引信所需的微型惯性测量组合等；二是为新武器系统研制提供技术储备，如微型航空器、微机器人、纳卫星和皮卫星和弹道武器等所需的各种类型的微系统。

目前研究重点应为：微型惯性器件（制导传感器）及微型惯性测量组合，可供民用，亦可用于精确打击武器和弹道修正引信；生物微系统和微分析仪器；信息微系统，用于通信、电子/信息等；微型动力系统，包括微推进器、微型电池等。此外，还有共性关键技术，纳米技术基础及应用，集成应用和微机理研究等。

(3) 重视基础技术，开展微机理研究，保持发展后劲

微制造技术水平低是我们微系统研究落后国外的主要原因之一。靠技术引进不可能从根本上缩短与国外先进水平的差距。只有坚实的基础技术支撑才能有技术创新，才会有自己的知识产权。基础技术是发展微系统技术的"重中之重"。今天的基础技术也将为明天的创新发展奠定基础。

微米/纳米技术基础及微机理研究是探索微系统发展的新思想、新概念、新原理、新方法等研究活动。我们工作中已感到某些基础研究的制约。重视基础研究，开展微机理研究实为当务之急，时不我待。为此，需发展介于宏观与微观之间的研究方法，例如宏微观力学、宏微观热力学等。还应注意机电热、固液气、物理和化学等相结合的交叉学科的研究方法。也需开展微机械特性和微摩擦学、微光学、微系统表征与计量方法等基础研究。探索各种可能的应用时，还需认真研究环境可能产生的损伤机理等。

(4) 统筹规划，发挥优势，大力协同

如何组织微系统科学研究和技术开发，是目前必须着手解决的问题。国家那么多机构、那么多计划，要力避重复，不要各自为战。国外大体上有三种模式：第一种是国家支持搞基础研究，工业界支持有关方面提出的专项开发计划，成立一个单位集中搞，如瑞士电子和微技术中心；第二种是由国家成立一个组织机构统筹安排，如日本的微机械中心，国家和企业的项目都由它选择承担单位并与之签订合同；第三种是美国的模式，就是各搞各的，谁有条件谁就搞。从我国的实际情况出发，统筹规划，发挥优势，大力协同为好。

历史上，美国曾为遏止半导体技术和工业下滑，并保持半导体技术坚实的工业基础，1987年由政府/工业界共同成立工业界和国防部

半导体制造技术集团（Sematech），把半导体工业界集合起来，共同开发半导体技术，由政府与工业界共同投资。主要任务是发展和增强半导体工艺技术与工艺设备及材料。工艺技术开发的任务为：1）研究开发高级半导体制造技术；2）在试验生产线上试验和验证这些技术；3）利用这些新开发的技术生产多种产品。Sematech强调"竞争前"的技术合作，即在产品研制前的技术开发阶段各成员公司通力合作，集中人力和财力，提高投资效率，避免在同一水平上重复；进入产品研制阶段再展开竞争。例如，成员公司在共同开发的 $0.35~\mu m$ 技术的基础上竞争DRAM产品。Sematech使美国半导体技术能力大大加强，半导体工业出现新的转机。其成功的关键在于：一是政府强有力的支持；二是技术开发形成合力、共同攻关。

MEMS 技术发展战略研究中需考虑的几个主要问题

(2000 年)

今年三月份，国家科技部开始着手组织研究 MEMS 技术及产业的发展问题，在广泛征求专家和有关部门意见的基础上，于五月十五日正式成立了 MEMS 技术的发展战略研究专家组，其任务是用半年左右的时间完成 MEMS 技术及产业的战略软科学研究。今天利用在上海召开 MEMS 学术会议的机会，与同志们探讨一下 MEMS 技术发展战略研究中应考虑的几个主要问题，目的是广泛听取大家的意见，把发展战略研究工作做好。下面根据科技部〔2000〕148 号文件精神，结合我国的实际情况，主要谈五个方面的问题。

1 MEMS 发展战略及指导思想

1.1 发展 MEMS 的重要意义及指导思想

微机电系统（MEMS）是在微电子技术的基础上发展而来、融合微电子技术和精密机械加工等多种微加工技术，并应用现代控制技术构成的微型系统。它的发展将对 21 世纪的人类生产和生活方式产生革命性的影响，并在未来高科技战争中起举足轻重的作用，成为关系国民经济发展和国家安全保障的战略高技术。

2000 年在 MEMS 学术会议上关于 MEMS 技术发展战略研究的讲话。

"发展高技术，实现产业化"，"在高技术领域占有一席之地"，这是微机电系统技术发展战略的指导思想。MEMS 技术和其他高技术一样，具有高度的竞争性和时效性，因此必须从一开始就抓紧、抓好。我们要抓住机遇，有所作为，探索发展中国 MEMS 技术的途径，迅速改变目前相对落后的状况，早日在 MEMS 技术领域有一席之地。

1.2 MEMS 技术发展战略

针对我国 MEMS 技术研究发展现状，应有重点地发展我国的 MEMS 技术，并尽快形成我国的 MEMS 产业。其发展战略拟考虑：

1) 集中决策，大力协同，以形成有市场竞争力的产业为中心，以国民经济发展和国家安全为目标，走技术创新和国际化的发展道路。

2) 本着"有所为，有所不为"的指导思想，重点发展国民经济和国家安全急需的产品和军民结合的两用技术，努力使 MEMS 的研究成果向商品化转化。

3) 跟踪世界微机电系统技术的发展方向，加强多学科的集成，特别是与微电子及软件产业、生物领域的结合。强化引进、消化、吸收和创新，努力实现技术跨越。

4) 加强前瞻性和创新性基础研究，占领技术的战略制高点，为 MEMS 的深入发展提供可持续的技术保障，以保证 MEMS 技术在高水平上持续发展。

5) 大力发展 MEMS 产业化，实现以产品应用促进技术的发展，实现 MEMS 技术的可持续发展。

6) 发展 MEMS 产业不能由国家包办，应遵从市场规律，充分利用社会力量。国家宜采用风险资金投入的方式，对有商品化前途的产品进行引导，同时鼓励多种资金投入，以股份制的方式调动各方力量共同发展 MEMS 产业。

1.3 发展战略研究中需要注意的几点

发展战略研究必须贯彻改革创新的精神。要从掌握好一切有利于 MEMS 技术发展的原则出发，深入研究改革创新的具体措施和办法，大力提倡探索、创新和冒险的精神，不人云亦云，不迁就现状。

发展战略研究要正确处理好重点与一般、明确目标和有限探索性目标的关系，有所为，有所不为。对看准了的目标，必须集中人力、物力和财力，搞好资源优化配置，尽快地出成果并形成产业，而对探索性的基础研究、应用研究项目可以采取放开一点、多一点的方式，但要注意适时筛选、集中，进行滚动式管理。

发展战略研究要坚持市场导向和应用导向的方针，虽然 863 计划不包括产业化，但 MEMS 发展战略研究，不仅要包括产业化，而且要以它为最终目标。

科学试验是探索未知，在成功之前遇到困难、失败是正常的，高技术具有高度的风险性，要容许失败，在失败中得到有益的经验和教训，激励人们勇攀科学高峰。

发展战略研究要充分认识到人才在技术和产业发展中的重要地位，要研究人才培养的机制和具体办法，在人才使用培养上创新。人才要在实践中培养，工作中识别，人要用其所长。例如，我们大家在发展战略研究中可以发现人才，为将来的专家委员会提供人选。

2　MEMS 技术发展的目标、方向、重点和技术发展方针

1) 发展某一产业、某一技术必须有明确的战略，首先必须有发展的目标、方向、重点和技术路线，从而为发展指明方向，否则产业和技术创新将无法衔接和持续，在分散、孤立的状态下进行，形不成合

力。从过去的经验看，集群的系统技术创新对产业和技术发展有重要的推动作用，这就体现了集中决策、大力协同的重要性。

当代最重要的科技方向是信息技术、生物技术及纳米技术等，最近华裔美籍科学家在南京大学的讲话中特别提到了微小型化问题，指出它是将来最重要的发展趋势之一。微机电技术和纳米技术是我们的发展方向，从国际市场技术分类及我们召开的几次座谈会和调查研究来看，可以考虑将生物微系统、信息微系统和传感器微系统作为我国MEMS技术发展的重点，它们在我国比较有条件，有可能较快地进入产业化，与此相关的微型动力系统也应作为重点，至于这些微系统中的哪些项目应首先进行突破，则应根据需求和条件来选择，应广泛听取大家的意见。

2) 任何一个微系统都是多种技术的集成，我们要密切关注其他学科、专业的发展和最新成就。从技术发展趋势看，技术之间的交叉融合产生的创新机会较多，高新技术都是如此，因此超越某一学科、某一专业、某一产业技术内部的交叉融合，更有利于技术进步，有利于创新。

产业技术创新的趋势是企业越来越重视从外部获取技术，一个企业，无论规模多大，也不能适应目前技术相互交叉、相互渗透、新技术层出不穷、技术生命周期短、开发风险加大的趋势。引进、消化、吸收、创新，洋为中用，符合产业技术发展规律，可以提高技术起点，加快创新和发展。

技术创新的扩散是使更多部门都用上，这对经济的贡献度比一般的技术创新贡献度更大。某个单位、某个小组及某个课题的技术创新有可能作出更大贡献。

3) 以需求为牵引，对基础研究、产品攻关和产业化工作统筹安排，技术上采取"目标产品带动关键技术，系统研究带动元器件开发，具体任务带动学科发展"，这是 MEMS 技术发展的方针。

发展 MEMS 技术，要科学界定开发研究、应用研究和基础研究的内涵和相互关系。我个人认为，开发研究是指攻克目标产品的关键技术，为产业化提供可靠的产品关键技术储备。应用研究到一定水平，一般在实现原理模型、探索到主要关键技术后即可转入开发研究。应用研究一是指探索新系统、新器件，弄清主要关键技术，构建出原理模型；二是对技术基础进行研究，对设计、材料、工艺、测试、控制等技术开展应用研究，为技术开发提供技术基础。基础研究一是指微机理研究，重点围绕应用研究中的微机理问题；二是指纳米技术研究。以上界定是否科学，请大家讨论。

4) 政府在促进技术进步的机制上的作用，关键在于制定方针、政策、提供技术基础设施和产业的共性技术，这项工作应有远见地尽早安排。

强化科技与经济的结合要贯彻始终，重点是当前的技术产业化，但对下一步及长远也要有部署，要注意产业的共性技术问题，要加强前瞻性、创新性基础研究，占领技术上的战略制高点，为持续发展做好接力，做好准备。为此要选择高水平的代表性的技术产品为对象，攻克技术难关，以促进共性技术和前瞻性、创新性基础研究的发展，其实质是为新产品做好技术先导。

3　面向需求、面向市场

根据指导思想，要实行"需求牵引，技术推动"的方针，只有需求才能通过产品牵引 MEMS 技术的发展，而 MEMS 技术的发展和提高将大力推动新产品的诞生。这就是技术上采取目标产品带动关键技术、系统研究带动器件开发、具体任务带动学科发展的方式，以最终建立我国的 MEMS 产业。

3.1 国际上的市场分析

近年来 MEMS 技术专利呈指数级上升，技术推动的力量在促使产业化的到来，SPC（System Planning Corporation）1999 年发布的题为 *Emerging Applications and Markets* 预测报告中关于 MEMS 技术市场的有关结论主要包括：1）多数产业观察家预测在下个 5 年之内 MEMS 销售额将有显著增长。2）MEMS 市场的宠儿包括：汽车、信息技术领域中的压力传感器、加速度计、微陀螺、打印机喷墨头及硬盘驱动器头。3）化学测试系统诸如芯片上的实验室（LOC）等将是潜在的市场应用。其他还包括微陀螺级的传感器、显示器及基于 MEMS 的 RF 通信器件。4）不同于半导体工业，MEMS 工业界没有搜集和散布 MEMS 市场信息的系统化的途径，使得许多从事 MEMS 业务的公司经常缺少必要的获得商业机会的市场信息。5）到目前为止，MEMS 的军事应用开发还很有限。众多公司已经认识到在这方面大有作为，特别是基于 MEMS 的惯性测量组合。

SPC 还指出了 MEMS 产业化可能遇到的一些问题：1）封装（MEMS 器件成本的 75% 或更多是投入在封装和测试上）；2）缺少标准化测试方法；3）将传感器和相关电路集成在同一个芯片上遇到的困难；4）传统的缩比法不适用于微观尺度。据 SPC 统计，1996 年全球 MEMS 市场销售额为 1 675～2 940 百万美元，而到 2003 年将为 6 500～11 540 百万美元，增长率为 20%～30%。图 1 和图 2 则分别为 SPC 提供的 1996 年与 2003 年不同类应用所占的市场销售份额。

3.2 市场研究的若干途径

市场研究需搞好几个结合，即军与民的结合；新产业与传统产业

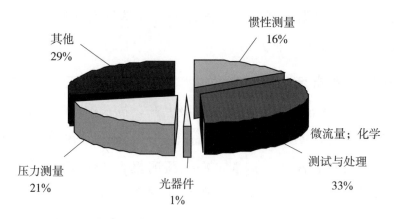

图 1 按技术分类 1996 年 MEMS 市场销售估计

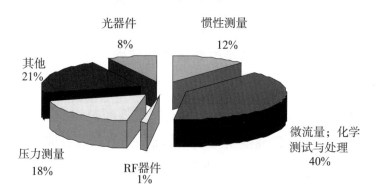

图 2 按技术分类 2003 年 MEMS 市场销售预测

改造之间的结合；国内与国外市场的结合；近期、中期和远期市场的结合。具体途径包括：1）了解各行业发展计划和需求。这是市场研究的基本要求，即所谓需求拉动，通过了解和掌握各行业的发展计划和需求，可以分析 MEMS 技术在这些行业中的切入点，从而有效地开发出富有针对性的技术和产品，可以有效地减少开发风险和成本，因而这种方法特别受到产业界的欢迎，例如韩国的三星、LG 公司等专门设立了研究客户发展计划和需求的部门。2）参与讨论提出技术发展可能性建议。开发者内部以及开发者与用户之间必须进行多次反复和有效的沟通，特别是在开发 MEMS 新的应用时，多方之间积极的协商和讨论有助于开拓思路，寻找到技术发展的正确途径。3) 调查研究替代已有产品的技术，这是一个很重要的途径，因为有关的产值预计主要通过替代产品来计算，已有产品通常已经占据了一定的市场份额，如果通过应用 MEMS 技术对现有产品或传统产业进行改造或替换，显然市场份额会有增无减，而且可以增强现有用户对应用新技术和新产品的信心。4) 向产业部门介绍可能有的技术创新项目，即所谓技术推动，产业部门为了保持和获得市场竞争的领先位置，具有高度的技术敏感性，因为技术推动所产生的市场效应一旦成功，则可以产生巨大的经济和社会效益。通过向产业界推广和介绍 MEMS 技术，可以迅速将真正具有应用前景的 MEMS 新技术和新产品推向市场，实现产业的跨越式成长。

3.3 抓好应用和试用是面向需求、开发市场的重要环节

MEMS 科学研究和技术开发不能停留在样品上，目标是产品，这是一个基本点。项目开发出来后，就要送去试用、应用，为下一步的改进提高和商品化铺路。这方面比较好的例子有：清华大学高钟毓教授与主题办办公室签了合同，向他们提供自行研制的 3 轴、±30 g、精度 1 mg 的微机械加速度计，年底以前争取试用，此外兵器 26 所准备将三条微电子工艺线中的一条改造成微机械加工线，年底可达项目要求；另一个例子是东南大学陆祖宏教授研制的生物芯片，目前组建了南京益来基因医学有限公司，以市场为导向，合理配置社会资源，实现技术优势向产品优势快速转化，优先推广个人识别芯片、PCR 基因芯片检测系统、促红细胞生成素（ELISA）检测试剂盒等。个人识别芯片将首先在公安系统进行推广，用于犯罪嫌疑人的 DNA 鉴定。PCR 基因芯片检测系统可以配合芯片同时进行推广，并可以推广到各

大科研院所以及医院等需要进行基因扩增研究的单位部门。促红细胞生成素检测试剂盒主要配合国内促红细胞生成素生产的迅猛发展来推广，用于病人促红细胞生成素指标的检查。以后再逐步推广感染性疾病基因芯片等其他产品。目前益来公司准备与南京医科大学、华西医科大学、江苏省公安厅联合开展个人识别工作；并与南京妇幼保健医院联系，建立南京产前筛查中心。

可能还有一些科研与应用结合得比较好的例子，限于时间关系这里不一一列举。从最近了解的情况看，生物芯片、信息微系统（主要是 RF MEMS 和光开关）与微传感器（主要是微加速度计和微陀螺）有可能是我国较早进入产业化的行业。

3.4 市场是检验产品的最佳场所

MEMS 产品必须用市场经济的办法来营销，市场是检验产品的最佳场所。不可能由国家来包分配，但各有关单位可以主动联合，利用各自特点，扬长避短，大力协同去开发产品，开拓市场，共占市场份额。即使是军品，由于数量有限，也只能通过竞争去争取。

4 对微米/纳米、微系统技术的几点认识

1) 微电子技术及其在其基础上发展起来的微系统技术是微米/纳米技术的主要方向。从工程技术的观点看，微米/纳米技术是一个发展过程，微米技术与纳米技术是有所区别又有联系的宽广技术领域，是指微米级（0.1～100 μm）到纳米级（0.1～100 nm）的材料、设计、制造、测量、控制和应用的技术。在技术上，从微米量级发展到纳米量级是一次飞跃。

2) 微制造技术是微系统、微机电系统技术的基础。微制造技术是微系统技术的基础，它包括：设计、材料、工艺、测试及控制技术，

以及微机理研究；微器件是微系统的基本组成单元，包括：微传感器、微执行器、微结构、信号处理与控制、接口和能源等；微系统建筑在上述基础上（见图3）。

由此可见，如果微制造技术上不去，一切都无从谈起，它是技术应用研究中的重中之重。

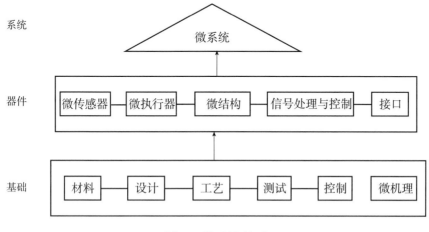

图 3 微系统基础

3）微系统技术是在半导体平面工艺的基础上发展起来的，但又不同于平面工艺。特别是三维体加工技术、应力释放、表面质量、尺寸精度与集成电路兼容工艺以及封装测试等后续工艺，没有稳定可靠的工艺，就制造不出高质量的批量产品。

4）从机理上看，微米量级的器件仍遵循宏观世界的规律，但又在不同程度上显示出微小事件的特征，而纳米尺度的现象和器件的特性与客观世界的规律有很大差异。要重视和加强微机理的研究，探索微观物质结构的机理和规律从而完善微系统，甚至可以探索到新原理、新现象和新规律。为此，需发展介于宏观与微观之间的研究方法，例如宏微观力学、宏微观热力学等。还应注意机电热、固液气、物理和化学等相结合的交叉学科的研究方法。也需开展微机械特性和微摩擦学、微光学、微系统表征与计量方法等基础研究。探索各种可能的应用时，还需认真研究环境可能产生的损伤机理等。

如果说微制造技术是微系统技术的技术基础，微机理可以看成是微系统技术的理论基础的重要组成部分。只有加强前瞻性和创新性基础研究，才有可能占据技术的战略制高点，使我国在这个领域占有一席之地。

5) 纳米技术是发展方向，要安排一定的基础研究工作。结合当前MEMS的发展需要，对特殊材料如薄膜材料的发展要引起重视。

6) 微系统技术研究涉及范围广，需要大力协同。由于投入大，决定了只可能搞几个各具特色的点，充分利用已有条件，钱要花在刀刃上，搞好资源优化配置，不搞重复建设。

5　集中决策、强化管理、大力协同

5.1　我国的发展现状和存在的问题

我国从 80 年代末开始了微型机电系统的研究，从跟踪国外研究开始，如硅微型压力传感器、微型电机、微型泵。十年来研究队伍逐步扩大，90 年代末已有 40 多个单位的 50 多个研究小组，在新原理微器件、通用微器件、新的工艺和测试技术，以及初步应用等方面取得了显著的进展。

微型机电系统研究在我国已形成了以下几个主要研究方向：

(1) 微型惯性器件和惯性测量组合（MIMU）

已研制出梳齿式电容等多种结构和原理的微型加速度计，器件精度达 mg 级，可抗 5 000 g 的冲击。掌握了 10^{-18} F 电容测量，设计了高动态力平衡回路，具有较高的零偏稳定性。已研制了振动轮式微机械陀螺、正交复合梁压阻微机械陀螺,开展了微型光波导陀螺的研究等。振动轮式微机械陀螺已有敏感信号,压阻微机械陀螺在大气中其角速度灵敏度为 0.22 V/[(°)·s]。利用引进的微加速度计和微陀螺,研制了 MIMU 样机。

(2) 微型传感器和致动器

已研制出压阻式、电容式、力平衡等多种原理、结构和性能的微型压力传感器及压力-温度复合传感器,微型压力传感器已有小批量生产。还研制出了微型谐振传感器、微型振动传感器、微型触觉传感器、微型真空力敏传感器、微型硅电容式麦克风、微型湿度传感器、隧穿红外探测器等。微驱动器方面已研制出静电微马达、电磁微马达、压电微马达、摇摆式微型电磁驱动器、微型电磁悬浮直线电机、电流变柔性微致动器等。直径 1mm 的电磁微马达输出力矩为 1.5 μNm,转速达 2 000 r/min,重 12.5 mg,可连续工作 1 小时。

(3) 微流量器件和系统

已研制出多种结构的热致动微泵、压电致动微泵和小型泵,微泵最大流量达 365 μL/min,最大背压达 118 cm 水柱;已研制出多种微型阀以及气体、液体微流量传感器等。研制出喷嘴直径为 20~50 μm 的电阻电热式微推进器微喷样机,其推力达 1.2 mN。

(4) 生物传感器、生物芯片和微仪器

研制了微电泳芯片、DNA 芯片、微光谱分析系统、微型机械光开关等。

(5) 微型机器人和微操作系统

研制了细小管道微机器人,能在直径为 10 mm 和 20 mm 的管道中爬行;6 自由度微操作平台;细胞操作系统等。

(6) 硅和非硅制造工艺

已具备了常规表面和体硅微细加工技术和条件，引进了高深宽比深层反应离子刻蚀设备、双面光刻机以及硅硅键合设备。利用北京高能所拥有同步辐射 X 线源，可进行简单的 LIGA 工艺加工。

制造工艺研究方面，开发了三维微机械加工新技术——DEM 技术，并引进了微注塑设备，可加工深宽比达 10 的金属和塑料结构；开发了形成硅多层微机械结构的"掩膜-无掩膜"腐蚀新技术；开发了硅/硅低温键合技术；开发了微电火花加工技术，能加工出直径小至 25 μm 的微小轴和直径为 50 μm 的微小孔。还开展了金刚石薄膜微机械加工技术、准分子激光直写加工、分子组装加工以及微构件组装等研究。

(7) 测试技术

开展了硅微结构材料的机械特性测试技术，微力、微扭矩测试技术，微小电容测试技术等的研究。

(8) 相关理论和技术

在微机械理论基础、建模仿真和 CAD 方面也开展了相应的研究。我国微系统和微米纳米技术研究的经费主要来自国家自然科学基金委、科技委和国防预研。90 年代初，基金委就支持了两项微系统方面的重点基金，接着科技部投入了一项微机电系统的攀登 B 计划，1996 年国防预研"九五"计划设立了微米纳米技术课题，1999 年科技部把集成微光机电系统列入重点。现在已有的有一定规模的微系统和微米纳米研究基地有：

• 清华大学微米纳米技术研究中心；

• 北京大学微米纳米加工国家级重点实验室；

• 电子部 13 所微系统小批量制造基地；

- 上海冶金所传感器国家重点实验室。

总的看来，经过几年的努力与发展，我国微系统技术与开发有了一定的基础。但目前国内大多数研究工作仍停留在实验室阶段，而且低水平重复的问题比较严重，一些有需求的研究成果并没有转化为生产力，基础研究薄弱，有的还缺乏必要的科研试验条件。这有待于进一步的资金和政策支持。

5.2 关于管理问题

（1）国外的管理模式

日本：日本通产省自1991年度开始实施为期10年、总投资250亿日元的"微型机械技术"大型研究开发计划。目前日本的国家和企业有关微系统的项目基本都统一由微机械中心安排。其特点是统一计划、分散实施，将主要的投资投向那些新的持续性的研究项目。其目的是引起企业对新技术的兴趣，同时打好共性技术基础。

瑞士：政府支持基础研究，产业界支持科研人员提出的开发构想。例如，成立瑞士电子和微技术中心CSEM，政府占总投资的30%，产业界占50%，技术成果或专利产生后，由新成立的公司进行产品的商品化运作，再将回收的资金重新投入研究开发，实现良性循环。这种管理模式值得我们借鉴。

欧盟（European Commission）：为推进其产业化进程制定了很多支持计划。主要包括 EUROPRACTICE 和 NEXUS（Network of Excellence in Multifunctional Microsystems）。NEXUS 主要是为了将用户和产品、技术的供应者连接起来，为微系统的产业化提供战略指导及联络更多的相关机构、组织。而 EUROPRACTICE 则具体帮助用户实现他们的产品，如为初学者提供培训，使他们了解CAD工具，帮助他们设计、生产小批量的产品。这些计划直接促进了欧洲已有微系统技术的产业化进程。

美国：美国国家关键技术计划把"微米级和纳米级制造"列为"在经济繁荣和国防安全两方面都至关重要的技术"。美国国家基金会把微米/纳米技术列为优先支持的项目。美国政府从1995年起对微系统方面的科研年投资达3 500万美元。美国国防部为保持连续的军事技术优势，十分重视发展微系统技术。克林顿总统的2001财年预算申请中给纳米技术的投资增加2.25亿美元，达4.95亿美元，并把国家纳米技术倡议作为最优先的科学技术倡议。1999年4月，DARPA又将它的电子科技办公室（Electronics Technology Office）更名为MTO（Microsystems Technology Office，微系统技术办公室）。这表明微系统已经不再仅仅是一项独立的研究计划，而是DARPA重点支持的有极强竞争力的核心基础技术。DARPA还支持对国内商业和学术研究用户建立定期的、共享MEMS制造服务。这种服务可以让几百个研究和工业用户节省费用和快速制造MEMS器件和系统。美国在MEMS技术上的投入巨大，管理上基本上是松散型，谁有钱谁搞。但在历史上，美国曾为遏止半导体技术和工业下滑，并保持半导体技术坚实的工业基础，1987年由政府/工业界共同成立工业界和国防部半导体制造技术集团（Sematech），把半导体工业界集合起来，共同开发半导体技术，由政府与工业界共同投资。主要任务是发展和增强半导体工艺技术与工艺设备及材料。工艺技术开发的任务为：1）研究开发高级半导体制造技术；2）在试验生产线上试验和验证这些技术；3）利用这些新开发的技术生产多种产品。Sematech强调"竞争前"的技术合作，即在产品研制前的技术开发阶段各成员公司通力合作，集中人力和财力，提高投资效率，避免在同一水平上重复；进入产品研制阶段再展开竞争。例如，成员公司在共同开发的0.35 μm 技术的基础上竞争 DRAM 产品。Sematech 使美国半导体技术能力大大加强，半导体工业出现新的转机。其成功的关键在于：一是政府强有力的支持；二是技术开发形成合力、共同攻关。Sematech 强调资源共享，合作前竞争的模式对我们是一个有益的启示。

（2）针对我国存在的问题，要做到全部集中不现实也无必要，但集中决策、大力协同则非常必要

首先要制定起指导作用的全国性的 MEMS 发展战略和发展规划，根据市场需求和科技工作达到的水平，对有市场竞争力的产品要支撑，对重要的开发研究要技术攻关和重要的应用研究及基础研究项目要支持。

项目管理采用国际规范化的指南发布体系，根据指南纲要和项目申请指南，向有优势的研究单位定向发布，由专家组进行评审。

5.3 初步设想

MEMS 技术发展战略软科学研究成果包括一份总报告和几份附件报告。总报告包含总体发展目标和政策措施，初步考虑附件报告包括：1）关于我国 MEMS 产业的现状、需求和发展规划；2）关于我国 MEMS 技术的开发研究（关键技术攻关）报告；3）关于我国 MEMS 技术的应用研究及基础研究报告。附件报告要以发展战略为指导，包括可操作的具体措施。工作方式采取专家组集体讨论、分工编写、汇总定稿的形式，年底以前完成报告，争取提前，以便明年正式启动发展计划。

初步工作计划安排为：六、七、八月份调查研究，听取各方意见，九月中下旬北京开会集中听取意见，研究讨论，各单位根据发展战略的总思路提出产业发展、开发研究关键技术攻关、应用研究、基础研究项目意见，特别是改革、创新的政策措施方面的意见。

关于微米/纳米技术的认识与思考

(2000 年 12 月)

1 前言

微米/纳米技术（Micro/Nano Technology）是一门新兴的、多学科交叉的研究领域，汇集了电子、机械、材料、制造、测量，以及物理、化学和生物等不同学科新生长出来的微小和微观领域的科学技术群体。当前，作为微米/纳米技术重要组成部分的微机电系统（Micro Electromechanical System，MEMS）和纳米技术已成了科学技术发展的热点、微小型化的重要发展方向。

MEMS 和纳米技术由于其对其他学科和技术的带动性和创新性，被认为是面向 21 世纪的新兴技术，对国民经济和国家安全有着重大的作用。纳米技术必将改变绝大多数人造物质及其构成部件的设计与制造方式，从而引发一场工业革命，成为未来国民经济新的增长点。微米/纳米技术将是国防科技与武器装备保持长远和持续发展，实现跨越式发展的一次新的机遇。所以说，微米/纳米技术将是 21 世纪科学和工程发展的战略领域，对人类健康、财富和安全的影响，至少会像本世纪抗生素、集成电路和人工聚合物的综合影响那样重要和深远。

目前，各发达国家政府为争夺 21 世纪微米/纳米技术的领先地位，纷纷制订了大型发展计划。美国国防高级研究计划局（DARPA）目前发展 MEMS 的计划主要有：微机电系统计划、微流体分子系统计划、微光机电系统（MOEMS）计划、高级能源技术计划、MEMS基超低成本/轻重量相控阵雷达技术计划和微机械传感器惯性导航系统（INS）计划等，并已公布了这些计划的76项研究项目。纳米技术方面，2000年2月美国政府公布了《美国国家纳米技术倡议》，并置于发展科学技术最优先的地位。计划2000年的经费为2.7亿美元，2001年上升为4.95亿美元，比2000年增长2.45亿美元。

日本的国家"微机械技术计划"，现处于第二阶段后期。目前研究工作集中在：1）系统化技术，制造4个实验微机械系统；2）功能器件技术，作动器、微接头电池组等；3）共用基础技术，控制、致动等；4）微机械技术和系统研究。2001年将实施为期7年的"纳米材料工程"计划，每年拟投资50亿日元。不久前日本政府拟定了"纳米技术"战略，希望通过开发纳米技术加强其产业的国际竞争力，促进经济发展。战略的三个发展重点是：加强开发5～10年后可以实用化和产业化的纳米技术；重视开发10～20年后的前瞻性技术；创新的基础研究。

欧洲微系统研究水平较高，且有其特色。法国仍在执行1997年制订的国家微技术和微系统工业发展计划，开发新材料、新颖设计和提高微加工能力的纳米仪表性能。法国2001年将投资8亿法郎建立微米/纳米技术园区，2005年建成。德国政府实施的1994—1999年的"微系统技术"计划中，政府每年资助6千万美元。计划包括：工业合作研究开发项目，发展系统技术、微系统和器件；工业原型样机开发和基础研究等。英国信息技术联合研究计划（JFIT）的合作研究计划中，将纳米技术作为一个计划项目列入。欧共体则把300多个与微系统相关的机构组织起来，组成"多功能微系统优秀网络"进行研究开发。

2 微米/纳米技术的概念与相互关系

对于微米/纳米技术，理查德·P. 范曼（Richard P. Feynman）1959 年以"实际上大有余地"为题的讲话，曾预见制造技术将沿着两个途径发展，一个是从宏观到微观，即 Top‑Down（由大到小）的途径，另一个是从最小构造模块的分子开始，进行物质构筑，即 Bottom‑Up（由小到大）的途径。历经 40 年的发展，特别是 80 年代以来，扫描隧道显微镜（STM）、能束制造技术及 X 射线光刻技术和分子工程与常规平面工艺密切结合这三个微观尺度制造领域的革命性成果，大大地推动了微米/纳米技术的发展，顺理成章地出现了微机电系统（MEMS）。我们现在说的微米/纳米技术是指包括集成电路、MEMS和纳米技术等的一系列技术。MEMS 是美国的叫法，日本称为微机械（Micro Machine），欧洲称为微系统（Micro System）。它们均以微小（Micro）为特征，有的强调机械特征，有的强调系统特征。MEMS、微机械和微系统的内涵略有差异，还可以包括微光学系统（Micro‑Opto‑Electro‑Mechanical‑Systems，MOEMS）。MEMS 的概念在发展中会得到逐渐扩充和修正。我国普遍采用微机电系统一词。

MEMS 是采用 Top‑Down 发展途径的用大机器做出小机器，用小机器又做出更小机器的产物。借助半导体批量制造技术可以制造许多宏观机械系统微米尺寸的 MEMS，从而在小型机械制造领域开始了一场革命。1987 年美国加州大学伯克利分校在集成电路工艺基础上，研制成功世界上第一台静电微马达，也是世界上最早的 MEMS。MEMS 的尺寸现在还没有明确规定，一般认为 MEMS 的尺度在微米到毫米范围。1 mm（10^{-3} m）～10 mm 范围的称微小型（mini）机械；1 μm（10^{-6} m）～1 mm 范围的称微型（micro）机械；1 nm（10^{-9} m，或 10 Å）～1 μm 范围的称纳米（nano）机械。根据当前的技术能力，尺寸在 1～10 mm 的小型机械也归为 MEMS，借助生物工程及分子组装技术实现的纳米机械或分子机械称为纳机电系统（Nano-Electro-Mechanical Systems，NEMS）。

Top-Down 发展途径通过各种微加工工艺将加工尺寸越做越小，达到纳米尺寸。日本电气公司（NEC）用离子束加工工艺已开发出厚度小于 0.1 μm 的微小零件。美国康奈尔大学正在开发纳机电系统，关键特征尺寸可以达到几百至几纳米。研究工作包括改进制造工艺，探索在纳米尺寸致动和探测运动的新方法。采用光刻法能在硅和其他材料上产生独立的物体，厚度和横向尺寸降到大约 20 nm。用类似工艺可制作接近分子尺寸的相应尺寸的通道和孔。NEMS 可能在传感器、医学诊断、显示和数据存储等应用方面带来革命性变化，会使个别双分子结构和功能试验成为可能。

目前，纳米技术还没有一个确切一致的定义。1999 年 9 月美国国家科学与技术委员会的纳米科学、工程和技术跨部门工作组提出的《纳米技术研究方向》认为，"纳米技术是通过在纳米长度尺寸，即在原子、分子和超分子结构水平上控制物质，创造和利用材料、器件和系统的技术。"2000 年 2 月的《美国国家纳米技术倡议》在纳米技术定义一节认为："纳米技术的本质在于以逐个原子的方式在分子层次上进行工作的能力，产生具有根本不同的新分子组织形式的大型结构。"根据对纳米技术内涵的理解，可以认为：纳米技术系一门在 0.1～100 nm 空间内，研究电子、原子和分子运动规律和特性的多学科交叉的高新技术。随着对纳米技术认识的提高，对纳米技术的定义会更确切。纳米技术的研究领域包括：纳米电子学、纳米材料学、纳米显微学、纳米机械学、纳米生物学和纳米制造等。

纳米技术是采用 Bottom-Up 发展途径研究重新排列原子的问题，使人们能按照自己的意志操纵单个原子，制造具有特定功能的结构。主要是通过在原子、分子和超分子层次上，对结构和器件进行控制和保持界面稳定性，以及在微米尺度和宏观尺度上保持这些"纳米结构"的完整性。纳米尺度下新的性质，不一定能够从大尺度下观测到的性质来预测。性质上最重要的变化并不是尺度数量级减小引起的，而是由于纳米尺度下固有的或变为起支配作用的新观察到的现象引起的，例如尺寸限制、界面现象的支配作用和量子力学。一旦能够控制特征尺寸，也就能够把材料性质和器件功能提高到超过我们目前知道的，乃至认为有可能的水平。减小结构尺寸，出现了诸如碳纳米管、量子线与点、薄膜、以 DNA 为基础的结构和激光发射器等具有独特性能的实物。只要我们能够发现并充分利用这些基本原理，这样一些新形式的材料与器件将预示着一个科学技术革命的时代。

当特有结构特征值介于孤立原子和大量分子间——即在 $10^{-9} \sim 10^{-7}$ m（1～100 nm）范围时，这些物体往往显示出与原子或分子大不相同的物理特性。例如，碳纳米管的强度是钢的 10 倍、重量却是钢的六分之一，并已证明纳米粒子能瞄准并杀死癌细胞。纳米尺寸系统具有成几百万倍地提高计算机效率的潜力。纳米技术实际上影响着每种人工制造物体的产品——从汽车、轮胎和计算机电路到医学与组织替代品，并导致发明仍属想象中的物体。最近，美国康奈尔大学研究人员还把用金属镍制成的螺旋桨接到三磷酸腺苷酶（ATP）分子的中轴上开发出分子马达。这种马达以每秒 3～4 圈的转速运转了 40 min，还表演了一个尘埃粒子先被旋转的螺旋器吸入，再被甩出的景像。

Top-Down 和 Bottom-Up 虽然是两条不同的发展途径，但用微加工方法可以做到纳米尺寸，用纳米技术同样可以做成器件和系统，仅仅是发展途径不同而已。再者，MEMS 的微机理研究实质上是研究纳米尺度下有别于宏观世界的物理现象，在这一点上 Top – Down 和 Bottom – Up 又合二而一了。

从微米/纳米技术现在的发展来看，我认为目前微米技术处于前沿、战略、现实的阶段，纳米技术则处于前瞻、战略和基础的阶段。现阶段我们开展纳米技术研究，根据可能投入的力量和从有利于目前和今后的发展来看，研究工作应首先考虑纳米材料和纳米操作（工艺）等方面。

3 广义制造技术是微米/纳米技术发展的关键

MEMS 研究不仅涉及元件和系统的设计、材料、制造、测试、控制、集成、能源以及与外界的联接等许多方面，还涉及微电子学、微机械学、微动力学、微流体学、微热力学、微摩擦学、微光学、材料学、物理学、化学、生物学等基础理论。但是制造技术，即微加工技术是实现微米/纳米器件或系统的关键。之所以到 1987 年才出现世界上第一台静电微马达，主要原因之一就是此前半导体加工技术及三个微观尺度制造领域还不足以为 MEMS 提供必需的加工工艺。微制造技术是在微电子平面工艺基础上发展起来的，在此基础上又发展了体硅和表面硅加工、键合和组装等技术。后来又先后有了深反应离子刻蚀（DRIE）、LIGA、准分子激光和光成形等多种工艺创新。微观尺度制造技术的演进与革命，是促成 MEMS 这一重大变化的推动力。没有微观尺度制造技术的演进与革命，MEMS 就不可能有今天的发展。然而，微加工工艺仍是制约 MEMS 发展的一个主要因素。

纳米技术研究开发同样需要拥有新型仪器设备，用以操纵（制造）和测试这些微型"纳米"结构件的特性。80 年代出现的实现范曼想象的那种能力的仪器设备，包括隧道扫描显微镜、原子力显微镜以及近场显微镜，为纳米结构测量和操作提供了"眼睛"和"手指"。同时计算机能力的发展，也提供了对纳米尺寸的材料特性进行复杂模拟的能力。这些新型设备和技术对纳米技术发展是一个极大的激励。纳米结构通过亚微尺度构件（理想的做法是自组合和自组装）提供了一种新的材料制造范例，它是一种"自下而上"而不是"自上而下"的超小型方法。

制造工具和加工工艺是开发微米/纳米技术的前提，德国开发LIGA技术的例子可以说明这一点。德国为走出微电子技术竞争失败的阴影，下大力气发展微米/纳米技术，他们看准了制造工艺的作用，研究开发了LIGA技术，为制造高深宽比器件和批量生产提供了有力的工具，推动了微米/纳米技术的发展。不仅如此，德国还把LIGA技术与准分子激光烧蚀结合使用的新加工工艺，准分子激光用于加工电镀模板初始三维结构的第一道工序。

MEMS除芯片制备外，很多后工序也很难、很复杂。某些集成电路制造中已解决的问题，在MEMS中却很难办，MEMS器件封装就是一个例子。集成电路是平面的、无可动部件，芯片封装相对容易一些，而MEMS芯片内有的含有可动部件，给封装工作带来很大困难。封装成了MEMS器件制造的瓶颈，极大地影响了成品率和经济效益，因此必须重视MEMS的封装问题。

从国际上开发MEMS的情况看，美国侧重在微电子技术的基础上，通过微芯片取得制造工艺的突破；日本则偏重从机械加工工艺实现微机械的制造，强调通过非光刻的传统机械线实现机械微型化，是一条用大机器造小机器，用小机器造微机器的途径；德国的特色是在LIGA工艺的应用上取得进展。这些国家的加工工艺各有特色，但均取得显著成效。总体来看，目前美国和日本处于微米/纳米技术领先地位。我们应在利用国外各种微加工工艺的基础上努力创新。

4 要有创新思维、创新方法，才能有所作为

微米/纳米技术的出现和发展是科学创新思维的结果，微观尺度制造技术的演进与革命，是促成这一重大变化的推动力。今天，尽管人们已经意识到 MEMS 和纳米技术有可能在未来数十年内引发一场工业革命，但在基础探索、发现，以及最终的制造方面面临的问题，无疑也是很艰难的。突破这些困难的关键是对微米/纳米技术要采用创新的研究方法、多学科的综合集成，以及不同的制造范例。

我们目前只是初步了解一些制作"设想的"纳米结构的原理，以及如何经济地制作纳米器件和系统。但即使是被制作出来了，那些纳米结构器件的物理/化学特性也只是刚开始被揭示出来，目前的微型器件都是以尺寸长度在 100 nm 以上的模型为基础的。在掌握纳米结构的物理/化学特性，以及在开发预期的控制它们的方法方面，取得的每一次重大进展，都将有可能促使我们将纳米结构装置和纳米器件设计、制作和组装成工作系统的能力大大提高一步。对纳米尺寸基本现象的研究手段及了解程度，目前还都是初步的，而且尚难对纳米技术的基础科学和应用科学划分明确的界线。

在具体的纳米技术得以出现以前，对许多学科领域，诸如物理学、化学、材料科学、电子工程，及其他一些学科的基础科学，都还必须进行充分揭示。当前对纳米科学进行的基本研究和探索中，遇到的一些普遍性的和跨学科的问题，就说明了这一技术的复杂性，具体为：1) 纳米结构装置，特别是在室温条件下，能达到怎样的新的且与众不同的量子特性？2) 相邻接的纳米结构间，界面区域的特性与大块物体

的特性有何区别？新的技术是如何利用这些特性的？3）纳米晶体和纳米棒中，原子表面的重建和排列是什么样的？是否有可能在纳米晶体内制作外延芯壳（core-shell）系统？4）能否通过人工合成单一长度和螺旋形碳纳米管，并将其精制成单独的分子产物？能否在一维纳米结构内重复制作异质结？5）通过对单分子特性的研究，能从中对复杂的聚合物、超分子和生物系统得到什么样的新认识？6）在出错率变得无法接受前，人们能在多大程度上根据复杂的设计程序，利用平行自组装技术控制纳米尺度元件的相对配置？7）是否有能控制先进器件应用必需的尺寸、形状、构成和表面状态的方法，用来经济地制作纳米结构？要回答这些问题就需要有创新的思想、方法，才能有所发现和发展。MEMS研究同样需要创新的思维和方法，没有创新就没有突破性进展。

微米/纳米技术发展至今还为时不长，还有较大的争取前沿地位的空间。科学上要有创新就必须加强基础研究，基础研究是技术创新发展的源泉，无论微电子还是激光都是在基础研究基础上创新发展的结果。MEMS或纳米技术的基础研究现在已得到国家和有关部门的重视，应该抓住当前这一有利时机努力工作，选择应重点开展的基础研究项目。微加工工艺在承袭现有方法的基础上，迫切需要结合我国国情发展创新的加工工艺。

另外，MEMS材料方面除硅材料外还应探索新材料，如聚酰亚胺、钨、镍、铜、金、石英、氧化锌、锆钛酸铅（PZT）、钛镍合金（TiNi）、砷化镓（GaAs）和类金刚石薄膜等。

5 结束语

我国开展微米纳米技术研究还是比较早的，至今已有10余年历史。原国家教委、原国家科委和原国防科工委等，先后组织了微机电系统的研究开发，制订了一些研究计划，目前国内有40多个单位的50多个研究小组在从事微机电系统的研究工作，开展纳米技术研究的单位也在日渐增加。这些工作主要集中在大学和研究所，其中以大学为多。经过十余年的研究开发，我国微机电系统研究已初步形成以下几个研究方向：（a）微型惯性器件和微型惯性测量组合（MIMU）；(b) 微型传感器和微执行器；(c) 微流量器件和系统；(d) 生物传感器、生物芯片和微操作系统；(e) 微机器人和 (f) 硅和非硅制造工艺。与此同时，也培养了一批微米/纳米技术研究人才。

国际上微米/纳米技术发展很快，新的研究成果不断涌现，初步应用取得了一些进展，并形成了一定的批量制造能力，产品开始占有市场。我国近年来研究工作虽然取得了显著成绩，但缺乏突破性的成果。造成我国微米/纳米技术研究进展滞后于发达国家的原因，大致上有如下几点：

1) 微米/纳米技术研究涉及众多学科和技术，综合集成性极强，而我们习惯于传统的单一学科的研究方法，对综合基础研究的认识和投入的力量远不及国外，这就制约了我们的研究进度，因此需进一步重视综合性基础研究工作。

2) 从全国来看，要努力创造条件办好开放实验室，科技人员要既懂设计又熟悉工艺制造，以有利于设计与工艺制造的紧密结合，缩短设计和加工、试验周期。这必将大大促进研制速度、早出成果。

3) 充分发挥现有加工基地的作用，解决加工能力欠缺，加工周期长的问题。在可能的情况下应实行资源共享和合作研究，充分发挥各部门的优势，弥补研究力量、设备等不足的问题。目前集成电路的发展，将设计、制造与封装分为三大段，分别在不同工厂里进行阶段性工作，实行专业化分工协作，这是值得我们学习的。

4）研究队伍整体力量不足，吸引和留住人才力度有待提高；与国外合作和联合研究少，难以进行各种深入交流。

5）研究经费力度不及发达国家，如何使有限经费集中用在研究工作上和保证研究人员应有的福利待遇有待进一步改进。

6）目前的MEMS研究计划应集中力量开发MEMS器件。我在1997年发表的《微机电系统的科学研究与技术开发》一文中就提出，"现阶段主要从微电子微细加工技术的创新应用入手，主要开发MIMU惯性器件"，现在看来这个意见还是必要的。因为器件是基础，我们要吸取已有的经验。利用国外器件进行应用研究，在没有国产器件前可以争得时间，但不能分散我们开发器件工作的力度。

上面谈的几点仅是目前个人对微米/纳米技术发展的认识与思考。鉴于微米/纳米技术学科内涵丰富、发展变化较快，需要不断学习、总结和提高，希望从事微米/纳米技术的同志能解放思想，群策群力，尽快把我国微米/纳米技术搞上去。

微米/纳米技术当前发展动向

(2001年10月)

近年来参加了几次微米/纳米技术方面的国际会议,也参观访问了国外一些科研生产单位,对目前国际上微米/纳米技术研发现状和发展前景有了一些新的认识,这里我结合当前微米/纳米技术发展动向,着重谈一下微机电系统技术发展过程中面临的热点和难点问题,供同志们研究参考。

1 目前微米/纳米技术发展平稳,但孕育着新的机遇和突破

总的来说,当前国际上微米/纳米技术处于平稳发展阶段,但孕育着新的机遇和突破。MEMS的研发工作仍主要集中在传感器、执行器、信息(包括光通信和无线通信)、生物医学等领域,并取得不同程度的进步,特别是信息领域,无论是光开关,还是射频开关的研究均取得了较大进展。目前典型的MEMS产品主要还是微压力传感器、硅微加速度传感器、喷墨打印头等,其市场年销售已达400亿美元。纳米技术的研发工作主要围绕纳米材料、纳米器件等方面。

但是,业界普遍预测的去年年底将实现MEMS光开关产业化的预期目标并未实现,这使人们对MEMS研发及产业化的前景产生了一些

2001年10月24日,在第五届全国微米/纳米技术学术会议上的讲话。

怀疑和悲观，同时也说明人们对 MEMS 技术发展中产生的困难估计不足。但我们认为，微米/纳米技术相对平稳的发展中正孕育着新的机遇和突破。主要表征为以下几点：

（1）发达国家和地区继续加大对微米/纳米技术研发的投入

微米/纳米技术作为一项军民两用的战略高科技，越来越引起各发达国家政府、研究机构和企业界的高度重视和支持。美国政府在今后五年中将投入 2 亿美元用于 MEMS 的研究与开发。美国国防高级研究计划局（DARPA）目前发展 MEMS 的计划主要有：微机电系统计划、微流体系统计划、微光机电系统（MOEMS）计划、高级能源技术计划、MEMS 基超低成本/轻重量相控阵雷达技术计划和微机械传感器惯性导航系统（INS）计划等，并已公布了这些计划的 76 项研究项目，2001 年又新增加 26 个项目。纳米技术方面，美国政府 2001 年 2 月公布了《美国国家纳米技术倡议》，将其置于发展科学技术最优先的地位。2001 年计划经费为 4.95 亿美元。

欧洲。德国政府从 1990 年开始每年对 MEMS 的投入均保持在 5 000 万美元的力度，2000 年至 2003 年是它们的第三期计划，计划包括：工业合作研究开发项目，发展系统技术、微系统和器件；工业原型样机开发和基础研究等，投资额度为 2 亿美元。法国 2001 年将投资 8 亿法郎建立微米/纳米技术园区，计划于 2005 年建成。欧共体还把 300 多个与微系统相关的机构组织起来，组成"多功能微系统优秀网络"（NEXUS）以促进研究开发。

亚洲地区。处于审批中的日本 MEMS 国家发展计划，今后 5 年的投资相当于过去 10 年投资的总和。前不久，日本政府制定了"纳米技术"战略，该战略的三个发展重点是：加强开发 5～10 年后能实用化和产业化的纳米技术；重视开发 10～20 年后的前瞻性技术和创新的基础研究。在亚洲地区，韩国除了在 1995 年开始实施为期 7 年的"国家 MEMS 计划"，又在 1999 年末启动了为期 10 年的"智能微系统计划"，中国台湾方面今后 3 年中对 MEMS 的投资将达 3 亿美元，并计划建立三个 MEMS 研究开发中心。此外值得注意的是，目前国际上 MEMS 研发中，政府投入仅占总投入的 30% 左右，众多的投资来自工业界或民间的创业基金，不少 MEMS 高技术企业也应运而生。

（2）知识和技术创新大量出现

国际上 MEMS 论文和专利数量每年按指数级迅速增加。以年初在瑞士召开的第 14 届 IEEE MEMS 2001 年会为例，共收到了 400 篇 MEMS 研究论文，而在 1989 年只有 50 多篇。值得注意的是，2001 年会议上宣读的 44 篇论文涉及的领域与 1989 年的会议上涉及的领域是相同的，显然 12 年之后同一领域中其知识和技术创新的成果与十多年前相比是不可同日而语了。大量的知识与技术创新为 MEMS 新产品的出现与产业化奠定了基础。一些光开关、RF-MEMS 器件已经完成样机研制阶段，正在攻克大规模生产中的成品率和可靠性问题，一旦获得突破，必将带动 IT 领域的新的技术革命。

（3）研发和生产工艺齐头并进

在传统 MEMS 制造技术由研究阶段向批量生产阶段转变的同时，SOI、高深宽刻蚀等许多新工艺的研究开发取得重大突破。至目前为止，MEMS 标准化工艺和设备在逐步完善，并已用于在美国、德国、瑞士和中国台湾等建立的批量生产 MEMS 专用线。

以我们访问过的以生产微电子和 MEMS 设备著称的德国 Karl SUSS 公司为例，该公司产品每年以 30% 的速度递增。值得注意的是，其产品的结构开始发生重大变化，即用于批量生产的 MEMS 设备超过了科研开发用 MEMS 设备数量，硅片尺寸由过去 4 寸变为现在的 8 寸。

(4) 呼吁加强微米/纳米技术教育与相关人才的培养

每次参加国际会议都可以强烈地感到世界各国特别是发达国家高度重视微米/纳米技术人才培养的氛围。日本、美国和欧洲陆续开展了微米/纳米技术专业的普及教育、学位教育和工程继续教育，以培养微米/纳米技术的多层次人才。其发展目标是既要满足迅速增加的微米/纳米技术领域的人才需求，还要为下一步微米/纳米技术的深入发展准备人才。其目的在于保证微米/纳米战略高技术的可持续发展和保持已有的优势地位。

从微米/纳米技术发展的总体来看，我认为：目前微米技术处于前沿、战略和现实的阶段；纳米技术则处于前瞻、战略和基础研究阶段。纳米器件进入实用阶段可能还需要 10～20 年的时间，具有诱人的发展前景。

2 当前 MEMS 发展面临的热点与难点

(1) 当前 MEMS 发展面临的热点

当前 MEMS 发展面临两个热点，热点之一是信息 MEMS 的研发成为人们关注的焦点；热点之二是国际上具有 Foundry 性质的研发基地或生产线建设显著增加，为 MEMS 产业化奠定基础。

①信息 MEMS 仍是当前 MEMS 研发的热点

其涉及的器件有两类，其一是已形成产业的硬盘读写头、喷墨打印头及 DMD；其二是目前正在研究应用在光通信和无线通信中的微型光开关、光开关阵列和 MEMS 射频开关（即所谓 RF MEMS）等，它们可以替代现代信息领域中所采用的庞大的光学器件和射频器件，实现系统的微小型化和性能的更新换代，已成为人们深入讨论的热点话题。第 7 届微机械高层会议上美国代表团提交的 5 篇论文中，有 3 篇是专门论述信息 MEMS 的发展的,涉及 MEMS 射频开关、MEMS 光开关和高密度存储。德国代表所报告的论文 *Development an Application of RF MEMS Technology* 中,展示了 Raytheon 研制的射频开关和以此为基础开发的 MEMS 移相器。射频开关的指标已达到:插入损耗小于 0.07 dB (40 GHz),隔离度大于 35 dB (40 GHz),开关时间小于 10 μs,开关电压 30~50 V。韩国 LG 公司侧重研究用于无线通信的 RF MEMS 器件,主要包括 MEMS 开关、可调电容、微型电感和谐振器。三星公司同样把信息 MEMS 作为重点发展方向之一,在其所列的 6 大研究方向中,MEMS 射频开关和 MEMS 光开关是重要的两大发展方向。新加坡则将 MEMS 光开关作为重点发展的方向之一。

②具有 Foundry 性质的研发基地或生产线建设显著增加,为 MEMS 产业化奠定了基础

微机电系统器件能否最终稳定发展,关键取决于它们能否成功替代传统器件得到广泛应用。做到这一点的决定因素是批量生产能力,没有批量生产能力就无法体现这些器件性能价格比的优势。发达国家在微器件批量生产方面已取得初步成效。据最新统计,目前全球共有 25 条 MEMS 加工线,主要集中在美国和欧洲。其分布为:美国 11 条,莱茵河地区 4 条,法国和瑞士 4 条,欧洲其他地区 5 条,中国台湾 1 条。其建设动向为:

一是 6 英寸线的比例增加,而 4 英寸线的比例缩小;二是投资来源发生根本变化,政府直接投资仅为总投资的 1/6;三是建设规模向两极发展。例如,与 1999 年相比,2000 年每条加工线上技术人员少于 25 人的加工线以及技术人员在 100~500 人之间的加工线均增加了,而技术人员在 25~50 人和 50~100 人之间的加工线减少了。

例如，瑞士 Colibrys S. A 公司是刚从母公司独立出的专门从事 MEMS Foundry 加工的公司，该公司开发了近十套成熟的标准工艺，对全欧洲服务。我国台湾地区建立了开放的公用实验室（工业技术研究院，英文简称 ITRI），其目的是促进 MEMS 研究与工业界的联系。此外有 3 个创业公司利用已有的 6~8 英寸线，建立规模量级的 MEMS Foundry，实现从半导体向 MEMS 制造的转化。其中一个公司还与 ITRI 签订了正式的合作协议，以充分利用 ITRI 的研究资源。此外，比利时的 IMEC 公司，利用现有条件，在 IC 研究线的基础上，增加相应的 MEMS 专用设备，扩充了 MEMS 研究能力，与国际合作共同开发 MEMS 技术和 MEMS 器件。

（2）当前 MEMS 发展面临的难点

当前 MEMS 发展面临的难点主要表现在：原型机的研究与产业化之间存在距离，即研究工作与产业化衔接不够。这一点在 MEMS 光开关的研究开发中得到了证明。去年，美国多个公司宣布将在年底推出商品化的光开关，但实际上只有郎讯公司拿出了试用样品，其他公司均未达到预期目标。而郎讯公司的光开关样品也因结构支撑部分的应力问题，导致镜片形变，至今尚未实现真正的产业化，据与会的美国代表预测，光开关的产业化仍需要一定的时间。这些现象表明，针对产业化的 MEMS 技术攻关研究工作需要深入开展。

当前 MEMS 产业化发展较为缓慢，原因是多方面的。正如其他技术一样，技术发展从研制、开发到产业化还有一个过程。我认为其中主要原因有以下两点：

一是微制造技术仍是制约 MEMS 发展的一个主要因素。例如标准工艺与工艺标准化问题，微机械与电路的单片集成问题以及 MEMS 封装问题等。

微制造技术是在微电子平面工艺基础上发展起来的，在此基础上又发展了体硅和表面硅加工、键合和组装等技术。后来又先后有了深反应离子刻蚀（DRIE）、LIGA、准分子激光和光成形等多种工艺创新。微观尺度制造技术的演进与革命，是促成 MEMS 这一重大变化的推动力。没有微观尺度制造技术的演进与革命，MEMS 就不可能有突破性发展。MEMS 除芯片制备外，很多后工序也很复杂。某些集成电路制造中已解决的问题，在 MEMS 中却很难办。例如 MEMS 的单片集成，由于机械部分与电路部分先后走不同的工艺，如何在工艺上处理以保证质量；再如 MEMS 器件封装也是一个新课题。集成电路是平面的、无可动部件，芯片封装相对容易一些，而 MEMS 芯片内有的含有可动部件，就给封装工作带来很大困难。柏林弗朗霍夫可靠性和微集成研究所的 Herbert Reichl 等教授经研究后认为，小批量生产及所需特殊的技术和材料将导致 MEMS 封装本身的成本占到整个微系统的70％。封装成了 MEMS 器件制造和产业化的一个瓶颈，极大地影响了 MEMS 产品成品率和经济效益，因此目前各国普遍开始重视 MEMS 的封装研究工作。

二是目前对 MEMS 产品质量及质量保证体系还关注不够，致使 MEMS 产品的质量处于不够稳定状态。尽管 MEMS 生产线产业是在 IC 生产线产业上发展来的，但两者还是有着显著区别的，MEMS 加工线必须满足客户多样化的需求，这就导致：技术适应性要强，不仅要有标准工艺，还要有工艺的标准化以保证质量。相对而言，MEMS 产品品种多，批量少。这样要想达到 IC 产业的质量标准和经济规模是很困难的，因此，MEMS 产业发展还有较长的路要走。这对我们来讲既是机遇，也是挑战。

从深层次来看，解决 MEMS 技术发展中面临的困难，特别需要强调创新思维和创新方法。MEMS及纳米技术的出现都是科学创新思维的结果，近年来这方面专利数的大幅度增加就是一个很好的证明。今天，尽管已经意识到MEMS和纳米技术有可能在未来数十年内引发一场工业革命，但在基础探索、发现，以及最终的制造方面的问题无疑也是十分艰难的。突破这些困难的关键是对微米/纳米技术要有创新的特别是根本性创新的研究方法、多学科的综合集成以及不同的制造范例。

3 把握机遇，促进我国微米/纳米技术新发展

前面我谈到当前国际上微米/纳米技术发展平稳，孕育着新的机遇和突破。我认为对中国而言，最大的机遇就是目前还没有哪一个国家已经垄断了微米/纳米技术的研发与产业，大家都处在"百舸争流"，你追我赶的状态，我们一定要抓住当前的有利时机，有所作为，争取脱颖而出。我国开展微机电系统技术研究至今已有10余年历史，这些工作主要集中在大学和研究所，其中以大学为多。总的看来，经过几年的努力与发展，我国微系统技术与开发有了一定的基础，初步形成以下几个研究方向：（a）微型惯性器件和微型惯性测量组合（MIMU）；（b）微型传感器和微执行器；（c）微流量器件和系统；（d）生物传感器、生物芯片和微操作系统；（e）微机器人和（f）硅和非硅制造工艺。与此同时，也培养了一批微米/纳米技术研究人才。但目前国内大多数研究工作仍停留在实验室阶段，科研成果尚未转化为生产力。令人鼓舞的是，"十五"伊始，国家有关部门都在制订微米/纳米技术的发展战略和计划，这是推动我国微米/纳米技术及产业发展的一个良好机遇，应牢牢把握。

加速我国微米/纳米技术研发的关键是要形成一个有利于创新和产

学研结合的机制。从我国当前的实际情况看,有些地区、有些单位正在作这方面的努力,并取得了一些成效。建议国家有关部门要重视研究微米/纳米技术发展战略,做好顶层设计,各个计划定位明确,避免重复,真正做到选择有优势的单位承担任务,鼓励产、学、研结合。

从全国来看,要努力创造条件办好开放实验室,设计人员要熟悉工艺制造,以利于设计与工艺制造的紧密结合,提高研制质量,早出成果;要充分发挥现有加工基地的作用,解决加工能力欠缺,加工周期长的问题;要利用国外器件进行应用研究以赢得时间,但是器件是基础,不能因此分散我们开发重要器件的力度,在器件和基础技术上要有创新和突破;要加强微米/纳米技术研究队伍及新生力量的培养,加强国际交流和合作研究。

上面谈的几点仅是个人的认识和思考,由于微米/纳米技术学科内涵丰富,与时俱进,需要不断学习和总结提高。本届微米/纳米技术大会为同志们提供了一个集思广益、深入交流的机会,希望大家一起来共同努力,促进我国微米/纳米技术的新发展。

抓住机遇，促进微米纳米技术新发展

(2003 年 2 月)

微米纳米技术对其他学科和技术具有带动性和创新性，被认为是面向 21 世纪的新兴技术乃至主导技术之一，它们对国民经济和国家安全的重要作用已被人们广泛认同和重视。这里我想与大家一起来探讨一下微米纳米技术之间相互融合、相互促进的问题，并简要谈一谈相应的工作方法。需要说明的是，微米纳米技术发展很快，这里谈的是个人在学习和工作中的一些思考和体会，水平有限，仅供同志们参考。

1 以实例看微米/纳米技术的迅速发展

首先，我想仅以微型发动机的发展和演变为例来说明微动力源及器件方面取得的进展及突破是多么令人鼓舞。

1987 年，美国加州大学伯克利分校在集成电路工艺基础上，成功研制了世界上第一台静电微电机，也是世界上最早的 MEMS 标志性器件。

继德国研究出小型电磁电机后，1999 年，上海交通大学研制成功直径为 1 mm、重 12.2 mg、世界上最轻的电磁型微电机，最大转速为每 18 000 r/min，输出力矩 1.5 μN·m。

2001 年 5 月，美国康奈尔大学的科学家研制出世界上第一台生物

本文发表于《纳米技术与精密工程》2003 年第 1 期。

分子纳米发动机。它由两部分组成：一部分用有机物充当发动机，另一部分用镍无机物充当螺旋桨。发动机螺旋桨长 750 nm，宽 150 nm。它由 ATP 合成酶驱动发动机运转。每加一次能量，纳米发动机可连续工作 1 小时。

2002 年 1 月，加利福尼亚大学研制出的分子发动机，整台设备直径 11 nm，高 11 nm，可带动比自身体积大十几倍的物体。其原理是 6 个合成 ATP 分子构成一个三冲程、三缸发动机，第七个分子构成中央的"轴心"。通过化学反应，使这 6 个 ATP 分子围绕轴心做有序活动。组装后的分子发动机每秒钟旋转 8 次而不会变形。

2003 年 7 月 24 日，加利福尼亚大学伯克利分校的 Alex Zettl 及其同事宣布研制出了迄今为止世界上最小的人造发动机。这种纳米尺度的发动机宽度仅有 500 nm，300 个这样的电子转子才相当于人类头发丝的直径大小。科学家们在一个硅晶片上附着一个多层的碳纳米管，然后再涂上一个 200 nm 左右的金原子涂层。通过对晶片特定区域的蚀刻之后，金属薄片就可以进行自由旋转。通过对晶片不同区域施加不同的电压，就可以控制该金属碟片的运动。另外重要的是它能够适应不同的工作环境，包括高度真空和强烈的化学条件。这一研究成果最终将广泛应用于微机电系统领域中的光学开关以及微流体芯片等领域，并为发动机带来全新面貌。

此外，今年 1 月 16 日《自然》杂志报道，美国哈佛大学成功开发出一种新型纳米激光器，细到人的头发丝的千分之一，可自动调控开关。将其安装于微芯片上，能增加计算机磁盘和光子计算机的信息存储量，加速信息技术的集成化发展。新型纳米激光器的技术关键就在于，它具备电子自动开关的性能，无须借助外力激活。去年 6 月，康奈尔大学和哈佛大学分别开发出具有开关功能的纳米晶体管，体积小至现在的百分之一，相当于单个分子。该技术可使计算机电路再缩小至 6 万分之一，造出仅有句号大小的超微计算机。9 月，威斯康星州大学在室温条件下通过操纵单个原子，研制出原子级的硅记忆材料，其存储信息的密度是目前光盘的 100 万倍。

从 16 年前的静电微电机到今天的分子发动机，从普通的激光器到新型的纳米激光器，我们可以充分感受到微米纳米科技的迅速发展，同时也激励我们进一步思考，从中得到启示，抓住机遇，从而大力促进我国微米纳米技术的新发展。

2 微米/纳米技术科学研究和技术开发

早在 1994 年酝酿召开我国首届全国微米纳米技术学术大会时，我们对会议的名称进行了反复斟酌，最终决定使用"微米纳米技术"这个名称，当时纳米科技远没有像今天这样炙手可热。时至今日，近十年过去了，现在看来"微米纳米技术"的提法应该说是正确的，是符合实际的。我们当时之所以把"微"和"纳"放在一起，就是既考虑了微米技术的前沿性和现实性，又着眼于纳米科技的前瞻性和基础性，两者都是具有战略性的科技工作，应兼顾基础研究、应用研究和开发研究的协调与可持续发展。大家都知晓，这两年纳米科技方兴未艾，成为影响未来人类生活的世界三大新科技之一，世界各国政府包括我们国家都加大了发展纳米科技的投入并制定了各种战略目标，其重要性毋庸置疑。但是也要注意到有一种倾向，那就是把微机电系统技术和纳米科技割裂甚至对立起来了，认为搞纳米科技是先进的，搞微技术或微机电系统技术是落伍了，我认为这不是发展微米纳米技术的正确观点。

实际上无论谈微米技术还是纳米技术，大家都会提到 Richard

Feynman，他 1959 年发表的题为"实际上大有余地"的讲话，科学地预见了制造技术将沿着两个途径发展，一个是从宏观到微观，即 Top-Down（由大到小）的途径，另一个是从最小构造模块的分子开始，进行物质构筑，即 Bottom-Up（由小到大）的途径。历经 40 年的发展，特别是 80 年代以来，扫描隧道显微镜（STM）、能束制造技术及 X 射线光刻技术和分子工程与常规平面工艺密切结合，这三个微观尺度制造领域的革命性成果，大大地推动了微米纳米技术的发展，顺理成章地出现了微机电系统（MEMS）。MEMS 主要是采用 Top-Down 发展途径的用大机器做出小机器，用小机器又做出更小机器的产物。借助半导体批量制造技术可以制造许多微米尺寸的 MEMS，从而在小型机械制造领域开始了一场革命。

纳米技术则主要是以 Bottom-Up 发展途径研究重新排列原子的问题，使人们能按照自己的意志操纵单个原子，制造具有特定功能的材料结构。通过在原子、分子和超分子层次上，对结构和器件进行控制和保持界面稳定性，以及在微尺度和宏尺度上保持这些"纳米结构"的完整性。需要注意的是，物质性质上的最重要的变化并非是由尺度数量级减小引起的，而是由于纳米尺度下固有的或变为起支配作用的新现象，例如界面现象的支配作用和量子效应等引起。一旦能够控制特征尺寸，也就能够把材料性质和器件功能提高到超过我们目前知道的水平。减小结构尺寸，出现了诸如碳纳米管、量子点与线、薄膜、以 DNA 为基础的结构和纳米尺度的激光发射器等等具有独特性能的实物。

Top-Down 和 Bottom-Up 虽然是两条不同的途径，但用微加工方法可以做到纳米尺寸，例如北京大学利用所发明的新型侧墙工艺制作出了 40 nm 宽的梁，中国科学院上海微系统与信息技术研究所也研制出 12 nm 厚的超薄悬臂梁，并据此观察到了硅杨氏模量的尺度效应，等等。反过来，用纳米技术同样可以做出器件和系统。在实际工作中我们应注意根据不同情况以及 Top-Down 和 Bottom-Up 这两种方法的特点，创造一种最佳结合的工艺方法。再者，微米技术的微机理研究实质上是研究从微米尺度到纳米尺度下有别于宏观世界的新现象，包括出现的各种新效应，如介观效应、表面效应和量子效应等等。当然，尺度的变化会引起物质结构性能从量变发展到质变，但从探索微小尺度的新机理这一点看，微米纳米技术属于殊途同归。

总体上看，微米纳米技术之间是一个相互交叉融合、相互促进的关系。目前如果要应用纳器件，就离不开微机电系统，离不开微电子。一个简单而形象的例子就是扫描原子显微镜 STM，它的探针最前面的部分是"纳"，后面就是"微"和"电"，三者集成在一起，紧密结合，协调工作。

具体从微米技术对纳米科技的作用和影响的角度看，一方面微米技术可以作为工具来访问和分析纳尺度的世界，典型的例子就是 80 年代出现的实现 Richard Feynman 想象的那种能力的仪器设备，包括隧道扫描显微镜、原子力显微镜以及近场显微镜，为纳米结构测量和操作提供了"眼睛"和"手指"。另一方面，过去十多年里积累发展起来的微米技术中的集成系统技术将发挥重要作用。对新的纳米器件，还有生物器件进行系统集成是微机电系统技术面临的主要挑战之一。通过利用微米技术进行集成，从而将基于纳效应的功能和特性转变成新的器件和系统。此外，微机电系统技术中的封装、交互以及集成技术将为纳米科学包括物理、化学、生物学和材料的研发提供帮助。今年早些时候召开的 Nanotech 2003 会议上，美国前 DARPA 的 MEMS 项目主管 Albert Pisano 在谈到 MEMS 与纳米技术的关系时，也发表了类似的看法，他认为"MEMS技术能担当纳米科学走向纳米技术的桥梁。"

从另一个角度看，MEMS同样需要利用纳米科学技术来寻求新的突破，纳米科技将为微米技术拓展敏感事物和认识事物的窗口。在这一方面纳米材料将会发挥重大作用。研究纳米材料很重要的一个目标是研究纳米材料的新颖的特殊性能，以开发出新的特殊功能的产品。简单地说，研究开发纳米材料不仅需要尺度上的表征，还包括各种性能的研究和挖掘，这是一项艰巨的工作。可以这样说，将哪一项包括纳米材料的物理、化学和生物学等内在特性弄清楚都将是重要的开拓性的贡献。最近科学家发现，如果碳纳米管的分子结构不同，它的光谱特性就不同，它的物理性能也有区别。当碳原子形成的六角形整齐地排列成管状时，其对电的反应就像金属一样。当六角形呈螺旋状排列时，其电特性就像半导体。还没有发现有其他任何材料能在分子结构不同时具有如此不同的特性。纳米材料在晶粒尺寸、表面与体内原子数比和晶粒形状等方面与一般材料有很大的不同，这些材料的奇异性能是由其本身尺度上的结构、特殊的界面和表面结构所决定的。这里面探索性的工作很多，如果把这些问题都搞清楚了，就有可能利用它们的内在特性和相互作用开发特定功能的器件产品。只有实现了纳米器件，物质的潜能才有可能被人们充分有效地利用。当然，纳米材料还有一些其他方面的应用，但更本质的更重要的应用体现在其内在特性的开发上。其次关于纳机电系统，它应是以超高灵敏度、超低功耗的纳器件为基础的应用系统，纳器件本身不应叫作系统，它是基本功能单元。

从国外的研发状况看，微米纳米技术相互融合已成为趋势和主流：一些基于微纳集成的科研计划和基础设施应运而生，比如：Los

Alamos and Sandia 国家实验室成立了集成纳米技术中心（CINT）；法国将在 2004 年正式建成微米纳米技术发明中心，德国微系统技术 2000 未来计划也将微纳接口列为未来重要领域。德国 Karlsruhe 研究中心将在其中期计划中将微系统计划和纳米技术计划合并到一个计划中，欧盟积极支持微米纳米技术和信息技术的集成和合并；美国国家科学基金会/商业部在"改善人类性能"的标题下倡议关于"合并技术"（纳米—生物—信息—认知）的讨论；越来越多的国际会议和展览将微纳结合作为重要主题，如今年 5 月 26—28 日在德国柏林召开的 NAMIX 第一届会议及今年 9 月 22—26 日即将在法国格勒纳布尔（Grenoble）召开的 MINATEC 2003 都将会议主题定为微纳集成和接口，明年在美国波士顿召开的 NANOTECH 2004 主题也覆盖纳米、生物及微技术等。

3 开展微米/纳米技术具体工作中应注意的几个问题

微米纳米技术是一门新兴的、多学科交叉的研究领域，汇集了电子、机械、材料、制造、检测以及物理、化学和生物等不同学科新生长出来的微小和微观领域的科学技术群体，是科学技术创新思维的结果，极富挑战性。一方面，它给我们提供一个非常大的机遇，就是创新的机遇，甚至是发现新规律的机遇；另一方面，尽管人们已经意识到微机电系统技术和纳米科技有可能在未来数十年内引发一场工业革命，但在基础探索、发现以及制造方法等方面将面临相当的艰难和风险，因为它是过去没有人从事过的新领域，这是客观的实际情况。突破这些困难的关键是对微米纳米技术要采用创新的研究方法，多学科综合集成，运用先进的检测和工艺手段以及不同的制造范例。现在据了解，国内有一些单位成立了微米纳米技术研究中心，这是件好事。

希望在实际工作中要重视微米纳米技术相互之间的关联性，把微米纳米技术工作紧密结合起来开展，这有利于创新，有利于赢得机遇。同时还要强调的是，研究设计工作必须与制造技术紧密相联系，这有利于微米纳米技术的结合，有利于它们相互促进。

此外，我们一定要注意加强国际国内的交流和合作。大家可能都有这样的体会，每次参加学术研讨会或相关的调研活动，都能够从别人那里得到一些借鉴或感悟，这肯定有利于我们开阔思路，有利于我们的科研工作。第六届全国微米纳米技术学术会议，无疑为从事微米纳米技术研究工作的专家学者提供了一个集思广益的平台，希望同志们抓住机遇，敢想敢为，为推动我国微米纳米技术的发展贡献力量。

微纳技术进展、趋势与建议

(2006 年 12 月)

1 产业发展趋势

近年来，MEMS 产业发展迅速，据著名的国际 MEMS 咨询机构约尔公司（Yole）分析，2005 年全球基于硅和石英器件的 MEMS 市场高达 50 亿美元，预计 2010 年将达到 100 亿美元（见图 1）。其产品主要包括：喷墨打印头、光 MEMS（主要是数字微镜器件）、压力传感器、惯性传感器、微流控器件、RF MEMS、硅麦克风及微型燃料电池，其中硅麦克风和微型燃料电池为新出现的产品（见图 2）。

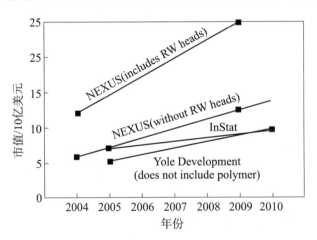

图 1 国际上 3 个 MEMS 咨询机构预测的 MEMS 市场趋势

本文发表于《纳米技术与精密工程》2006 年第 4 卷第 4 期。根据作者 2006 年在第八届中国微米纳米技术学术会议上的讲话整理而成。

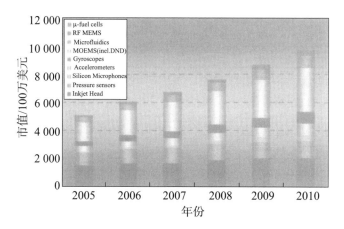

图 2 2005—2010 年全球 MEMS 市场产品分类

从 MEMS 市场分类的角度，需要重点关注应用在生物领域中的微流控市场、恶劣环境中应用的 MEMS 市场、光 MEMS 市场、RF MEMS 市场、惯性 MEMS 市场和移动电话中的 MEMS 市场等。

1.1 微流控/生命科学

探知生命的奥秘是人类一直努力的方向，微纳米技术的进步为这一领域的发展提供了机会。微流控科技在生物领域的应用是近年来微纳米技术最活跃的方向之一，具有降低分析成本、缩短反应时间、增加精度、多功能集成等优点。据统计，2005 年微流控市场已达 9 亿美元，预计到 2010 年将达到 20 亿美元。

目前，仅有少数微流控器件实现了商业化（主要是药物筛选或用于科学研究），大部分处于原理样机阶段。

微流控器件是在各种不同的衬底材料上制造完成的，硅衬底主要用于药物发明和流体配制；聚合物衬底则可以用于药物配制、诊断、实验室分析、生态学和农业食品等（见图 3）。

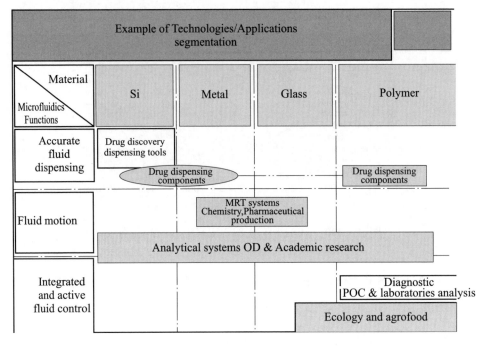

图 3 微流控功能/衬底材料

在聚合物衬底上加工微流控器件的主要方法有：1）阳模/阴模制造，方法有光刻、激光消融、电火花加工、微机械加工、电铸；2）原型制造，方法有激光消融、微型立体光刻、软光刻、纳米压印；3）批量制造，方法有基于压模的制造技术、热模压、微注塑、轧制、直接等离子体腐蚀。后处理技术是发展中的瓶颈技术，包括激光焊接、热键合、超声键合、粘接、打孔和表面改性（见图4）。

1.2 恶劣环境中的 MEMS

MEMS 器件通常的使用环境：温度在 $-60\sim150\ ℃$ 之间；而且是在非化学腐蚀和可控压力环境下。使用新型材料如碳化硅、绝缘体上硅（SOI）等，可以使 MEMS 在高温、低温、辐射、化学腐蚀环境中使用，这也是 MEMS 近年来出现的新的市场。据统计，恶劣环境中

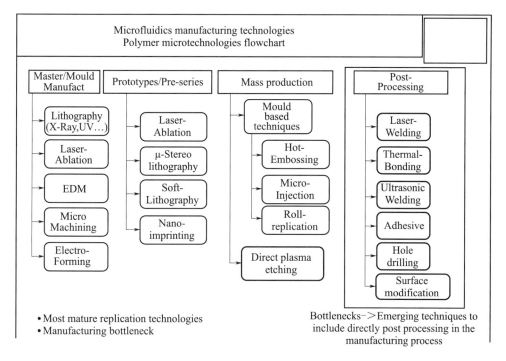

图 4 微流控器件制造技术流程

MEMS 市场 2005 年约为 1 亿美元，2010 年将超过 5 亿美元，其应用领域主要包括汽车、航空、空间、工业、石油钻探等（见图 5）。

以航空应用为例，过去空客 A320 中仅使用了 20 个压力传感器，而现在 A380 中就使用了 200 个，从而提高了飞机的可靠性、降低了维护成本。

1.3 光 MEMS

光 MEMS 的主要应用领域包括显示、红外成像、光谱仪、条形码读出、无掩膜光刻、自适应光学、指纹传感器和头盔显示器等。

据统计，光 MEMS 市场 2005 年约为 14 亿美元，预计 2010 年将达到 32 亿美元。其中最大的市场仍然是数字光处理投影仪、数字光处理电视和微型热辐射计。其他器件包括光互连、光衰减器件、微型光

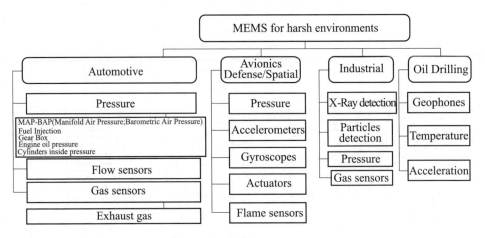

图 5 MEMS 在恶劣环境中的应用领域

谱仪等。需要注意的是，新出现的产品中包括了干涉调制显示仪（iMoD）（见图 6）。

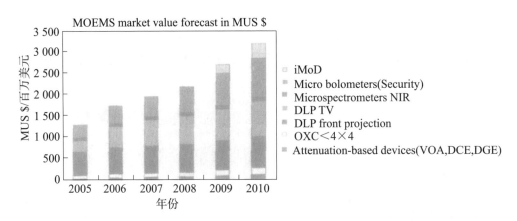

图 6 光 MEMS 市场趋势

1.4 RF MEMS

RF MEMS 的主要市场有移动电话、宽带局域网、消费类与信息技术、基站、微波通信、汽车雷达、仪器、卫星及相控阵等。

据统计，RF MEMS 2005 年市场达 2 亿美元，预计 2009 年接近 12 亿美元，主要市场是移动电话和消费类与信息技术产品，如图 7 所示。

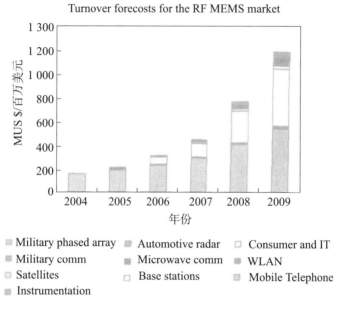

图 7　RF MEMS 市场趋势

1.5　惯性 MEMS

惯性 MEMS 主要应用领域包括军事、航空/航天、汽车、消费类与信息技术。2004 年加速度计和陀螺市场均为 4 亿美元，估计 2009 年将均达到 7 亿美元左右，其中消费类市场发展较快，年增长率在 20％～30％，如图 8 所示。

MEMS 陀螺应用领域比较广泛，对量程和精度分别有不同要求，例如，在手机中的应用指标是低精度、宽量程。未来在汽车和军事领域将有较快增长，如图 9 所示。

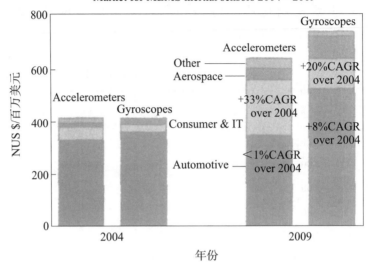

图 8　惯性 MEMS 市场趋势

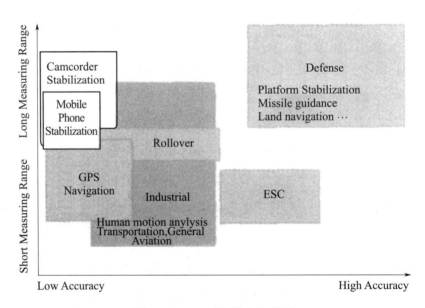

图 9　MEMS 陀螺应用领域

1.6 移动电话中的 MEMS

移动电话中的 MEMS 器件主要有薄膜体声谐振器、硅麦克风、加速度计、自动聚焦部件、微型燃料电池、多频振荡器、嵌入式 RF MEMS、图像稳定化器件等。2004 年市场接近 1 亿美元，主要是薄膜体声谐振器，2008 年将接近 6 亿美元，如图 10 所示。

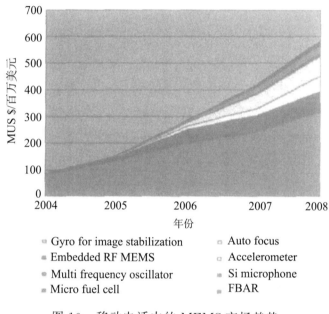

图 10　移动电话中的 MEMS 市场趋势

由于中国占全球移动电话产量的 1/3，是世界上最大移动电话生产国，因此国际公司比较重视在中国的发展。例如，国际最大的麦克风制造公司娄氏公司（Knowles Acoustics）在苏州的制造厂已经开始量产，以巩固该公司在全球 MEMS 硅麦克风生产领域的领先地位。

2006 年 5 月，国际最大的移动电话振荡器制造公司硅时公司（Silicon Times）开始在上海销售 MEMS 振荡器。

2 最近的一些进展

国际微纳技术发展迅速，应用领域不断扩大，同时产业进程加快。下面介绍一些国际上比较新的研究与开发的项目，这些项目的进展主要得益于微纳制造技术的进步。

研究项目（主要在大学进行）包括人工复眼、陀螺诊断癌症、碳纳米管热辐射计、微机械脉冲磁强计。

技术开发（主要在公司进行）包括纳米飞行器、电润湿显示、碳纳米管热沉、药物定点释放。

2.1 人工复眼

按照光学成像原理，动物眼睛主要分为两类：一类是照相眼系统（如人类）；另一类是复眼系统（如昆虫）。蜜蜂的眼睛是由几万个相同集成光学单元沿弯曲的表面排列而成，每个单元吸收很小角度的入射光，整个阵列则形成宽场的视角。蜜蜂人工复眼的单元则由微透镜、聚合物锥、波导芯、波导包层和光探测器构成，如图11所示。

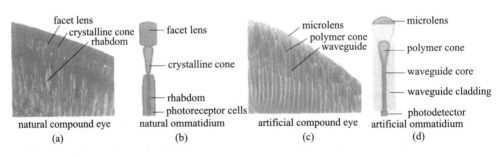

图 11 蜜蜂自然复眼和人工复眼构成的比较

最近，加州大学伯克利分校研究人员在 SCIENCE 杂志发表论文，宣布利用微纳制造技术研制出了蜜蜂复眼的曲面阵列光学元件。他们主要利用了波导的微透镜辅助自写技术和光敏聚合物树脂中两种交联机制。第一步,当 UV 能量超过聚合物发生光聚合反应阈值能量时,聚合物折射率会增加;第二步,通过加热引起的交联反应会使聚合物折射率减小,而 UV 曝光过的折射率不变(见图 12)。通过这两步过程,就可以在聚合物中形成波导芯和波导包层(自写工艺)。

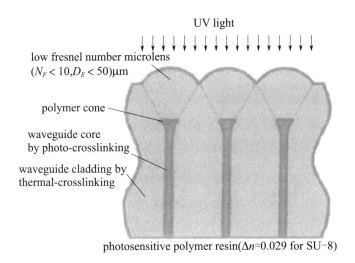

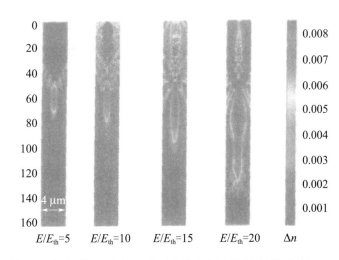

图 12 波导的微透镜辅助自写过程与折射率的分布

制备人工复眼的微纳加工过程如图 13 所示。

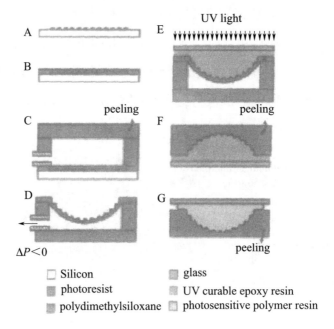

图 13 制备人工复眼的制造流程

A—用光刻胶制备微透镜模板；B—第一层 PDMS（聚二甲基硅氧烷）压膜（22 μm）；
C—PDMS 键合；D—PDMS 变形；E—注入 SU-8 胶，UV-光互联

从实验结果看，人工复眼在物理尺寸、光学特性等方面与自然复眼非常接近，它在数据读写、医疗诊断、监视成像等方面具有潜在的应用前景。

2.2 陀螺诊断癌症

最近，英国纽卡斯尔大学发明了一种直径小于 0.1 mm 的陀螺仪圆盘，这种微型陀螺仪可以用来探测由癌细胞分泌出来的特定蛋白质。当这些蛋白质绑定一种 DNA 通过陀螺仪的圆盘表面时，圆盘就能探测到。欧盟已经为这项生物传感器研究提供了 1 200 万欧元的资金，争取在未来 4 年内把这项技术应用于临床治疗，检测的范围也会扩大到传染病的检测，如肺结核等。由于是新立项的课题，详细的原理还

未公开。

2.3 碳纳米管热辐射计

最近，美国加州大学河滨分校的研究小组在 SCIENCE 杂志发表论文，称单壁式碳纳米管会因光照而导致导电性改变，这一特性在真空中格外明显（见图 14）。这项特性可能有助于碳纳米管在红外线热像传感器、光谱仪及红外天文学等领域应用。

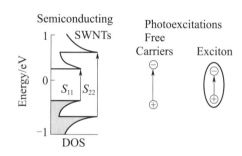

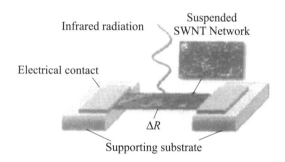

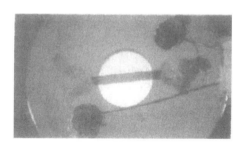

图 14　单壁碳纳米管受辐射的过程

单壁碳纳米管薄膜电阻率的改变主要来自材料温度的变化,但此响应通常因热能透过基材耗散而受到限制。因此唯有将薄膜悬空并在真空下操作,才能观察到完全的响应,由于此响应极大,加上单壁碳纳米管在红外波段的吸收很强,因此它很适合用来制作红外线传感器。从公布的性能来看,低温下(50 K,真空)的性能:吸收系数比碲镉汞高 1 个数量级;吸收区域在紫外线到远红外;由于质量在纳克量级,因此为低热容。常温下(330~100 K)其电阻温度系数与二氧化钒在相同数量级,因此有可能成为热释电探测器、非晶硅阵列等低成本探测器的替代品。

2.4 微机械脉冲磁强计

精确测量脉冲磁场是一项富有挑战性的工作。过去,霍尔器件或磁阻器件虽然体积很小,但仅能测量较小的磁场,且受温度影响较大。贝尔实验室最近在 SCIENCE 杂志发表论文,说明用 MEMS 技术研制出的磁强计可以测量强度为 60 T、脉宽 100 ms 的磁场。其主要优势是适合于局部区域的磁体测量,因尺寸小,有更高的谐振频率,因此能够实现快速响应(见图 15)。

2.5 纳米飞行器

众所周知,20 世纪 90 年代比较著名的微型飞行器是"黑寡妇",其主要性能是:翼展为 152.4 cm,质量 80 g;它有 2 个微处理器、命令接收器、数据传输器、两轴磁姿态传感器、MEMS 压力传感器、速度陀螺仪、马达、微涡轮喷气发动机。

2006 年 5 月,DARPA 授予洛克希德·马丁公司一项总金额为 170 万美元、为期 10 个月的合同,开发一种用于城市作战环境、可在室内或室外搜集军事情报的纳米飞行器(NAV)。这种 NAV 尺寸和形

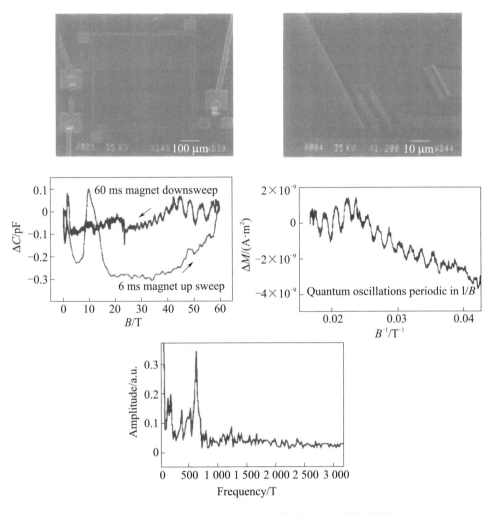

图 15 硅表面微机械加工的磁强计照片及测试结果

状都与枫树种子相近，机长约 3.8 cm，最大起飞重量约 10 g。它采用的单叶片式机翼除用来控制升力和俯仰外，还嵌入了遥测、通信、导航、制导用摄像机和电池。该机可搭载一个重约 1.98 g 的传感器任务模块，并利用化学能（超微型火箭）在悬停状态下将其发射到 1 005 m 处。

在典型作战使用情况下，该机由 1 名士兵发射，后者可通过嵌入机翼的制导用摄像机观察飞行路径，以引导飞机飞向目标，就像枫树种子一样，该机在飞行中会旋转，但其摄像机将稳定地保持前视，并不断地将图像传输到控制员的掌上显示器，在投射出任务载荷后，操作员可控制它返航。当系统成熟后，该机将装备 1 台简单的自动驾驶仪，从而具备有限的自主飞行能力。

值得注意的是，上述的纳米飞行器的形状和枫树的种子相近，和我们熟知的"黑寡妇"的形状完全不同，具有仿生的特点。

2.6 电润湿显示

2003 年，飞利浦公司 Hayes 等人在 NATURE 杂志发表论文，提出电润湿显示原理，通过施加在水和顶部或底部电极之间的电势差，可以独立并可逆地开关每一油层，因此引起光学性质的变化，进而又由这些光学性质的变化转换成视觉的变化，如图 16 所示。

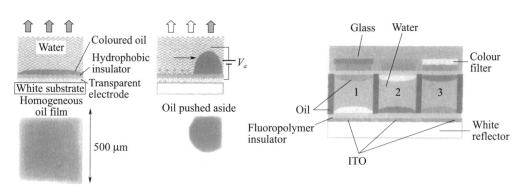

图 16　电润湿显示原理

不久前，新创始集团（New Venture Partners）与荷兰菲利浦公司共同成立了名为利科维斯塔（Liquavista）的一家新公司。2006 年 6 月，该公司在旧金山展示了计算器大小般的电润湿显示原理样机，如图 17 所示。电润湿显示是一种超薄显示科技，在未来可将其用于数字随身听、手机、手表以及其他也需要高亮度及迅速更新速率的便携式产品,电润湿显示这项技术据称能比反射式显示技术提供更好的亮度及更快的响应时间。

图 17 电润湿显示原理样机

2.7 碳纳米管热沉

碳纳米管(CNT)有许多奇异的特性。最近,日本富士通公司利用碳纳米管制造的热沉就利用了碳纳米管具有极高的热导率这样的特性。

热导率[单位:W/(m·K)]数值分别是:碳纳米管为 1 400;铜为 400;金为 300。

目前大功率芯片为了散热,将芯片紧贴在散热基板上,但是由于输入/输出引线又有寄生引线电感,因此最好用倒装焊技术,将芯片倒装在基板上,但是倒扣芯片的电极必须通过金属电极与基板相连,因此又存在散热问题,富士通公司利用碳纳米管制造的热沉就很好解决了这个问题,如图 18 所示。

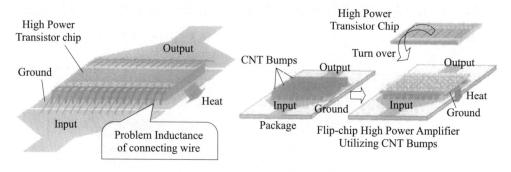

图 18　碳纳米管热沉示意

2.8　药物定点释放

美国吉普瑞克斯（CHIPRX）公司的药物定点释放系统已经投放市场，该结构中含有电池、控制电路、生物传感器、药物储藏室、药物释放孔、生物兼容的防渗透膜等，如图 19 所示。

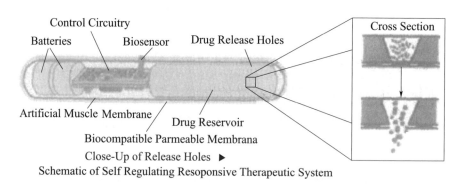

图 19　药物定点释放系统示意

3　问题与建议

虽然微纳技术处于高速发展中，但是仍有一些基础性的问题急需解决，例如、标准、测试与表征、表面性能控制、微纳尺度互连、多学科交叉、基础研究与应用研究结合等等。

3.1 标准

标准的等级一般分为：国际、区域和工业共同体标准等。

建立标准的目的是降低加工厂、设计者、设备制造商及终端用户之间的自由贸易壁垒，而且用户要求 MEMS 便宜、互换性好、容易集成到其他系统。目前已经有一些组织在推动标准化的建设，例如，欧洲电技术标准化委员会（CEMELEC）、德国标准化协会（DIN）、电气电子工程师协会（IEEE）、国际电技术委员会（IEC）、美国国家标准与技术研究所（NIST）等等。但是，MEMS 标准的建立仍存在不少技术困难，主要包括相互竞争的制造技术、不同的设计程式及多种多样的封装技术。

3.2 测试与表征

由于尺度变小，材料及其相关特性表征非常困难。有关纳米结构的测试和微弱信号的检测等问题一直困扰着研究人员。

例如，关于纳米悬臂梁的检测，IBM 的 Fritz 等人在 SCIENCE 杂志上发表论文，提出将生物、化学分子转移到悬臂梁进行测量，其基本原理是先在悬臂梁上涂覆一层有机材料，它会与被探测物质发生作用，从而引起悬臂梁变形，通过测量悬臂梁偏转、应力或谐振频率的改变就可测量生物/化学分子的重量。目前，常用的一些检测悬臂梁变形的方法还包括光学、压阻、压电、电容及隧道电流等。

美国西北大学的研究者最近在 SCIENCE 杂志上发表论文，将金属氧化物半导体场效应晶体管（MOSFET）嵌入到悬臂梁根部，可以探测 5～10 nm 的偏转。对维生素的测量结果：偏转灵敏度与光学方法相当；比其他原理高 1～2 个数量级。

3.3 表面性能

尺度缩小引起表面积/体积比增加，表面在决定材料的性能中起主要作用。

一方面可利用表面性能研制传感器；另一方面微纳材料尤其是纳米材料，其基本的性能（如力学、热学、电学、磁学、光学等性能）与表面状况密切相关，如何进行可控制备材料的表面状况研究面临着挑战。

Harvard 大学研究者在 SCIENCE 杂志上发表论文，利用硅纳米线表面与化学物质的作用，改变了表面电荷状态，进而改变了硅纳米线的电导，通过测量电导可确定 pH 浓度。例如，可探测 pH 浓度到 pM（10^{-12} mol）的浓度范围。

3.4 微纳尺度互连

将纳米结构与微米结构互连后，才能与宏观世界联系起来，但这是一件极具挑战性的课题，第 20 届 IEEE MEMS 会议（2007 年）已将微纳尺度互连作为一个方向进行征文，目前的主要方法是先做电极，后做纳米线。

3.5 学科交叉

以人体药物释放系统为例，研发这种系统需要微电子、生物、化学、传感器、机械等学科人才的协作。学科交叉是进行"跨学科"的研究活动，学科交叉点往往就是科学新的生长点和新的科学前沿，有利于解决人类面临的重大科学问题和社会问题。

学科交叉归根到底就是不同专业人才之间的合作，只有排除了非学术因素（职业道德、利益分配等）的影响，才能实现。

3.6 基础研究与应用研究紧密结合，有利于相互促进发展

基础研究是以认识自然现象、揭示客观规律为主要目的的探索性研究工作；基础研究是技术创新，特别是核心技术产生的主要源泉。

基础研究的重大突破，往往能够极大地推动新技术的发展，带动新兴产业的崛起。微纳科技发展的趋势在于，已逐步从单纯满足科学家对自然现象和规律认识的兴趣，转向更加注重服务于人类社会发展的需要。经济社会发展需求对基础研究的推动力已经大大超过单纯的科学自身发展的吸引力。应用研究在获得知识的过程中具有特定的应用目的；应用研究直接推动科技进步，使生产过程合理，效率提高，成本降低，不断涌现出新产品。

在微纳科技发展中，学科间交叉融合、相互渗透的趋势日益明显，研究对象的复杂性不断增强，研究成果转化为产品的周期越来越短，速度越来越快。因此，基础研究与应用研究紧密结合，有利于相互促进发展，本文中列举的脉冲磁强计、电润湿显示、碳纳米管热沉等就是基础研究与应用研究相互促进发展的实例。

微米纳米科学技术发展及产业化启示

(2007 年 12 月)

摘 要 近年来,微米纳米科学技术发展迅速,在产业化方面已经取得了不少进展。本文从发展现状和趋势上,结合我国具体情况,阐述了以下观点:开展创新工艺和设计方法研究,提高微机电(MEMS)应用和产业化能力;开展基础研究,提高 MEMS 器件和系统的水平;加强微米/纳米科学技术成果的结合。最后对今后研究工作提出了建议。

关键词 微机电系统;纳米科技;发展现状与趋势

1 近年来微米纳米科技发展的状况和趋势

1.1 微机电器件与系统

2006 年微机电(MEMS)技术发展迅速,图 1 为 2006 年 MEMS 各类器件和系统的份额图。其中打印头(Inkjet heads)仍然占据着主导地位,之后依次为压力及惯性传感器、微光机电等。微机电技术在未来仍然是发展热点,知名 IT 专栏作家约翰·德瑞克在 2006 年末对 2007 年全球 IT 发展趋势做五大预测时指出:MEMS 依然是半导体市场最热门的发展方向。

本文发表于《纳米技术与精密工程》2007 年第 5 卷第 4 期。合作作者:丁衡高,朱荣。

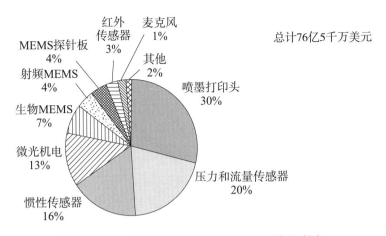

图 1　2006 年 MEMS 各类器件和系统的份额

MEMS 应用和产业化发展始终是人们关注的重点。图 2 为 2006 年世界 30 强 MEMS 生产企业的年收益，其中德州仪器（TI）公司以数字光处理器件（DLP）芯片产量高居 30 强的榜首。TI 公司的 DLP 是为高清晰度显示设备研制的，其核心是一种数字微镜器件（DMD）光半导体芯片。由于 DLP 技术实现了商业化，该技术的应用也越来越广泛，目前，全球已经有将近 40 家著名的电视和放映设备厂商开始采用 DLP 子系统。LG 公司也在韩国电子展览会上展示了一款采用 DLP 子系统的 132 cm（52 英寸）彩电。MEMS 器件正在加速向具有信号处理功能的微传感器芯片，以及能够完成独立功能的"片上系统"（微系统）方向发展，DLP 芯片就是一种微系统。

早期的 MEMS 器件性能不佳，无法实用化。但随着各国在制作工艺、封装等方面的尝试和努力，价格便宜、小型化、性能可接受的产品正逐渐走入人们的生活。

目前，MEMS 器件应用最成功、数量最大的产业当属汽车工业（见图 3）。现代汽车采用的安全气囊、防抱死制动系统（ABS）、电喷控制、转向控制和防盗器等系统都使用了大量的 MEMS 器件。为了防

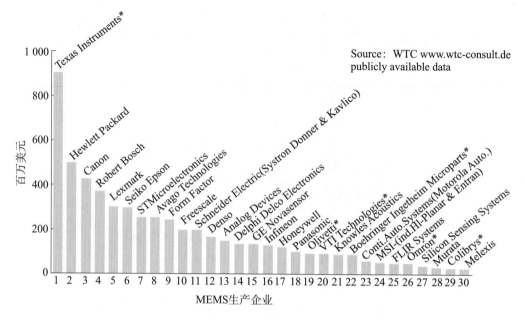

图 2　2006 年世界 30 强 MEMS 生产企业的年收益

止汽车紧急刹车时发生方向失控和翻车事故，目前各汽车制造公司除了装备 ABS 系统之外，又研制出电子稳定程序（ESP）系统与 ABS 系统配合使用。发生紧急刹车情况时，这一系统可以在几微秒之内对每个车轮进行制动，以稳定车辆的行车方向。

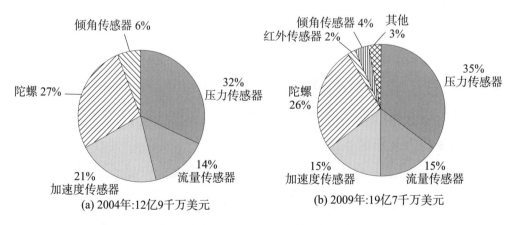

图 3　全球车用 MEMS 传感器份额

第二大应用为消费电子市场。以下主要领域正在推动 MEMS 传感器在消费电子中的应用，它们是：数字电视、打印机、游戏机、手机、笔记本、数码相机等便携设备中的硬盘保护；利用倾斜度的变化在新出现的复杂手持设备中实现多种功能界面（游戏和浏览网页）；为实现精确导航而使用的导航推算系统；比压电解决方案更经济省电的静态和动态成像设备中的防抖功能。许多公司都看到了 MEMS 传感器在消费电子市场的潜力并开始加大推广力度。任天堂公司已经将意法半导体（ST）公司的三轴加速度传感器用在新型家庭游戏控制台中，而耐克公司更是推陈出新，将基于 Z 轴加速度计的步程计功能与软件结合后，推出了一款具有热量计算功能的跑鞋。此外，针对手机中小规模使用，ST 还创新性地推出了将存储器、控制器和 MEMS 传感器集成在一起的 MMC 卡，帮助手机厂商加快产品的上市时间。拥有超过 25 年 MEMS 传感器开发和生产经验的飞思卡尔公司也不例外，该公司不久前推出了两款双轴加速度计和一款三轴加速度计，其灵敏度值可通过引脚选择进行调节，设计人员能够在（1.5～10）g 的加速度范围之内任意选择 X、Y 和 Z 轴感测的组合。据称，这三款器件瞄准的便是那些需要检测由于坠落、倾斜、移动、定位、撞击或振动产生微小变化的低成本消费电子产品。德国博世传感器（Bosch Sensortec GmbH）于 2007 年发布了他们的便携设备用三轴加速度传感器"SMB380"（见图 4），确定应用于手机，SMB380 外形尺寸为 3 mm × 3 mm × 0.9 mm，消耗电流在待机时为 1 μA，工作时为 200 μA，为目前世界上最小尺寸的三轴加速度传感器。

在 MEMS 应用过程中，世界各国也在不断努力进行新产品、新技术的开发，以适应不同需求。图 5 为世界 10 强 MEMS 研究和生产基地。这些正在研发或计划研发的 MEMS 新器件和系统应该引起我们的

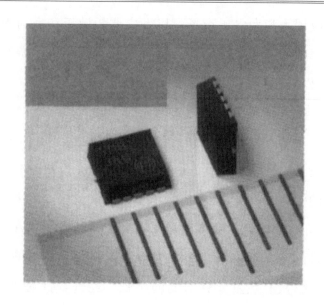

图 4　Bosch Sensortec GmbH 的"SMB380"三轴加速度计

高度重视,它们可能是未来应用市场的宠儿,具有较大的潜能和发展空间。据《新科学家》杂志报道,美国 NASA 正在开发一种小型而便宜的"智能气象气球",充满氦气,重 130 g,像沙滩充气球一般大小,用来监测火箭或飞船发射前夕发射场区的气象状况。这项研究的起因源于数月前一艘飞船在发射前夕突遭一场冰雹的袭击,导致油箱严重受损。"智能气象气球"的研发采用了"智能尘埃"技术;所谓"智能尘埃"就是可以感知环境并传输数据的 MEMS 器件。"智能气象气球"以太阳能作为能源,携带了温度、压力、湿度及 GPS 等传感器,配备了无线发射器,它们以组网方式形成系统工作,气球一旦漂离场区过远,还可以通过卫星链路发回数据。当然正在研发的新器件和系统还远不只于此,各国在以下方面也正在开展研究,这些成果不久将进入未来市场:1) 生物传感器系统,如欧洲 SPOT－NOSED 计划研究模拟人类和动物味觉感知的纳米生物传感器;2) 生物致动系统,如英国

Portsmouth 大学开发的一种基于 DNA 的转换器，这是世界上第一个生物纳米技术致动器；3）微能源，如基于人体体温的热电式微型发电机和振动式微型发电机等；4）RF MEMS，如美国 Rockwell Collins 公司开发的超宽带、多通道接收机和光谱传感器。

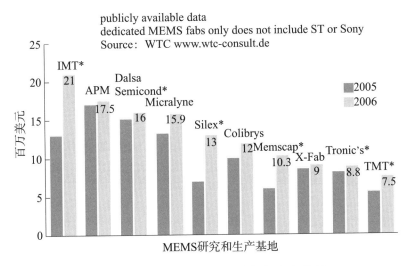

图 5　世界 10 强 MEMS 研究和生产基地

1.2　纳米科技

纳米科技作为 21 世纪的一个重要新兴科技领域，在理论与实践上正经历着高速发展。大量新型纳米材料与器件不断被开发出来，并在信息、生物医学、能源、国防以及人们日常生活的各个领域中展现出前所未有的应用前景。

我们可以从论文发表及专利申请数量上看出各国在纳米科技发展上的热点方向：欧盟在光学和光电材料、有机电子学和光电学、磁性材料、仿生材料、纳米生物材料、超导体、复合材料、医学材料、智能材料等方面有较大的研究力度；美国在纳米材料、医学生物材料、复合材料、功能材料等方面发表的论文较多，应用研究多在半导体芯片、癌症诊断、光学新材料和生物分子追踪等领域；中国在传统材料、纳米材料及其应用、隧道显微镜分析和单原子操纵等方面发表论文较多，主要以金属和无机物非金属纳米材料为主，约占80%，高分子和化学合成材料也是一个重要方面，而在纳米电子学、纳米器件和纳米生物医学研究方面与发达国家有明显差距。中国论文被引频次方面与发达国家相比仍有差距。据相关资料报道，2000—2006年，各国纳米技术发明专利数量均有较大幅度的增长，中国纳米技术专利申请数已排世界第3位，纳米技术专利各主要申请国所占比例如图6所示。国外专利研究侧重纳米有机物材料的应用，而且主要应用在催化剂、光学器件、半导体和导电材料、磁性材料和显示器等高技术含量的电子信息领域，中国在纳米涂料、纳米橡胶等领域应用较多。

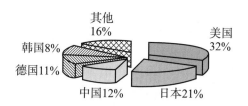

图6　各国纳米技术专利申请比例分布

纳米科技已成为科技发展的制高点，并成为许多国家提升核心竞争力的战略选择。2006年我国发布了《国家中长期科学和技术发展规划纲要》，对未来的纳米科技的发展作出了总体部署，包括实施纳米科学技术研究重大科学研究计划，重点研究纳米材料的可控制备、自组装和功能化，纳米材料的结构、优异特性及其调控机制，纳加工与集成原理，概念性和原理性纳器件，纳电子学，纳米生物学和纳米医学，分子聚集体和生物分子的光、电、磁学性质及信息传递，单分子行为与操纵，分子机器的设计组装与调控，纳米尺度表征与度量学，纳米材料和纳米技术在能源、环境、信息、医药等领域的应用；在重点领域及其优先主题中开展纳米生物药物释放系统的研究；在前沿技术中明确在纳米科学研究的基础上发展纳米材料与器件，发展微纳机电系统、微纳制造等。

2 观点与建议

从微米纳米科技的发展状况上看，目前的工作应注意支持有潜力、有应用前景的创新科学技术研究，以下3个问题应重点考虑。

2.1 开展创新工艺和设计方法研究，提高 MEMS 应用和产业化能力

市场需要便宜、可靠、性能好、久经考验以及易于扩大产量的 MEMS 器件，因而为提高我国 MEMS 和纳米器件的应用和产业化进程，我们应该致力于开发具有自主知识产权的创新制造工艺和设计，建立有特色的标准，提高竞争力。国外为抢占 MEMS 市场，都在努力开发新工艺，如拥有 20.32 cm（8英寸）晶圆量产 MEMS 传感器能力的 ST 公司面向加速度传感器和压力传感器这两个 MEMS 技术分支，开发出了分别名为 Thelma（微致动器及加速计用厚外延层）和 VenSen（压力和动力感应器用工艺）的标准生产工艺平台，为满足多元化的市场需求，提供适当的应用灵活性。ST 公司的专利还包括在 MEMS 传感器小型化道路中起到关键作用的单硅片制造技术。常规技术利用在玻璃底板或者硅晶元上键合硅薄膜形成空腔，这一程序阻碍了传感器体积的进一步缩小。而 ST 公司的全硅（full-silicon）技术无需键合就能够在单硅片上实现更小的空腔。这使腔体宽度能够从 100 μm 迅速缩小到 1 μm 左右，而表面薄膜也从 50 μm 降到了 10 μm，硅片内腔的减小为外围尺寸的减小带来可能。据悉，在标准技术下，

传感器尺寸为 2 mm×1 mm；采用全硅技术后，可以缩小到 1 mm× 300 μm。借助上述工艺，ST 公司已经将其 MEMS 传感器产品成功推进到第 4 代产品，随着集成度的不断提高，芯片尺寸也以超过摩尔定律的速度迅速缩小。由于目前仍然没有解决利用 COMS 工艺来实现 MEMS 器件的标准化制造，为了使 MEMS 传感器与模拟或数字电路集成在一起，将整个系统的功能完全集成在一个半导体封装中的 SiP (system in package) 成为必由之路。2005 年，ST 公司率先推出了采用小型化的 LGA（基板格栅阵列）封装的 MEMS 传感器，进一步降低了该器件的普及成本。

其他公司也在为实现 MEMS 的标准生产工艺而努力。Akustica 公司最近推出了全球首款基于 MEMS 技术的单芯片数字扬声器，利用 CMOS 标准制造工艺来实现 MEMS 结构。Dalas 半导体公司也在积极开发针对不同 MEMS 器件的模块化 CMOS 工艺。Bosch 公司也针对其新 MEMS 压力传感器（见图 7，3 μA，尺寸为 5 mm×5 mm× 1.6 mm）开发了 APSM（Advanced Porous Silicon Membrane）工艺，采用全硅表面加工，实现了与 IC 电路的集成。

图 7　Bosch 公司的 MEMS 压力传感器

我国科技人员也正在朝着自主创新设计和工艺研究方向努力，如东南大学国家专用集成电路系统工程技术研究中心[1-3]根据市场需求开发了基于低成本体硅工艺的 PDP 驱动芯片，该芯片针对长虹 PDP 模组和荫罩式 PDP 模组对驱动芯片的要求，联合芯片用户四川长虹和南京华显，同时联合芯片制造厂家无锡华润上华共同开发，制造工艺的自主研发有助于降低产品成本。东南大学的基于自主开发的高低压兼容体硅工艺的 100 V 192 路输出 PDP 列驱动芯片成本在 5 元人民币左右，比 ST 等公司用外延工艺设计的同类产品成本低 30%，极大地提高了市场竞争力。自主创新设计与制造工艺研究还可以使我们掌握核心技术的知识产权，可以在较长时间内发挥较高的市场竞争力，并有助于建立产品标准化。

相比于 MEMS 技术，微电子技术发展至今，虽然目前大部分加工技术已趋于成熟，但在未来仍然可以考虑利用新机理、新效应、新结构（如纳米功能结构），以进一步提高微电子器件性能、减小体积和功耗。例如，美国科研人员在 2006 年开发出一种特殊纳米芯片制造工艺技术，可使碳纳米晶体管在集成电路上"听从安排"，在硅片上垂直生长出单个碳纳米管，这突破了碳纳米管计算机研发的"瓶颈"。碳纳米晶体管应用于大型集成电路，有助于研发超级运算速度和低能耗的微处理器，预计其运算速度将比目前看好的下一代硅芯片要快 10 倍，且耗能更少。又如三星电子公司 2006 年发布了 40 nm 32G "NAND"型闪存，在全球率先开始了 50 nm DRAM 内存的生产，其容量达到 1 Gbit，并在国内外申请了 51 项专利。与三星电子 2005 年开发的 60 nm 工艺相比，其工作效率提高了 55%。

2.2 开展基础研究，提高 MEMS 器件和系统的水平

MEMS 器件和系统发展至今，虽然有体积小、重量轻、功耗低、

可批量生产、成本低、可与电子线路集成等优点,但却被定位在精度较低、性能不高的水平。其主要原因是加工精度不够、材料和结构的尺度和表面效应影响、微观摩擦和噪声等。为提高 MEMS 器件和系统水平,必须加强基础理论和微观机理研究,研究微结构尺度效应、微结构表面效应、微观摩擦机理、传热机理、误差效应和材料性能等,研究多物理场和多材料耦合效应,研究微观多种力效应,进行结构和系统的误差源分析和消除方法研究,并采用新材料和新检测手段,利用新型的物理-化学-力学-电磁现象/效应,要在前沿技术和基础研究上发展 MEMS 和纳米器件,发展微纳机电系统和微纳制造等。

从基础研究来说,纳米结构的出现,把人们对纳米材料出现的基本物理效应的认识不断引向深入。无序堆积而成的块体材料,由于颗粒之间界面结构的复杂性,很难把量子尺寸效应和表面效应对奇特理化效应的机理搞清楚,而纳米结构可以把纳米材料的基本单元(纳米微粒、纳米丝、纳米棒等)分离开来,这就使研究单个纳米结构单元的行为、特性成为可能,更重要的是人们可以通过各种手段对纳米材料基本单元的表面进行控制,这就使我们有可能通过调制纳米结构中纳米基本单元之间的间距,进一步认识它们之间的耦合效应。因此,纳米结构出现的新现象、新规律有利于人们为构筑纳米材料体系的理论框架奠定基础。

纳米结构新现象、新规律的研究也有助于产生新应用和新纳米器件,将纳米技术转化为生产力。如相比于分子束外延(MBE)方法制备的量子点,胶体量子点具有尺寸均一等优点[4-5],但目前对其电效应的研究尚不充分。东南大学先进光子学中心现已制备出光电性能优良的胶体量子点电致发光器件,该器件可用来制备平板显示器件。目前正研究量子点电致发光处于光学微谐振腔中的光电性质,以求实现电泵浦量子点激光[6-7]。此外，在电场作用下，量子点呈现量子限制（quantum-confined）的 Stark 效应，利用这种 Stark 效应，可以制备出可调光学滤光片。

要实现纳米功能器件和系统的应用，必须解决的一个关键问题是保证纳米材料（如纳米颗粒、纳米管、纳米线等）的性能稳定性、一致性、可用性、可靠性等。这些问题可通过进一步深入研究纳米粒子的力、电及其转换机理，从中掌握控制纳米粒子的生长和运动模式等，如研究粒子化学键掌握对纳米材料进行电学控制的方法，如研究纳米粒子在多力作用下的运动模式探索出纳米材料的力学特性。只有通过对材料内部粒子的力电控制方法研究，才能从根本上解决纳米材料性能的稳定性、一致性和可靠性等问题，才有可能构建出性能优良和可用的纳米器件和系统。

2.3 加强微米/纳米科学技术成果的结合

随着 MEMS 技术和纳米科技的发展，将两者结合的呼声越来越高。目前双方存在的问题有两方面。

1) 在纳米领域，为探索极小尺度现象，目前多数采用如电镜等大型设备和仪器进行观测，而要实现纳米功能器件，必须在结构中考虑有效的微观/宏观转换窗口，即可以将纳效应转换为宏观信息量（如电信号）并输出，与 MEMS 结构的结合，可以有效地解决纳米效应与电信号的转换和输出问题，可以实现纳机电功能器件和系统（NEMS）。

2) MEMS 器件和系统一般具有较为完备的信息转换和输出功能，能够完成信息的自获取、自转换和自传输，但由于受敏感材料性能和结构的限制，目前器件改善工作很难进一步深入，缺少突破性的创新成果。将纳米科技与 MEMS 技术结合，能够突破目前的两尴尬局面。

虽然纳米材料和结构目前仍然存在着如一致性和可靠性等问题，

但开展 NEMS 器件的基础研究还是十分必要的，有助于促进 NEMS 器件的研究进程，为后续 NEMS 器件的应用打下基础。

从工艺结合角度上看，纳米结构的制备有 3 条可能的技术路线：自上而下（top‐down）、自下而上（bottom‐up）以及混合型（hybrid）3 类加工方式。所谓自上而下是指从体材料出发，利用薄膜生长和纳米光刻技术（电子束光刻、X 光光刻等）制备纳米结构和器件。该方式的主要优点是工艺技术相对成熟，但这一技术路线要求使用精密和昂贵的设备，成本较高。自下而上的方式是从原子分子出发自组织生长出所需要的纳米材料与纳米结构，这就要求在材料的生长过程中就对它们的结构、组分、形状、大小和位置进行人为的控制，从而直接生长出具有所需要的结构和性能的纳米器件。这种加工方式充分利用了材料自身的特点，可以实现对器件结构参数的精确控制，但同样存在效率相对较低的问题；此外，要实现功能器件，电转换和电输出是关键，目前通过 bottom‐up 制备形成的纳米结构还不能解决信号的自输出问题。混合方式则结合了 top‐down 与 bottom‐up 二者的优点，采用纳米功能材料与微米结构相结合，既可以在 MEMS 结构（如微电极、微梁等）中控制生长出纳米材料，也可以利用组装和集成技术在 MEMS 结构中组合纳米材料，其显著成果是综合了微机电加工和纳米材料制备的先进技术，能够实现多种多样复杂的功能结构，既利用了纳米材料优良的性能，同时通过与微结构（如微电极等）的结合实现信号的自检测和自输出，构成完整的 NEMS 功能器件。

图 8 为采用 3 种方式制作出的纳米结构，图 8（a）为北京大学微电子学研究所采用聚焦离子束（FIB）刻蚀技术加工出的纳米螺旋[8]，厚度为 100～200 nm，螺旋直径为 500～700 nm；图 8（b）为美国 Georgia 理工学院的王中林研究小组采用气向沉积法制备出的 ZnO 纳

米带螺旋[9]，纳米带宽度约 30 nm，螺旋直径约 500 nm；图 8（c）为清华大学微米纳米技术研究中心采用可控电泳技术将 ZnO 纳米线组装在微电极上构成的 NEMS 梁结构[10]，纳米线宽 100~500 nm，厚度为 10~200 nm。

对微米/纳米结合构成的 NEMS 器件的功能性研究要注意选择合适的模型，虽然许多尺寸稍大的纳米结构仍然可以采用在 MEMS 器件中普遍使用的经典连续介质模型，但当材料或结构的特征尺寸小于 100 nm 时，某些材料的机电特性会发生全新的变化，如碳纳米管的杨氏模量和电学特性会随着结构尺寸的变化而变化，这些纳米效应及其表征虽然可以用基于原子力学模型、分子动力学、统计力学等物理分析手段进行研究，但一般面向应用的 NEMS 结构所含原子分子数往往过于庞大且行为复杂，远远超出了采用如原子模型模拟研究的能力范围，因此，必须发展跨尺度（原子-纳米-微米-宏观）的分析和计算方法，研究以原子模型为基础的连续介质力学模型。如在研究多壁碳纳米管或多壁碳纳米管束的复杂变形行为时，一般需要对单壁管采用连续介质模型，而管间相互作用采用基于原子相互作用 Lennard–Jones 模型[11]；又如，研究力电耦合的碳纳米管变形时，可以对变形采用连续介质模型，对电子行为用量子力学模型进行分析[12]。这种混合处理方法可以较好地解决计算复杂性和计算精度的矛盾，完全有能力再现大量已经观察到的纳米材料体系复杂行为，并预报新的重要现象，发现新的规律。

综上所述，只有将纳米科技（如纳效应、纳特性）与 MEMS 的创新成果相结合，才能在未来的微/纳器件与系统的发展中取得突破性成果。

(a) FIB加工的纳米螺旋　　(b) 气向沉积法制备的纳米带螺旋

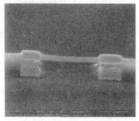

(c) 电泳组装构成的NEMS梁

图 8　不同方式加工出的纳米结构

3　结论

基于以上分析,为促进微米/纳米技术的健康长足发展,今后的工作还需要改善,以下为几点建议:

1) 重点支持有自主创新的加工和设计技术,形成自己特色;

2) 引导纳米科技与 MEMS 技术相结合,加强双方的技术交流和技术合作;

3) 加强基础研究,开展基础实验,通过基础理论和机理研究,从本质上解决 MEMS 器件和系统的应用问题,提升 MEMS 器件和系统水平;

4) 统一协调管理纳米科技和 MEMS 技术的资金投入,加强宏观规划和引导。

文中的分析、观点和建议希望供大家研究和参考。

参 考 文 献

[1] Sun W F, Shi L X, Sun Z L, et al. High-voltage power IC technology with nVDMOS, RESURF pLDMOS, and novel level shift circuit for PDP scan-driver [J]. IEEE Trans on Electron Devices, 2006 (4): 891-896.

[2] Sun W F, Wu J H, Yi Y B, et al. High-voltage power integrated circuit technology using bulk-silicon for plasma display panels data driver IC [J]. Microelectronic Engineering, 2004, 71: 112-118.

[3] Sun W F, Shi L X. Improving the yield and reliability of the bulk-silicon HV-CMOS by adding a P-well [J]. Microelectronic Reliability, 2005, 45: 185-190.

[4] Zhang J Y, Wang X Y, Xiao M, et al. Lattice contraction in free-standing CdSe nanocrystals [J]. Applied Physics Letters, 2002, 81: 2076-2078.

[5] Zhang J Y, Yu William W. Formation of CdTe nanostructures with dot, rod, and tetrapod shapes [J]. Applied Physics Letters, 2006, 89: 123108.1-123108.3.

[6] Zhang J Y, Wang X Y, Xiao M, et al. Modified spontaneous emission of CdTe quantum dots inside a photonic crystal [J]. Optics Letters, 2003, 28: 1430-1432.

[7] Zhang J Y, Ye Y H, Wang X Y, et al. Coupling between semiconductor quantum dots and two-dimensional surface plasmons [J]. Physical Review B (Rapid Communications), 2005, 72: 201306.

[8] Xia L, Wu W G, Xu J, et al. 3D Nanohelix fabrication and 3D nanometer assembly by focused ion beam stress-introducing technique [C] //The 19th IEEE International Conference on Micro Electro Mechanical Systems. Istanbul, Turkey, 2006: 118-121.

[9] Kong X Y, Wang Z L. Spontaneous polarization-induced nanohelixes, nanosprings, and nanorings of piezoelectric nanobelts [J]. Nano Letters, 2003, 3 (12): 1625-1631.

[10] Wang D Q, Zhu R, Zhou Z Y, et al. Controlled assembly of zinc oxide nanowires using dielectrophoresis [J]. Applied Physics Letters, 2007, 90: 103110.1-3.

[11] Liu J Z, Zheng Q S, Wang L F, et al. Mechanical properties of single-walled carbon

nanotube bundles as bulk materials [J]. J Mech Phys Solids, 2005, 53: 123-142.

[12] Li G, Aluru N R. Hybrid techniques for electrostatic analysis of nanoelectromechanical systems [J]. J Appl Phys, 2004, 96: 2221-2231.

Inspiration on Development and Industrialization of Micro/Nano Science and Technology

DING Heng-gao, ZHU Rong

(Department of Precision Instruments and Mechanology, Tsinghua University, Beijing 100084, China)

Abstract With development of mico/nano science and technology, great progresses are made in industrialization as well. Some viewpoints were presented in this paper according to Chinese current status and trend in development: motivate studies on innovative fabrication and design to enhance our capability on application and industrialization of micro-electro-mechanical system (MEMS); motivate fundamental research for improving the level of component and system of MEMS; motivate the integration of micro/nano science and technology. Finally, some advice were given for the future works.

Keywords Micro-electro-mechanical system; Nano science and technology; Status and trend in development

MEMS 器件研制与产业化

(2009 年 10 月)

1994 年我撰文提出开展微米纳米技术研究工作的意见，其后一直关注这项工作，与此同时参加了一些 MEMS 器件的研制。国内 MEMS 器件的立项研究差不多经历了 3 个五年计划，目前从技术上看，静态指标和环境考核均取得了较大进步。但是，从实验室样品到产业化的产品，还需要一个大跨越。产业化过程所付出的代价会远远超过实验室样品研制过程。回顾和总结我们的研制经历，初步有以下几点认识。

1 研制 MEMS 器件需要标准工艺

MEMS 技术是一个多学科交叉的新技术领域，涉及多种学科知识。在研制 MEMS 器件的时候，遇到了基础理论、设计与仿真、材料特性（机、电、光、热……）、微加工、集成化、封装和微系统测量等一系列问题。总的来说，这些问题可以分为设计、加工和检测三大类，其中加工技术的标准化是至关重要的。

普遍来说，研制一种 MEMS 器件的设计流程包括总体系统设计（含仿真）、结构设计、电路设计、工艺设计、封装设计和可靠性设计等环节，并通过工艺流程和系统测试来实现和验证。在此过程中我们

本文是 2009 年 10 月 12 日丁衡高院士在"第三届微米纳米'技术创新与产业化'国际研讨会（重庆）"上的发言提纲。

目前遇到的主要问题是，在不少情况下，无法从测试实验数据直接分析出产生性能偏差的原因。这是因为 MEMS 器件研制过程的各个环节均存在一些问题没有完全弄清。譬如说，总体设计理念尚不科学、数学模型过于理想化、结构设计不尽合理、加工误差大且重复性不好、检测与控制电路存在漂移等等。由于各个环节均不同程度存在问题，因而难以进行准确的对比分析，甚至有时会产生误导性结论。

经过多年的实践，我认为 MEMS 结构加工的标准工艺是研制工作的关键环节，是验证设计思想的保证。根据标准工艺搞设计、进行结构加工，可以获得足够数量的器件结构。通过对这些有精度、有一致性的结构进行测试，才能够发现规律性的问题，才能为设计提供正确的指导性思路。

2 研制 MEMS 器件需要开发特殊工艺

复杂的 MEMS 器件机械结构，需要多种工艺技术。仅依靠目前现有的 MEMS 工艺，还不能保证器件性能和可靠性达到要求。比如解决微机械振动陀螺线性化问题，需要考虑微机械梁结构本身的力学非线性、电容检测非线性和封装因素造成的非线性等几个方面。再如电容式微加速度传感器的分辨率问题，不能只考虑芯片检测电容本身，还要考虑结构中、引线中和接口电路中的寄生电容效应的影响，以及 $C-V$ 变换等信号处理环节引入的噪声等等。

因此，针对不同类型的 MEMS 器件，应根据其机械结构、信号转换方式和封装形式的需要开发适合这些器件的特殊工艺。比如说为了解决 MEMS 器件的粘连和漂移问题，国外开发了一种特殊钝化工艺，使器件结构表面形成低能量的、钝化的保护层。

3 ASIC 电路的重要性

MEMS 器件的一个重要组成是测控电路，它要完成信号的检测、处理和传输。测控电路的体积与成本是制约 MEMS 器件产业化的一个重要因素。因此，通过 MEMS 微机械部分与 ASIC 的结合，可以大批量制造集成化的 MEMS 器件。

在 MEMS 器件装配前和完成装配后，均涉及必要的测试、标定和调整工作。

采用分别在独立晶圆上制造 MEMS 微机械部分和 ASIC 的技术，可以使两者的测试工作在装配前完成。装配后的 MEMS 器件仍然需要测试，在使用过程中还需视具体情况进行标定和调整。因此，ASIC 电路除了具备基本的信号转换和处理功能外，还要具备器件的自检测功能、自标定功能和自补偿功能。

科学地解决好工艺和散热等问题后，单片集成化技术将机械部分与 ASIC 集成在一个片子上，是人们追求的目标。

4 圆片级封装的重要性

封装技术不仅决定了 MEMS 器件的可靠性、长期稳定性和成本，在一些器件中还决定了性能。MEMS 器件通常都有可动部分或"悬挂"结构。根据 MEMS 器件的种类和用途，需要设计专用的封装结构，总的来说包括气密性封装（含真空封装）和非气密性封装两大类。MEMS 封装经历了器件级封装和圆片级封装两个阶段。圆片级封装使原来后道工艺转变成高效率的前道工艺，是 MEMS 乃至半导体封装的发展趋势，它的全部工艺流程都可以在超净间里完成，并且一个圆片上有数十个、数百个甚至更多的器件，这大大提高了 MEMS 器件封装的效率和成品率，从而降低了成本、减小了体积。

5　MEMS 的批生产问题

目前，国内大部分 MEMS 实验室的装备还达不到中试水平，离产业化有较大的距离，所研制的器件大部分也达不到量产的要求。很多工艺技术还停留在理论和小量试验件层面上，无法检验其量产的可行性。因此，要尽快建立以批生产厂为主体、市场为导向、产学研相结合的技术创新体系，学习先进的 MEMS 批量生产技术。有助于我们建立自己的 MEMS 产业化能力。

经过十多年的积累，我们在技术和人才两方面都具备了一定实力，国内外市场需求和市场规模每年都在增长。弄清小量试验件与大批量产品的区别，进一步统一认识，科学解决好我们目前存在的问题，在未来几年实现我国 MEMS 器件的产业化是必然的结果。

在"MEMS 在机械与运载工程领域的应用研究"咨询项目启动会上的讲话

（2012 年 5 月）

今天这么多老熟人新朋友在一起认真思考过去的工作，探讨高端微机电系统在机械与运载工程领域的应用研究及产业化问题，展望 MEMS 未来的发展前景，我觉得很有意义。刚才听了三位同志的专题发言和同志们的讨论后，我想首先提这样一个问题，那就是多学科交叉的项目应如何发展？它的成果如何产业化？我们刚刚谈到了 MEMS 分类的问题，是把它划分到机械类？还是电子元器件类？多学科交叉的项目怎么划分？我觉得我们首先要解决这个认识问题，我们现在强调高新技术的发展与交叉科学发展的特点，因此，以前的分类观点要改变，要适应高新技术的发展。

我刚刚跟闻库同志谈了一下，他问我这样一个问题，我们国家把 MEMS 放到电子元器件一类，它是不是电子元器件？怎么认识这个问题？MEMS 是应用电子元器件工艺发展起来的，但它本质上不是电子元器件，MEMS 最大的特征就是它是由机械、光学、电子学、声学、生物学等多门技术构成的微器件和系统，一般说来，MEMS 主要包括

本文根据讲话录音整理而成。

微传感器、微执行器和微系统。这里强调一下它们的集成性和系统性，即使 MEMS 传感器也要和 ASIC 集成，成为有特点的微传感系统。随着 MEMS 的发展，其系统性和智能化水平也在不断提高，具有闭环控制、自检测、自标定等智能化的特点，而电子元器件没有这样的特点。由于对 MEMS 的特点没有搞清楚，所以就简单地把它归类到电子元器件，这样就从"认识上""行政上"阻碍了 MEMS 的发展。

我们现在应该清醒地看到 MEMS 技术发展的形势，从世界上来看，从国内的需求来看，MEMS 发展的形势是怎样的呢？昨天我看了最新的一期杂志里的一篇文章《一万亿美元的 MEMS 市场》，作者讲"再经过十年的发展，MEMS 市场将发展到一万亿美元"，他是怎么分析的呢？我看了之后觉得这篇报道里有一些很重要的观点值得我们思考，作者认为现在 MEMS 技术到了突飞猛进发展的时候了。文章中讲到由于信息通信、计算机、传感等技术的发展，促进了 MEMS 技术在气象方面，石油勘探和制造方面，物流方面，智能高速公路的管理方面，地震、海啸的预测方面，智能家庭方面，智能人体监控等方面应用发展的可能，加上我们有了研制 MEMS 的基础和经验，可以加速研制的进程。现在 MEMS 的产值是一千亿美元，预计在十年以后将达到一万亿美元，大家想想这是怎样一个份量！这篇文章中有一个很宝贵的介绍，我是第一次看到，这个表中举了许多个"第一"的例子：第一个压力传感器、第一个加速度计、第一个气体敏感器、第一个阀门、第一个喷射器、第一个光显示器、第一个生化传感器、第一个 RF MEMS、第一个微继电器、第一个振荡器，统计数据表明，这些"第一"从有概念开始到形成产品的时间平均为 28 年。随着开发经验的积累和科技进步，今后的产品化周期将会大大缩短。文章中谈到 MEMS 的建模与仿真技术及标准工艺的水平是缩短研制周期的重要环节。今天面临的形势就是这么大的市场需求，如果能够加速研制进度的话，十年以后 MEMS 产业的市场将达到一万亿美元，这里还未考虑军品的需求。应用需求太多，我们如何适应？这就是 MEMS 面临的发展形势。

针对这个形势，我再举一个国内的例子，前几天我到重庆开会，有一个叫王金山的同志，他让我到他那里看一看。据他回忆，2002 年我们一起在法国考察，在火车上，他问我 MEMS 产业他应该如何起步？我说："比较可以快速见效的，不是从研制 MEMS 开始，而是要从组装开始搞应用系统。"经过十年的发展，他创建的金山科技公司去年的产值达到 12.8 亿（含配套设备），公司总共才 600 多人，主要是制造智能药丸，共四种：一种查直肠、一种查小肠、一种查胃、一种查食道。他的产品现在已占了国内市场 95% 的份额，在欧洲的市场打败了以色列，现在还要在重庆江津建新厂，在国外建所搞 MEMS 研究开发。李克强同志曾参观过该公司并称赞道："这才是高技术！"说实话，我们没有想象到这个项目发展得这么快，这个例子说明，十年以后 MEMS 产业将达到一万亿的产值是完全有可能的，现在在中国的市场上已经看出一些苗头来了，这样的成功范例将会越来越多。

另外还有一个问题就是在十多年前我们研究 MEMS 发展战略的时候就提到过：要不要搞替代，当时我们提出替代与创新相结合，要并举。我举一个例子：1979 年我到奥地利去参观了一个叫 AVL 的研究所，它是研究弹道的，主要研究弹头着靶时的瞬间状态的，用闪光测速的方法和高速摄像的方法来记录弹头着靶的瞬间；同时该研究所还研究发动机的检测评估，很多国家的发动机拿到这里检测后，出一个包括发动机每一个环节损耗的性能评估报告，AVL 的本领是自己能做各种各样的传感器来适应发动机的测量，通过多年的测量积累了对发动机评估的关键参数。前不久我跟航天、航空的同志都说过，现在我们面临非常关键的一个问题是：我们的研究手段必须要上去，如果研究手段上不去，就出不了好成果。这个问题正好说明需求牵引与技术推动相结合来发展的重要性。现在新的火箭发动机中有很多问题需要解决，因为它是高压、高温，如要测涡轮泵回收的燃料的再燃烧效率，这一套系统是一个关键问题，我想在这些问题上面，我们的同志应该多交流，在这些方面启发思路，不仅能够提高我们科研的水平，而且也能够带动MEMS的发展，要坚持"需求牵引与技术推动相结合"。需求引领发展，应用促进发展。

刚刚姜澄宇同志提出了七条意见，我认为这七条中最重要的，或者说我们在搞调研，听取方方面面的意见时，最重要的就是：开展MEMS在中国的发展战略的研究。要把发展战略的研究摆在第一位来考虑。根据国内外的形势和目前我国的研究基础与研究条件，我们怎样来发展？过去在武器装备的发展方面，我们提出了"以整体效益为中心，需求牵引与技术推动相结合，选择跟踪，重点突破"的发展战略，还有高起点，但要遵循科研工作的规律，那就是必须要先做好基础研究、应用研究之后进行技术开发，这个规律不能违背，不能为了快，就马上搞技术开发。对于MEMS来说，我们到底应该遵循什么样的发展战略呢？从许多资料可以查到战略的基本含义，指全局性的、关系成败的谋略、方针和对策。我认为MEMS发展战略研究的重要内容就是要确定MEMS分阶段达到的近、中、远期目标是什么，以及实现这些目标要采取哪些措施。我们在制定发展战略的时候要紧密结合这两个方面来研究。

关于咨询报告，我认为有五个主要方面的问题需要了解、研究和思考：

1) MEMS 的用途和国内外 MEMS 发展的形势。国外已经将 MEMS 提到"第三次产业革命"的高度了，十年之后 MEMS 在全世界将达到一万亿美元的产值，那么我们国家将来能占多少份额？在报告中我们首先要把这个形势讲清楚。另外我们还要列举一些 MEMS 军用和民用中的例子，来说明 MEMS 能够达到的水平与展望，包括现在是什么水平，预期将达到什么水平。这一部分要有震撼性，让人们看到后要有震动，产生紧迫感，我觉得我们现在缺乏紧迫感。我们经常讲要转变经济发展方式，在很多领域如果应用 MEMS 技术可能将大大提高生产率和科研生产水平，甚至给生产和生活带来重大变革。报告中要把这一部分讲清楚。

2) 报告中要讲清楚 MEMS 的发展战略。把形势和任务联系起来，是我们多年来正确的思考方法。面对目前的形势和我们的任务，要根据发展战略，讲一下我们分阶段的目标，特别是近期的目标和下一步要搞的一些主要项目，以及这些项目预期能起到的作用。

3) 报告中要讲到产、学、研怎么结合的问题。要提出一条有中国特色的产业链，关于产业链现在有两种说法，一种说法是从科研开始、从源头开始的产业链，另外一种说法是从工程设计开始的产业链。我们国家不同于其他国家的是：我国企业里没有研究所，国外很多大的企业里都有研究所，是从源头开始的产业链，我们的产业链与国外不同。另一个重要问题，我们在前面已多次提过，就是产、学、研和用的结合，特别是对生产和生活影响重大的应用，我们要认真地进行调研，否则我们的产业链是不完善的、没有生命力的。因为需求引领发展，应用促进发展。

4) 要很好地研究关于微纳结合的问题。微纳结合的问题是基础研究问题，我们现在所搞的 MEMS 基本上都是经典的一些东西，还没跑出那个范围。而现在精彩之处在于纳尺度范围内的一些现象怎么为我们所了解、所用。最近，我们发现有一个很重要的纳尺度范围内的现象，就是通过纳尺度内的电子传输效应，我们可以研究出很多新的东西。上海交大的同志正在做这方面的研究，最近正在做比较精确的实验，现在初步实验结果的效率比经典的提高 3~4 个数量级。在纳尺度范围内的一些现象才是精彩所在，才是我们 MEMS 创新的源泉。但是现在存在一些机制上的问题，搞纳米研究的侧重于材料的问题，但材料中的某些向往的物理特性忽略掉了，怎么从一开始就要有微纳的结合？这个问题我们应该好好地研究，学术交流是一方面，这是必不可少的，另一方面可不可以将相近的课题结合到一起。美国在理论性的研究方面可以公开，应用性的研究不公开。但是就纳米技术来讲，如果我们认识到这一点的话，还是可以学到一些东西的。我们一定要认真地思考一下微纳结合的问题，只有这样，才能够真正发挥微观世界的威力。科学发展上的几次革命，还有技术革命、产业革命，从科学革命的角度来说，现在人类认识客观世界的趋势表明对宏观和微观的事物认识得不够，还有很多东西有待于我们去探索，宏观的就是宇宙，到现在还在争论到底有没有暗物质、暗能量，它是什么还没有弄清楚；微观方面纳尺度内的问题也是一样。我们讲微纳结合，我认为只有微纳结合才有生命力，MEMS 才有生命力，才可能走出具有中国特色的 MEMS 之路。

MEMS 的多学科性以及跨学科的合作研究，会有新的发现、应用和生长点，例如 MEMS 和生化的结合等等。我们已经强调过多次，这里就不在此展开讨论了。

5) 关于人才培养的问题。我觉得在人才培养这个问题上我们要好好思考一下，我认为我们要十分重视我们已有团队内的人才水平的提高,这是基础,不能片面地只盯着引进人才。我不反对引进人才,但引进的人才要能够适应当前中国的环境。我们已有团队人才的培养有一个基本的方法,那就是带着任务出国学习考察。另一方面我们要十分注意设备操作、维修、管理等方面人才的培养,如果我们在这方面不重视,就出不了好的成果。我们需要一些新的设备,一些要引进,但我们也不能忽视了自己的创造,需要自己想办法研制一些新的仪器设备来提高我们的研制水平。作为个人来讲必须承认我们的知识面是比较窄的,必须要学会用集成的方法综合各方面的技术来解决我们的测试问题,这决定了我们的研究水平。

我们的讨论还将继续,以上是这次讨论发言的提纲,我没有展开细说,大家共同来讨论研究。一共讲了五点,第一点是 MEMS 当前的形势,第二点是 MEMS 的发展战略及任务,第三、四、五及前言部分是为实现发展战略的策略建议及一些主要工作。

三十年不断发展的 MEMS 惯性传感器

(2023 年 10 月)

1 回顾

1994 年 11 月中旬,首届"全国纳米科学与技术学术会议"在北京香山召开,会上从我国实际出发,在考虑近期需求,又兼顾长远发展的基础上,确定了专有学术名词"微米纳米技术",这是符合实际的,既考虑了微米技术的前沿性和现实性,着眼于纳米科技的前瞻性和基础性,同时兼顾了基础研究、应用研究、开发研究和技术开发的可持续发展。在此之前,中央领导同志对开展微米纳米技术研究作了重要批示,肯定了微米纳米技术将在未来国民经济和国家安全建设中发挥重大影响。

"九五"期间,把研制高性能 MEMS 惯性传感器作为发展军用微米纳米技术的战略,紧紧围绕 MIMU 的需求开展技术攻关,带动了国内相关单位建立研究队伍、设立相关实验室、建设工艺制造平台,突破了若干关键技术,奠定了技术发展基础。

三十年弹指一挥间,迅速发展的应用需求牵引微米纳米技术不断创新。MEMS 惯性传感器已从实验室探索研究走向工程应用,技术和

本文发表于《导航与控制》2023 年第 22 卷第 4 期。收入本书时有部分修改,特别总结了主要经验。

本文是根据作者多年来与惯性技术界有关同志多次讨论整理而成。

产品都取得了巨大进步，在国防、车载导航、物联网、高端装备和消费类电子等方面得到了广泛应用。

2 发展成就

2.1 高性能 MEMS 惯性传感器产品已推广应用

当前，全球高性能 MEMS 惯性传感器产品的市场集中度较高，市场份额主要被 Honeywell、ADI、Northrop Grumman 等巨头占据，大约占据50%以上份额。国产高性能 MEMS 陀螺及加速度计的核心指标可以与国际大公司 Honeywell 对标，国内已可以生产，解决了装备应用问题。

从"九五"开始，在国家有关部门的推动和需求牵引下，大学和研究所纷纷设立研究队伍，围绕高性能 MEMS 陀螺和 MEMS 加速度计开展研究。MEMS 陀螺探索研究的主要方案有：质量块振动陀螺、音叉陀螺、四质量陀螺、环形陀螺等；MEMS 加速度计探索研究的主要方案有：跷跷板式、三明治式、梳齿式和谐振梁式等；采用 Top-Downd 的方式，以体硅工艺为主制造陀螺和加速度计敏感结构。

起步阶段的研究，在比较困难的条件下展开，经费、设备、经验、人才等比较匮乏，但是研究人员克服种种困难，逐步建立了仿真设计流程、制造工艺平台和测试评估实验室，具备了开展研究的相关条件和能力。后来，通过技术引进等手段，建立了 6 in 硅基标准制造工艺平台，提高了加工水平，保证了诸如深宽比、垂直度、底部钻蚀等机械参数的精度，进而大幅度提升了设计和工艺水平。2011年年底，国内开发出基于陶瓷外壳的 MEMS+ASIC 两片 SiP 封装的陀螺和加速度计。2019年，高性能 MEMS 陀螺和 MEMS 加速度计实现规模量产。至此，国内掌握了 MEMS 惯性传感器设计、制造、封装和测试等主要技术环节，具备了开发高性能 MEMS 惯性传感器产品的能力。

2.2　率先将 MEMS 惯性传感器应用于智能辅助驾驶

自 2009 年谷歌 Waymo 研发无人驾驶车开始，全球自动驾驶产业随之起步。2016 年国内企业开始发力，大力发展具有智能辅助驾驶功能的新能源汽车。在获取车辆位置信息的技术方法和传感器方案方面，不同厂商之间出现了发展路径上的差异。以特斯拉为代表的纯视觉方案（NOA），采用车载摄像头为主要感知设备，结合算法确定车辆的位置信息；以小鹏汽车为代表的"中国方案"，即多传感器融合方案（NGP），采用车载摄像头、车载超声波雷达、卫星惯性组合导航系统等多种感知设备，实现多源融合定位；还有正在研发中的车路协同方案（V2X），将车载的部分感知能力移到道路侧，通过 V2X 实现车辆位置信息的获取和交互。

智能辅助驾驶和自动驾驶对定位系统的基本要求为高精度、高可靠性、高可用性，同时需要满足功能安全的要求。仅依靠 GNSS 定位，在复杂环境和极端天气的情况下存在风险，而惯性导航则成为有效的安全冗余。MEMS 惯性传感器在体积、重量、功耗、价格、寿命等方面与激光陀螺和光纤陀螺相比，具有无可替代的巨大优势。然而，MEMS 惯性传感器的零偏稳定性指标与激光陀螺和光纤陀螺相比，差距较大。因此，几乎没有人认为 MEMS 惯性传感器可以应用于无人驾驶系统中感知车辆的位置信息。很多企业采用光纤惯导和卫星进行组合，在车上进行试验，取得了较好的结果。但是，高昂的价格、较低的生产效率和较长的生产周期使得这个方案没有办法在车上大规模应用。

在国内造车新势力，特别是小鹏汽车智能辅助驾驶方案的强力牵引下，国内企业敢为天下先，率先尝试 MEMS 惯性传感器＋GNSS 组合上车试验，取得了令人满意的结果。车载高精度定位 P-Box 对于位置精度、安全性以及成本把控和量产交付能力有很高的要求，其主要难点在于：1) MEMS 惯性传感器芯片的设计、制造、封装和批量测试技术；2) MEMS 惯性传感器的精度标定和算法；3) 成熟的大规模量产能力；4) AECQ100 车规级器件认证和 ISO 26262 车规功能安全认证。在保证产品质量和产品安全性的前提下，车载 P-Box 的量产交付能力建设是实现规模化上车应用的核心关键。在此之前，国内还没有哪个单位进行过类似的尝试。为了满足车企的交付需求，车载 P-Box 定点企业不得不转变思路，以数字化为基础，整合供应链、制造、交付、物流、资金周转和现金管理等体系，形成高效的数字化业务流程，生产过程所有数据实时采集上云，构建了高效的数字化交付体系。

国内企业将 MEMS 6 轴集成 IMU 应用于智能辅助驾驶，在车载定位这个细分赛道，蹚出了一条高水平有特色的新路，走在了国际前列。

2.3 消费级 MEMS 惯性传感器崭露头角

我国作为全球最大的电子产品生产基地，正消耗着全球四分之一的 MEMS 传感器，需求和市场是巨大的。但是，目前我国大部分 MEMS 传感器仍然依赖进口。Bosch、ST、TDK 等 IDM 大公司拥有雄厚的技术实力和资金实力，自己设计传感器，自己生产传感器晶圆，他们的产品无论在价格还是性能等方面具有巨大的优势。

2006 年起，我国陆续出台措施，引导 MEMS 传感器行业稳步发展。在政策和市场的双重加持下，涌现出了一批优秀的国产 MEMS 传感器企业，多家公司成功上市和上市辅导。国内已有提供专业 MEMS 代工服务的公司，也有提供封装和测试代工的企业，具备了 MEMS 结构芯片的加工、封装和测试能力，但是一致性、重复性、良率等还不能满足产品竞争的要求。

消费领域同样存在产业化中供应链的安全问题，国产化替代的大背景下，系统厂商对于 MEMS 陀螺和 MEMS 加速度计产品的需求量巨大，国内企业迎来了新的发展机遇。

3　主要经验

3.1　当初把高性能 MEMS 惯性传感器作为发展军用微米纳米技术的战略是正确的

20 世纪 80 年代末，以 MEMS 技术为代表的微米纳米新技术在西方发达国家兴起。通过分析国外发展情况，结合我国实际需求，在"九五"期间提出把研制高性能 MEMS 惯性传感器作为发展军用微米纳米技术的战略，紧紧围绕 MIMU 的需求开展技术攻关。经过近 30 年的艰苦努力，现在国内完全掌握了高性能 MEMS 惯性传感器设计、制造、封装和测试等主要技术环节，具备了高性能 MEMS 惯性传感器的批量生产能力，军用 MEMS 惯性传感器产品水平已基本上和国外相当，产品基本实现自主可控。如今，高性能 MEMS 惯性传感器和 MIMU 被公认为是微米纳米技术领域最核心和最重要的器件之一，有着重要的地位和作用。

3.2　MEMS 惯性传感器的颠覆性主要体现在创新应用，它不仅大大提高了传统装备的制导化和智能化水平，还催生了大量新装备和新产品的涌现

MEMS 惯性传感器的微型化、芯片化、低成本和可大批量制造的特点颠覆了传统惯性仪表的概念，惯性仪表从昂贵复杂的精密仪器变成了高度集化的芯片。它的出现使传统廉价的弹药如航弹、炮弹甚至是子弹成为具备制导能力的智能弹药，使汽车也可以自主导航定位，实现自动驾驶。MEMS 惯性传感器还催生了各种微小型无人机、无人

车、无人艇和人形机器人等大量新装备和新产品的涌现。MEMS 惯性传感器的颠覆性和重要价值主要体现在其他惯性传感器不能发挥作用（用不起，放不下）的创新应用领域，和其他惯性传感器的关系主要是互补而不是相互替代。

3.3 高性能 MEMS 惯性传感器的大批量推广应用是靠持续的技术突破和敢为天下先的创新精神

芯片式单轴 MEMS 加速度计和芯片式单轴 MEMS 陀螺与其他惯性传感器相比，体积、重量、成本等减小了 1~2 个量级。但是要组成 MIMU 和惯性系统还是需要和其他惯性传感器一样将 3 个加速度计和 3 个陀螺通过立体组装起来才行。尽管 MEMS 惯性传感器实现了芯片化、可大批量生产，但是 MIMU 和惯性系统还是需要复杂的立体组装，不能大批量生产，成本和体积不能进一步降低。在保持精度不降低的情况下，实现六轴 MEMS 惯性传感器的平面单片集成，直接做成芯片式 MIMU 是技术突破的方向。借鉴消费级 MIMU 单片集成的工艺路线，结合高性能 MEMS 惯性传感器的理论和设计方法，国内解决了这一难题。要实现 MIMU 和惯导系统大规模低成本量产还要解决自动化制造、自动化测试和自动化全温标定等技术难题，要实现在汽车上批量应用还要解决车规认证和功能安全认证等可靠性技术难题。国内企业在国际上率先实现了高性能六轴 MEMS IMU 在智能辅助驾驶上的大批量应用，是持续技术突破和敢于创新的结果。

4 未来发展

4.1 L3 及以上级自动驾驶需要安全可靠、低成本、高效能的 MEMS 惯性传感器

目前，全球自动驾驶渗透率情况以 L1、L2 级为主，L3~L5 级渗

透率较低。国内乘用车市场自动驾驶技术以 L2 级为主，L3 级尚未落地。根据国际前沿科技咨询机构（ICV）预测，2023—2027 年全球自动驾驶渗透率 L2 及以上级呈现增加的趋势，其中 L2/L2＋级预计 2027 年渗透率达 58%，L3 级预计 2027 年渗透率达 25%。

从 L2 级到 L3 级，自动驾驶的安全性非常突出，譬如对自动驾驶车辆进行测试，15 万公里测试能够发现 99.9% 的问题，但是剩余 0.1% 的问题可能在 15 亿公里都未能发现和解决。这个 0.1% 乘上每年上路的几亿台车，那就是天文数字。因此，如何让自动驾驶汽车比飞机更安全？具有功能安全的、具有 99.999 9% 可靠性的传感器必不可少。

价格在 30 万元以上的乘用车市场占有率很有限，因此乘用车高精度定位产品的装车率在 1.8% 左右。未来高精度定位产品一定会往 20 万元乃至 10 万元的车渗透，那么巨大的成本压力就会随之而来。在不影响安全和质量的前提下，MEMS 惯性传感器降本是必然的，价格会非常低，而且性能和可靠性要求还非常高。

4.2 人形机器人打开了 MEMS 惯性传感器的成长空间

长期来看，根据麦肯锡公司预测，全球人形机器人市场空间可达 120 万亿元，是一个崭新且空间庞大的蓝海市场。据艾瑞咨询公司预测，2021 年到 2025 年国内智能机器人市场规模的年平均复合增长率将达到 40%，2025 年国内智能机器人市场规模接近千亿元。这将带动 MEMS 惯性传感器的需求量不断增长。

MEMS 惯性传感器可以获取人形机器人的角速度和加速度数据，通过 MEMS IMU 可监测人形机器人的实时状态、位置信息以及运动轨迹，维持人形机器人完成走、跑、蹲等动作的姿态平衡。一台人形机器人就要采用多个 MEMS IMU，市场空间广阔。MEMS IMU 与其他传感器融合，如立体声摄像机、关节编码器、力扭矩传感器、足部接触传感器等，实现数据互补，估计姿态足的质心位置、速度、方向、角速率和角动量，共同进行机器人状态反馈并完成下一步动作，应用于机器人下蹲起立、前后行走、上下楼梯、回避障碍等场景。

4.3 MEMS 惯性传感器朝着集成、融合、智能方向迈进

在高性能和低成本的需求牵引下，MEMS 惯性传感器的主要性能指标零偏稳定性将达到 0.05（°）/h，并且实现多轴集成，即三轴陀螺和三轴加速度计集成在一起，同时与卫星芯片将逐渐走向融合。影响卡尔曼滤波算法精度的时间更新和量测更新，将随着集成化而减小数据延时，提高定位精度。随着集成化程度越来越高，产品的成本将更具竞争力，芯片集成或许成为产品终极形态。先进的封装技术，特别是 3D 堆叠封装技术，可以将多个芯片组合封装，可在有限的体积内集成更多的传感芯片，实现更复杂更强大的功能。

MEMS 惯性传感器的零偏精度和标度误差直接影响航迹推算的精度。因此，对惯性传感器的误差分析和补偿是提高定位精度的主要方法。随着 ASIC 技术的不断进步和成熟，传感器的信号检测与处理电路、闭环控制电路和计算单元将高度集成在一起，误差补偿算法、自校准、自标定以及功能安全算法都将在传感器芯片层运行。

随着 MEMS 惯性传感器性能指标不断提升，体积和功耗不断减小，成本和价格不断降低，其应用领域将不断扩大。在牢牢占据消费电子市场的同时，将紧跟 AI 的发展步伐在智能汽车、人形机器人、无人机、无人系统中得到广泛应用。国内的高校、研究所和创新企业要勇于把握在竞争中创新发展的机遇，不仅研制、生产出高水平有特色的国产 MEMS 惯性传感器，还应在国际市场占有一席之地。

科学技术综述

微型惯性测量组合的关键技术

(1996年2月)

摘 要 采用微米/纳米技术的微型惯性测量组合具有低成本、长寿命和高可靠性等优点，本文论述了国外这项技术的发展动向，提出了微硅及微型光学惯性器件关键技术的研究任务。

0 引言

在微米/纳米技术发展的推动下，微硅加速度计（Micromachined Silicon Accelerometer，MSA），微硅音叉陀螺仪（Micro Silicon Tuning Fork Gyro，MSG），具有集成光路的闭环光纤陀螺仪（IFOG）在国外已经研究成功，并将逐步应用于惯性技术的各个领域，它们构成的微型惯性测量组合（Micro Inertial Measurement Unit，MIMU）具有低成本、长寿命和高可靠性等优点，因而必将使惯性技术发生巨大的变革。

1977年，美国Stanford大学等单位开始在硅片上采用微加工工艺批量生产MSA和MSG。进入90年代，德国和日本也已生产并使用这些微硅惯性仪器。

1964年，作者曾提出应重视光学陀螺的研制。国外在1982年，

本文发表于《仪器仪表学报》1996年第17卷第1期。合作作者：丁衡高，章燕申，马新宇，张斌。

研制成功环形激光陀螺（Ring Laser Gyro，RLG），已成为产品，得到广泛的应用。1979年，美、日等国开始研制干涉型光纤陀螺仪（IFOG），其中采用了多功能集成光学芯片（Multiple Function Integrated Optic Chip，MIOC）。至1993年，各种IFOG和谐振式光纤陀螺仪（Resonant FOG，RFOG）均已研制成功，目前精度从0.05（°）/s到0.01（°）/h，已在许多领域得到应用。但IFOG体积和重量较大，成本较高，而RFOG较IFOG体积小、造价低，具有明显的优点。既然RFOG已经研制成功，那么采用包括光波导谐振器的MIOC来取代RFOG中的光纤谐振器应是RFOG技术今后的发展方向。

早在1978年，美国Northrop公司开始探讨在硅片上制造光波导的技术，并于1991年研制出样机，精度为10（°）/h，称为微型光学陀螺仪（Micro Optic Gyro，MOG）。MOG和RFOG的工作原理完全相同，实质上是一种环形激光陀螺仪，只是采用了无源谐振器。可以认为MOG和RFOG将成为新一代微型RLG，目前采用Brillouin散射激励（Stimulated Brillouin Scattering，SBS）的RLG已经研制成功，精度达到了10（°）/h，这种RLG也可称作Brillouin RLG（BRLG）。

可见，集成光学技术作为微米/纳米技术的一个重要组成部分，已经在微型陀螺仪中得到了广泛的应用，其中MOG和微硅惯性仪器MSA、MSG等有许多共同点，在本文中，将讨论它们的技术关键，提出解决的途径。

1 微硅惯性仪器

1.1 微硅加速度计

美国Draper Lab、Rockwell、Litton等公司均生产静电力平衡式MSA。在平板硅片上做出检测质量（摆片），由两根硅挠性梁连接在框架结构上,如图1所示。检测质量、挠性梁及框架都由同一片单晶硅片经过各向异性刻蚀制成。摆片厚度比框架薄,构成与电极之间的间隙,形成三端差动电容。差动电容既用来敏感摆片相对电极的运动,又构成施加再平衡静电力的电极。Draper Lab MSA 的芯片尺寸为 300 μm×600 μm。德国 LITEF 公司类似的产品为 B-290 TRIAD 型三轴 MSA,其零偏稳定性为 250 μg,逐日稳定性为 500 μg。

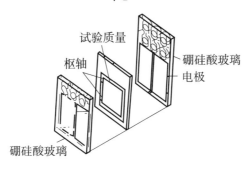

图 1　微硅加速度计原理图

1.2　微硅陀螺仪

MSG 的类型分角振动和线振动两种,前者为框架式,后者为音叉式。

(1) 框架式 MSG

Draper Lab 于 1989 年研制出这种 MSG,它由内外两层框架组成,如图 2 所示。内框架上有惯性质量,外框架在交变静电力驱动下绕枢轴振动。当外框架以小角度振动时,内框架就能敏感绕框架平面法线方向的输入角速度,内外框架运动的测量和控制都由电极进行。通过电极加矩保持力矩平衡,MSG 中芯片的长度为 600 μm,零偏稳定性约为 50 (°)/h,能承受 80 g 的环境条件。

(2) 音叉式 MSG

Draper Lab 和 Rockwell 公司在 1994 年研制成功这种 MSG,如

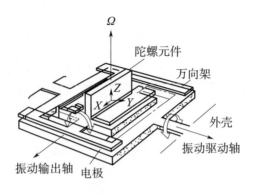

图 2　框架式 MSG 原理图

图 3 所示，它采用梳状结构的音叉产生大幅度振动，以提高陀螺的灵敏度。它的电子线路包括驱动惯性质量的自激谐振回路和测量电路两部分。这里的技术关键是微电容(10^{-18} F)测量和微型机械结构的设计与计算方法。这种 MSG 的标度因数为 40 mV/（rad/s），线性度为 0.2%，带宽为 60 Hz，零偏稳定性为 100（°）/h。

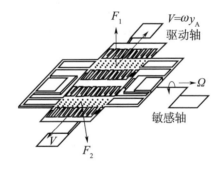

图 3　音叉式 MSG 原理

1.3　微硅加速度计陀螺仪

Northrop 公司正在研制采用玻璃-硅-玻璃三层结构构成的两个加速度计和两个抖动机构，由挠性的四边形装置连接起来。对称的抖动模式使抖动装置的线动量和角动量互相抵消。初步试验证明，预期加速度计的零偏稳定性为 1 mg，陀螺的零偏稳定性为 20（°）/h。

1.4 微硅惯性测量组合

1994 年 Draper Lab 公司研制的 MIMU 包括三只 MSG、三只 MSA 和附加的电子线路。陀螺的零偏稳定性为 10（°）/h，加速度计的零偏稳定性为 250 μg。

2 微型光学陀螺仪

2.1 结构

MOG 的组成部分为：光波导无源谐振器，声表面波移频器，引入和引出激光的耦合器，半导体激光器和光检测器等。它在结构上的特点是把各种光路、声表面波器件和控制电路集成到同一块硅片上，如图 4 所示，下面分别讨论其关键技术和解决的途径。

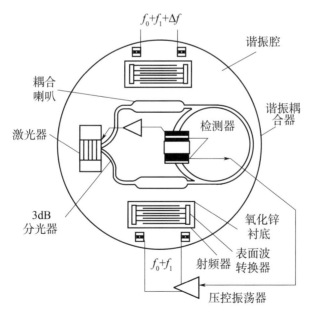

图 4 MOG 结构原理图

2.2 光波导无源谐振器

它是 MOG 的核心部件,其性能指标主要是品质因数。品质因数取决于光路损耗,包括耦合损耗和环形光波导本身损耗,关键技术在于降低环形光波导的损耗。

光波导基片材料的选择是设计的首要问题,可以选择的材料为硅和 $LiNbO_3$。在集中考虑波导的光特性和集成化程度情况下,硅材料优于 $LiNbO_3$,因此应选择硅基片。国外在硅基片上已制成损耗 < 0.1 dB/cm 的光波导。工艺上,光波导有离子注入扩散以及电子刻蚀等方法。在设计上,正确选择光波导环的直径是重要的,当曲率半径为厘米量级时,光波导的损耗较大(主要是辐射损耗)。目前硅基片上光波导环形谐振器的品质因数 F 约为 15,MOG 的要求应 > 100。图 5 为 MOG 的分辨率与波导损耗的理论曲线。由此可见,如何减小光波导损耗是 MOG 研制的重大技术关键。

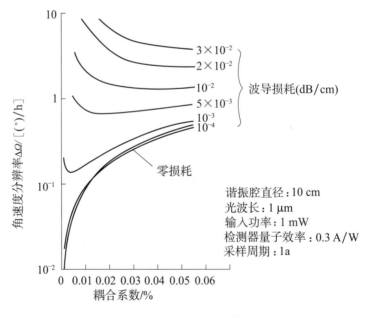

图 5 波导损耗与 MOG 分辨率的关系

光波导谐振器性能的测试是另一项关键技术，包括折射率、厚度和传输损耗等参量的测量，需要研制多种专用测量仪器。

2.3 声表面波移频器

在 MOG 中，采用两个 SAW 移频器分别实现跟踪顺时针（CW）和逆时针（CCW）两束光谐振频率的闭环控制，为此 SAW 移频器成为 MOG 的技术关键。如图 6 所示，SAW 移频器的原理是声光效应，使用蒸镀法在硅基片上形成压电材料薄膜，然后，采用激励压电薄膜使之产生表面波，并和光源给出的激光混频，以实现频率控制。这种 SAW 移频器保证了 MOG 的微型化和精度。

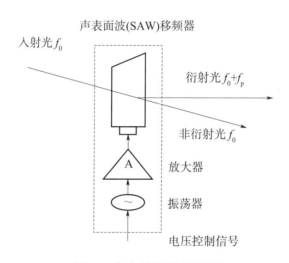

图 6 声光移频器的原理

声光移频器的性能指标为调制速率和频率稳定性。为了保证调制速率，应采用声光品质因数大、声吸收小、介质分布均匀、声速温度稳定性好的材料。此外，在梳齿形换能器设计上，应正确设计梳齿的间隔、延迟线带宽和换能器的长度，以避免声反射和声电再生效应，同时减少电极数，降低工艺的复杂性。

为了保证 SAW 的频率稳定度，应采用噪声系数低、输出饱和功率大、声孔径长的放大器，以提高等效品质因数。针对长期频率稳定性，应采用较厚的基片以减小应力松弛，同时用不同的电极材料，或在基片与电极之间加入阻挡层，以减弱基片和电极之间的化学作用。

2.4 耦合器

在 MOG 中采用环形谐振器，必须通过耦合器来实现光波的引入和引出，这是另一个技术关键。为此，在设计中应建立相位配合的偏振光电磁场，以提高耦合效率，常用的方法有以下四种：1) 对接耦合；2) 微棱镜耦合；3) 光栅耦合；4) 两个光波导互相耦合。应当采用后两种耦合器，它们的耦合效率较高。

2.5 激光器

理论分析表明，在环形谐振器的周长为 50 cm 的情况下，如果激光器的线宽为 8 kHz，则 MOG 的精度为 10^{-6} rad/s [0.2（°）/h]，但是如果其线宽为 5 MHz，则精度仅为 10^{-4} rad/s [20（°）/h]。由此可见，窄谱线的半导体激光器是提高 MOG 精度的一项技术关键。为了减少吸收损耗，光源波长应大于波导材料吸收边的波长。为此建议采用 1.55 μm 波长的光源。此外，光强需要稳定，应采用制冷电路，减小温度变化对输出功率的影响。当前，各种功率和波长的激光器已比较成熟。这里另一项技术关键是把光源集成在同一块硅基片上。

2.6 分光器及检测器

分光器的作用是将单一光源的激光分成两束，分别耦合入环形波导，形成 CW 和 CCW 方向的两束光。由于经分光器后两束反向传输光的光强不同，将造成 MOG 的零偏，因此这里的技术关键是保证分光光强的对称度。

检测器的性能为高频截止、非线性和噪声等。在选择检测器时，应尽量增大其量子效率，同时提高检测器的响应速度。现在，集成光学检测器的量子效率可达 0.4～0.6 A/W。

3 结束语

从 1995 年 5 月到 1995 年 7 月，在有关微米/纳米技术研究和博士生培养的几次座谈会上，作者提出，微米/纳米技术的研究应与国防和国民经济的需要相结合，建议在清华大学建立 MIMU 的实验研究装置，开展关键技术的研究。本文是座谈会的总结，文中提出的关键技术和解决途径只是初步的，它们将在今后的研究工作中加以深化。

国外 MIMU 的研制经验表明，我国应当选择 MSA、MSG 和 MOG 作为研制目标。应在较短的时间内建立 MIMU 的原理样机，为此应充分开展国际技术合作，包括选用国外器件，在有条件的情况下，利用国外设备研制关键器件，在国内组装等。同时，需要加强应用技术的基础理论研究，在实验研究的基础上，搞透基本原理和工程设计方法，并探讨解决关键技术的途径。

参 考 文 献

［1］ Lawrence A W. The Micro - Optic Gyro, Symposium Gyro Technology, Stuttgart 1983, Providing an inexpensive gyro for the navigation mass market, 1991.

［2］ Hafen M, E Handrich, et al. Micromachined Silicon Accelerometer B-290, Symposium Gyro Technology, 1994.

［3］ 丁衡高. 惯性技术文集［M］. 北京：国防工业出版社，1994.

［4］ Barbour N M, Elwell J M, et al. Inertial instruments: where to now?, I Saint Petersburg International Conference on Gyroscopic Technology, 25-26, May, 1994.

Key Technologies of Micro Inertial Measurement Unit

DING Heng‑gao ZHANG Yan‑shen MA Xin‑yu ZHANG Bin

(Tsinghua University)

Abstract Adopting the micro/nano technology, the Micro Inertial Measurement Unit has many advantages such as low cost, long life and high reliability. The paper analyzes the key technologies in its worldwide development and its future tendency. Some research tasks concerning these key technologies for the Micro Silicon and Micro Optical inertial devices are proposed.

微机电系统技术的实际应用——微型仪器

(2000 年 4 月)

摘　要　微机电系统被普遍认为是一项面向 21 世纪的可以广泛应用的新兴技术，作为其实际应用的微型仪器，具有传统仪器无可比拟的优势。从微型仪器入手，分别介绍了其概念、特点和市场应用，并在最后阐述了对微型仪器的几点认识。

关键词　微机电系统；微型仪器

1　微机电系统

一般说来，微机电系统（Micro Electronic Mechanical System，MEMS）是指可以用微电子等批量加工工艺制造的集微机械与微电子等部件于一体的部件或结构[1]，它可以分成多个独立的功能单元，输入的物理或化学信号由传感器转换为电信号，经过信号处理后，通过执行器与外界作用。我们习惯上把含有光学部件的装置也包括在 MEMS 内。这些部件结构单独或集成在一起，可以在微观层次感受、控制和作动物体，并产生类似宏观的作用效果。

MEMS 的特点和优点是显而易见的：体积小、重量轻、性能稳定；通过 IC 等工艺可批量生产、成本低，性能一致性好；功耗低；谐振频率高、响应时间短；综合集成度高、附加值高；具有多种能量转

本文发表于《国防科技大学学报》第 22 卷第 2 期。合作作者：丁衡高，袁祖武。

化、传输等功能的效应，包括力、热、声、磁及化学、生物能等。

鉴于上述特性和优点，MEMS 自 80 年代中末期发展至今一直受到世界各发达国家的广泛重视，被认为是一项面向 21 世纪可以广泛应用的新兴技术。目前 MEMS 已从实验室探索走向产业化轨道[2]，据美国 MCNC（北卡罗来纳微电子中心）MEMS 技术应用中心预测，当前 MEMS 业界的年增长率是 10%～20%，预期 2001 年将有高于 80 亿美元的 MEMS 潜在市场。而其中的大量产品主要包括汽车加速度计，压力、化学、流量传感器等，以及微光谱仪、一次性血压计等微型仪器（micro‑instrument）产品。

2 微型仪器

关于微型仪器，目前尚未见到有关对其概念的确切定义和描述。可以认为，微型仪器实际上就是具有仪器化功能的 MEMS 产品，是 MEMS 技术的实际应用，它具有一般仪器具有的监测（monitoring）、测量（measurement）、分析（analysis）、诊断（diagnostics）、控制（control）、作动（actuate）等功能，是一种新型的智能结构。其基本结构模式为微传感器＋信号和数据处理电路（含控制软件）＋外显示器或信号输出或微作动器等。随着微电子技术的发展，有些微型仪器的基本结构已经集成在芯片上，所以国外也开始将微型仪器称为芯片上的仪器（instrument on chip 或 chip‑sized micro‑instrument）。

从技术和工艺上来说，微型仪器技术涵盖了微小尺度的传感器和作动器的设计、材料合成、微型机械加工、装配、总成和封装及微专用集成电路等一系列微型工程技术，它是 MEMS 技术与微电子技术综合集成的产物。

下面我们举几个微型仪器的例子加以说明。

2.1 微光谱仪

德国 Karlsruhe 研究中心的微结构技术研究所和公司合作每年生产数千台微光谱分析仪,如图 1 所示,利用 LIGA 工艺制作,在 400～1 100 nm 范围内的分辨率为 7 nm,用于传输、反射和萤光测量;用于微分析系统的微型泵尺寸为 7 mm×10 mm,用模铸和薄膜技术,热致动线圈厚 250 nm,功率为 100 mW。

图 1 微光谱分析仪

2.2 MEMS 显示器

美国的 Texas Instruments 公司从 80 年代初就着手研究用于投影显示装置的数字驱动微镜阵列芯片(Digital Micromirror Device,DMD),已成功地演示了利用 768×576 像素的 DMD 芯片的彩色电视投影仪,并研制出 2 048×1 152 像素的 DMD 芯片样机。所研制的 DMD 芯片利用硅表面微加工工艺制作,其部分放大扫描电镜相片如图 2 所示,一个微镜的尺寸仅 16 μm×16 μm。图中有九个微镜,中间的一个去掉了最上层的反射镜,可看见下面的支撑机构。微镜通过支撑柱和扭曲梁悬于基片上,每个微镜下面都有驱动电极,在下电极与微镜间加一定的电压，静电引力使微镜倾斜，入射光线被反射到镜头上、投影到屏幕上，未加电压的微镜处的光线反射到镜头外，高速驱动微镜使每点产生明暗，投影出图像。

图 2　DMD 芯片部分 SEM 相片

2.3　微型化学传感器及微型黏度计

我们介绍伯克利微仪器公司（BMI）设计并推向市场的成熟的微型仪器产品[3]，包括测量空气中化学浓度的化学气体传感器 BMC200（图 3）及测量流体密度和黏度的微型黏度计 BMV105（图 4）。

图 3　化学气体传感器 BMC200　　图 4　微型黏度计 BMV105

BMC200 和 BMV105 均采用了一个独特的获得专利的称为 FPW

(Flexural-plate-wave)的硅微机械传感器技术（原理见图5）。包括一个微硅的MEMS芯片（尺寸为3.5 mm×7.6 mm），芯片上有一个约0.5 mm宽、6 mm长、3 μm厚的微机械薄膜，该薄膜由氮化硅层、金属层与压电氧化锌层构成。薄膜上集成了两个用来发射和接收超声波的交互数字式传感器。通过测量薄膜之间传递的声波的波速和振幅，微型黏度计BMV105则可确定液体的黏度和密度。对化学浓度传感器BMV200来说，当传感器放在化学气体中时，一些化学成分被聚合薄膜吸收，增加了薄膜的体积，降低了波速，从而测出化学浓度。

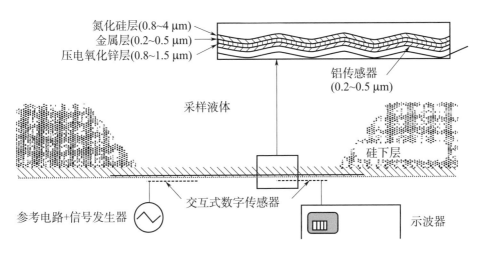

图5　FPW微硅机械微传感器原理示意图

与传统的类似的仪器相比，这些微型仪器的优势是无可争议的：体积更小、重量更轻、效率更高、价格更低，可以提供高精度的实时测量信息。这些信息可以用来有效地进行过程控制、调整、报警，或为其他自动监测和控制设备提供前馈或反馈，微型仪器既可以单独作为工作单元实现独立的功能，又可以作为在线的过程控制单元。

3 微型仪器的应用

从市场应用需求来看,微型仪器的市场大体可以分为以下几类:

(1) 环境科学

微型仪器在环境监测、分析和处理方面大有作为。它们主要是由化学传感器、生物传感器和数据处理系统组成的微型测量和分析设备。这些微型仪器可用来检测气体和液体的化学成分,检测核、生物、化学物质和有毒物品,其优势在于体积小、价格低、低功耗、易携带,市场前景广阔。

(2) 航天及星球探索

美国国家航空航天局(NASA)目前正制订一项雄心勃勃的微型仪器技术开发计划,主要目的是发展适合21世纪的小型、低价、高性能航天器,利用MEMS技术对航天器有效载荷和某些机电部件进行微型化,极大地缩小各种科学仪器和传感器的体积和重量,其结果是提高了功能密度,因此JPL称这些微型仪器将是新的微型实验室的心脏。它们主要包括:火星微登陆器、微加速度计、微磁强计、微湿度计、微气象站、微地震仪、微集成相机、微成像光谱仪、微推进器等等。

(3) 生物医疗

生物医疗领域是微型仪器比较活跃的领域之一,目前已研制的微型仪器包括:

• 一次性血压计,1995年Motorola与IC Sensor销出了2 000万只;

• "灵巧药丸"(Smart Pill),实际上是包含传感器、储药囊和微压力泵的微型仪器,可注入人体,并在人体内部的精确部位施放精确剂量的药物；

• 监测皮肤温度的集成无线微型仪器，它可用来远程监测皮肤温度，并将监测的信息以 350 Hz 的载波频率传送到接收器，在尺寸为 4.6 mm×6.8 mm 的芯片上集成了传感器的电子线路和微天线；

此外，用于医疗手术的微型仪器还有诸如用于修补血管，可替代心脏旁路手术的导管、假体单元、人造器官以及研制中的可清除血管壁沉淀物、攻击癌细胞的血管纳米机等。

（4）汽车工业

现代汽车的电子技术含量以每年 10% 的比例增长，而传感器的含量则以每年 20% 的比例增长，其中主要包括微加速度传感器及微压力传感器等微型仪器。将微型压力传感器置于轮胎中，可用来保持轮胎适当压力，避免充气过量或不足，仅此一项就可节油 10%。利用微加速度传感器可监控撞车，并为防撞气囊提供紧急充气信号，减少车内人员伤害。据美国系统规划公司估计，1998 年将有 1 500 万辆汽车要装备防撞气囊，加上压力传感器的需要，将形成巨大的市场需求。鉴于此，世界上各传感器公司纷纷加大投入，美国 Lucas NovaSensor 公司建成了一条年生产 600 万个加速度传感器的生产线，Rosemountg 公司的压力传感器的装货量达 4 亿美元，Motorola 公司则宣称将建立投资达 10 亿美元的汽车传感器产业。

（5）军事应用

微型仪器在军事上的应用具有深远意义，它或许就是未来战场的明星。如将袖珍式质谱仪作为有害化学战剂报警传感器安装在微型车辆或无人驾驶机上，就能在战地搜寻化学武器。利用各种微型传感器做成的草杆、树叶，上面装有微型电子侦察仪、微照相机等，将它们大量撒向战场后，在空中、地面、水上就能随时随地搜集情报。将传感器和灵巧武器散布在战场上，形成危险区，阻挠敌方步兵和坦克。在武器平台的蒙皮中植入传感元件、作动元件和微处理控制系统构成智能蒙皮，可用于预警、隐身和通信。美国弹道导弹防御局正在为导弹预警卫星和天基防御系统平台研制含有多种传感器的智能蒙皮。美国空军莱特实验室正在进行结构化天线的研究。演示实验表明，这种天线在气动特性、信息传输、结构重量与体积方面，都优于同样功能的普通天线。美海军则重点研究舰艇用智能蒙皮，以提高舰艇的减噪和隐身性能。

目前，世界上一些国家正在对蟑螂、蜜蜂、蚕蛾等昆虫进行试验，利用昆虫的生理结构和特性，进行人造昆虫的研制，这可算是一种特别的生物微型仪器，如人造光控甲壳虫，其底面积比一枚直径为 25 mm 的硬币还小。将光电池分别放在人造甲壳虫的身上，当光电池吸收外部的光能时，通过光电转换产生电流，驱动甲壳虫运动。所有这些精巧的微型武器，由于设计新颖，制作简便，造价低廉，可上至天空，下至地缝，无处不在，无处不入，能完成大型武器不能完成的多种任务，因此受到各国军方的极大关注。

除了上述的领域外，微型仪器还在工业控制、工业检测，玩具、服装等行业具有较强的需求背景，这里不再一一赘述。

4 对微型仪器的几点认识

通过对微型仪器的性能、特点和应用情况进行分析，我们有以下几点认识：

1) 微型仪器是 MEMS 技术的实际应用，是 MEMS 技术与微电子技术综合集成成果之一，它具有一般 MEMS 器件或 MEMS 系统特有的优点和性能，更为主要的是大大提高了仪器的功能密度和性能价格比。

2) 微型仪器技术中的核心技术之一是微型传感器技术，采用各种新原理、新概念的各类微型传感器，是实现微型仪器的关键，也是我们实现仪器小型化微型化的必要条件。此外，由于有的微型仪器上既有固定的部件又有活动的部件，而且采用诸如生物或化学活化剂之类的特殊材料，因此还需要认真重视微装配、微封装、微能源等问题。

3) 从广义上看，微型仪器还是一种新型的智能结构。目前国际上对智能结构的研究尚处在基础研究和实验阶段，今后的研究方向将包括：(a) 研制低能耗、大应变量、高稳定和长寿命的作动器材料；(b) 研制耐高温、低成本、易与基体材料融合的光传感器；(c) 研制可植入基体材料中的高性能微电子器件；(d) 新的结构控制技术；(e) 开发智能结构的设计、制造技术等等。

4) 微型仪器具有广阔的市场应用前景，并且新产品层出不穷，我们应密切跟踪国外发展动态，在研究和开发上采取联合攻关，成果共享的方式，加快研制进度，在21世纪MEMS产业化的道路上迈出我们自己坚实的步伐。

参 考 文 献

[1] 丁衡高. 微机电系统的科学研究与技术开发 [J]. 清华大学学报，1997，37 (9)：1-5.
[2] The 5th World Micromachine Summit' 99 [C], Glasgow.
[3] http://www.berkeleymicro.com; http://www.mcnc.org; http://trimmer.net.

The Typical Application of MEMS — Micro - Instruments

DING Heng - gao, YUAN Zu - wu

(General Equipment Headquarters of PLA, Beijing 100034, China)

Abstract　MEMS is widely considered as one of booming new technologies, and as its typical application, micro - instruments have unparalled advantage compared with traditional instruments, this article outlined the concept, characteristics and its application, some personal viewpoints about micro - instruments are forwarded.

Keywords　MEMS (Micro Electronic Mechanical System); Micro - instrument

微惯性仪表技术的研究与发展

(2001年12月)

摘　要　论述了当前国内外微惯性仪表设计和制造技术的研究与发展，强调了对相关基础理论研究的重要性，提出了计算机集成微制造单元的概念。

关键词　微惯性仪表；微机械陀螺；计算机集成微制造单元

1　引言

微惯性仪表，包括微机械加速度计、微机械陀螺和微惯性测量组合（MIMU），是一类重要的微机电系统（MEMS）。早在1994年就有人提出了发展微机电系统从抓微惯性仪表入手的设想。经过多年的探索与研究，微惯性仪表技术从无到有，已取得长足的发展。微惯性仪表技术的基础是微制造技术。从广义角度考虑，微制造技术包括设计、材料、工艺、测试、封装和微机理。研究理论涉及微电子学、微机械学、微动力学、微流体学、微热力学、微摩擦学、微光学、材料学，以及物理学、化学等基础理论。

2　微机理研究

一般来说，微机械按其尺寸可分为宏观微机械（$\geqslant 1\ \mu m$）、介观

本文发表于《中国惯性技术学报》2001年第9卷第4期。合作作者：丁衡高，王寿荣，黄庆安，裘安萍，万德钧，周百令，苏岩。

微机械(10 nm～1 μm)及原子和分子机械（＜10 nm）。微惯性仪表大多在微米量级，属宏观微机械范畴。其工作原理、设计理论等大体上可以沿袭传统的理论。但是，从一般技术到微米技术，从宏到微，这两者之间并不是简单的大系统小型化。随着微结构尺寸的不断缩小，构件可承受的外载荷和体积力变得次要，而构件间的摩擦力和其他表面力成为影响性能的主要因素。微型机械的力学系统特征和材料特征大致有这样几个方面：

(1) 力的尺寸效应

在微小尺寸领域，与特征尺寸的高次方成比例的惯性力、电磁力 (L^3) 等的作用相应减小，而与尺寸的低次方成比例的粘性力、弹性力 (L^2)、表面张力 (L^1)、静电力 (L^0) 等的作用相对增大，这也就是 MEMS 常用静电力致动的理由。

(2) 表面效应

随着尺寸的减小，表面积 (L^2) 与体积 (L^3) 之比相对增大，因而热传导、化学反应等加速，表面间的摩擦阻力显著增大。

(3) 制造误差影响

对于微结构，制造误差与结构尺寸之比相对增大，同时，由于微型机械往往是一次加工腐蚀成形，一般不进行调试与修正，这样，微结构的运动特性受制造误差的影响较大，再加上残余应力、弹性变形等的影响，使得运动精确度成为微结构研究的关键问题。这也是目前微惯性仪表成品率低的重要原因。

(4) 材料特性的变化

随着微结构尺寸减小，材料内部缺陷减少，材料的机械特性显著改变。对于硅材料，硅晶体存在各向异性，其机械特性会沿不同的晶向各异，还会随掺杂浓度的变化而变化。例如，其弹性模量 E 会随着掺杂离子浓度的增加而增加，当硼离子浓度达到 1×10^{20} cm^{-3} 时，E 增加 30% 左右。

对于微小尺寸领域的阻尼研究也是一个非常重要的问题，这主要有两方面的原因：一方面，特征尺寸达到微米量级后，常用于宏观物体的理论已经不能适用于微观物体。例如，在宏观物体间，人们早已总结出 Amontons 摩擦定律：$F = \mu N$，其中摩擦力 F 与载荷 N 成正比，其比例系数 μ（即摩擦系数）是常数，与接触面积无关。然而在微马达的研制过程中，人们却发现，减小转子与衬底之间的接触面积，可使微马达在较低电场强度下旋转。因此，转子与衬底之间的表面力是摩擦力的主要部分，这与经典的摩擦定律相去甚远。另一方面，在微结构中，由于尺寸效应的影响，使得表面阻尼力的作用远超过体积力，而微结构的动力能源很小，因此，降低阻尼以节约能耗也是一个关键问题。

研究表明：在不同的环境压力下，作用在微结构上的阻尼机理是不同的。可将气体压力分为三个区域。第一个区域是低真空压力区域，该区域大约在 10～100 Pa 以下，该区域固体材料表面的气体覆盖率大大减小，此时气体阻尼可以忽略不计，振动结构阻尼极大地依赖于表面积与容积之比、表面性质、材料特性以及加工工艺，即摩擦阻尼和材料阻尼。第二区域大约延伸至 10^3 Pa，该区域气体对振动结构的阻尼主要是单个气体分子与振动结构碰撞时的动量交换产生的。此时，气体分子之间的相互作用力很小，甚至没有相互作用力。第三区域压力较高，以致气体分子之间有较强的相互作用，从而粘性阻尼决定了振动结构的形态。为了提高微机械陀螺仪的性能，陀螺仪最好工作在第一个压力区域，但为了长时间保持陀螺仪性能不变，目前国内外不少单位是让陀螺仪工作在第二个压力区域。

3 微惯性仪表设计技术

微惯性仪表，一般主要是指微机械陀螺仪和微机械加速度计。从结构上讲，微机械陀螺仪比较复杂。以微机械陀螺仪为例，从工作原理上讲，它跟经典陀螺仪一样，即当绕驱动轴做高频振动的微陀螺仪敏感角速度时，其检测质量内各质点的哥氏加速度会对输出轴形成哥氏惯性力矩，从而产生输出信号。其本质是各质点敏感角速度产生哥氏加速度。哥氏加速度的表达式为

$$a_K = 2\Omega \times V_r \tag{1}$$

式中，Ω 为被测量的角速度；V_r 为质点自身驱动运动的切向速度。由式（1）可见，提高 a_K 的途径在于增大 V_r。当检测质量以频率做驱动振动，即 $x = X_A \sin(2\pi f_d t)$ 时，$V_r = 2\pi f_d X_A \cos(2\pi f_d t)$。可见，提高驱动振幅 X_A 及驱动频率 f_d 都可以提高 a_K 的数值。

但是，由于微陀螺的输出信号非常微弱，为了提高响应幅值，总是希望陀螺仪工作在共振状态，且品质因数 Q 的数值要足够大。过高的驱动频率会使驱动模态和检测模态的刚度值过大，从而使陀螺仪灵敏度降低。因此，从微陀螺仪结构设计的角度考虑，提高微陀螺仪分辨率的途径在于提高陀螺仪的驱动振幅和品质因数。此外，在不影响灵敏度的前提下，适当提高驱动频率也是可取的。

由于微惯性器件是一次成型的，尺寸误差对器件性能的影响较大，必须对器件进行精确的计算和仿真。因此，在微惯性仪表的设计中，对器件的有限元分析和仿真是必不可少的，需要进行模态分析、响应特性分析、温度场分析和静电场分析等。

模态分析是结构动力学的重要分支，通过模态分析，可以确定结构的模态参数，包括固有频率、振型、阻尼、刚度和质量等。微机械

陀螺通常都包含驱动模态和检测模态。陀螺仪驱动模态和检测模态的固有频率之差对陀螺仪检测灵敏度有很大影响。因此，如何在设计阶段精确确定其模态参数就显得非常重要。

响应特性分析包括谐响应分析和瞬态响应分析。

谐响应分析是用于确定结构在承受随时间按正弦规律变化的载荷时的稳态响应的一种技术。微陀螺都有驱动和检测两个模态，一般情况下，使两个模态的固有频率完全一致比较困难，总存在一个差值。这就面临着如何调整驱动频率，使输出响应达最大值的问题。运用有限元谐响应分析法可以较好地解决这个问题。

瞬态响应分析则可仿真陀螺仪在大加速度冲击时的变形以及所承受的应力情况，是研究微惯性器件抗过载特性的必要手段。

硅微机械振动陀螺仪大多采用静电驱动、电容检测的方式。通过静电场分析，可以分析计算静电驱动力矩、电容信号器的电场分布和阻尼孔的排列等问题。

硅材料是一种热敏材料，它的许多特性，如机械特性、物理特性等，受温度影响较大。当温度发生变化时，不但机械结构会变形，而且材料的弹性模量、拉伸强度、残余内应力、破坏韧性、疲劳强度等性能也会发生变化，有些甚至是质的变化。同时，硅微惯性仪表体积都很小，实际工作时通常被封装在一个很狭小的空间，环境温度的变化和电子元器件的发热量不容忽视。另外，在硅微机械加工过程中，浓硼扩散、键合、腐蚀等都会在硅结构内部产生应力和变形，从而对结构本身和输出精度都有影响。因此，需要对硅微惯性器件进行温度场分析。

MENS 技术的特点是多种学科的相互交叉，各种能量域，包括机械、电磁、热、流体等相互作用，使 MEMS 的仿真与建模越来越复杂。因此，微惯性仪表设计分析中遇到的最大挑战将是多能量域和多物理场的耦合分析。需要针对各个域的特点，寻找相应的建模方法和算法。

4 微惯性仪表制造工艺技术

微惯性仪表制造工艺是在微电子加工工艺基础上发展起来的。其中表面微机械加工技术主要是在基片上淀积或生长多晶硅层来制作微机械结构。由于与 IC 工艺兼容，因此可以将传感器与集成电路做在一块基片上。ADI 公司生产的 ADXL 系列叉指式加速度计，检测质量厚度为 2 μm，加速度计敏感元件部分边长 1 mm，信号处理电路分布于四周，整个芯片尺寸约 3 mm×3 mm。美国桑地亚国家实验室（Sandia National Lab.）与伯克利传感器执行器研究中心（the Berkeley Sensor Actuator Center）及 ADI 公司合作，利用模块化、单片集成微机电系统工艺（M³EMS 工艺），先在晶片上腐蚀浅槽，在槽底生长多晶硅来制作微机械，槽中充填牺牲层，进行高温退火，然后再制作电子线路，开发出单片集成的三维 MIMU。该工艺的槽深约 6 μm，利用该工艺，可以将传感器结构和电子线路集成在同一块芯片上。但由于是采用表面微机械加工技术，加工的检测质量的厚度都比较薄，制约了微惯性器件精度和灵敏度的提高。

微惯性器件的第二种加工工艺是基于深反应离子刻蚀（DRIE）技术的体硅加工工艺。比较典型的是硅/玻璃体硅溶解薄片法。该工艺包括硅片工艺、衬底玻璃工艺和组合片工艺。硅片工艺包括两次光刻，一次浓硼扩散。首先，在单晶硅片上进行浓硼扩散，以形成一定厚度的浓硼自停止层；然后，利用掩膜板在硅片上刻蚀出键合凸台；最后，采用 ICP 高深宽比干法刻蚀工艺，刻蚀出陀螺仪机械结构。衬底玻璃工艺包括一次光刻、一次金属化。首先，在 7740 玻璃上腐蚀出浅槽；然后，溅射 Ti－Pt－Au 多层金属；最后，用超声波剥离多余的金属层。组合片工艺是将加工好的硅片和玻璃片进行清洗、对准和静电键合。利用该工艺可以加工出 20～30 μm 厚的微结构，从而使仪表的精度和灵敏度都有所提高。但由于和 IC 工艺不兼容，因此只能采用混合集成电路。目前国内研制的振动轮式硅微陀螺仪都采用这种加工工艺来制作。

第三种是 LIGA 加工工艺，即 X 光同步辐射光刻、电铸及微压塑工艺。采用这种方法，结合牺牲层技术，可以制作大高宽比的可活动微结构。LIGA 工艺极大地增强了微结构加工能力，它所得到的微结构机械性能好，传感器的灵敏度高。但由于需要有同步辐射光源，因此成本较高。LIGA 技术作为一项有希望的技术，应给予一定的重视。

从某种意义上说，制造工艺是发展微惯性仪表技术的关键。微惯性仪表的特点之一是能够批量制造，因而成本可以降得很低。但是，批生产的前提是制造工艺必须稳定、可靠、规范，否则将成为制约微惯性仪表技术发展的瓶颈。

微惯性仪表制造工艺的一项主要技术是封装技术。由于微机械陀螺仪只有工作在一定的真空度下，才能满足性能要求。因此，如何进行真空封装，并保持一定的稳定性，是微机械陀螺仪从实验室走向实际应用的关键工序，必须引起足够的重视。

5 计算机集成微制造单元

目前在 IC 工业中，有很多 CAD 系统可用于集成电路的设计，极大地提高了设计效率。但它们中的大部分却不适用于 MEMS 的计算机辅助设计。由于 MEMS 在结构和制造工艺上的特殊性，需要对

MEMS 的三维结构进行多物理场的计算机辅助分析,并通过计算机对工艺和掩膜过程进行模拟,以使设计人员在设计阶段就能进行方案的比较和验证,并充分考虑工艺的变化对器件性能的影响,在制造之前能够充分检验工艺及掩膜的有效性。

图 1 所示是一个计算机集成微制造单元（Computer Integrated Micro Manufacture Unit，CIMMU）的框图。一个完整的 MEMS 必然是微机械敏感器件与微电子线路的集成组合系统,因此在 MEMSCAD 当中应当包括信号处理电路的计算机辅助设计和仿真,以及 MEMS 系统的仿真。

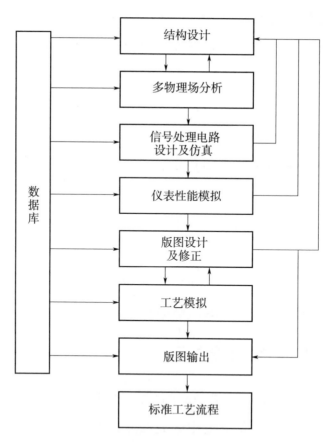

图 1　计算机集成微制造单元

MEMS的制造过程不仅改变结构的几何轮廓,还改变材料的性质。这种材料性质的改变将会影响结构的特性,必须建立相应的材料特性数据库,并可根据工艺流程将材料特性自动插入三维几何模型中。

当实现MEMS工艺流程的标准化和计算机控制后,一个计算机集成微制造单元的框图就如图1所示。

6 结束语

总之,我们应总结国内外对微惯性仪表的研究与发展状况,找出影响我国微惯性仪表技术发展的主要症结和关键所在,以期取得更快的发展和更大的成就。

参 考 文 献

[1] 丁衡高. 面向21世纪的军民两用技术——微米/纳米技术[J]. 中国惯性技术学报,1995,3(2).

[2] 丁衡高. 微机电系统的科学研究与技术开发[J]. 清华大学学报,1997,(9).

[3] 周兆英,叶雄英,等. 微型系统与微型制造技术[J]. 微米纳米科学与技术,1996,2(1).

[4] 王寿荣. 硅微型惯性器件理论及应用[M]. 南京:东南大学出版社,2000.

[5] 裘安萍. 硅微型机械振动陀螺仪结构设计技术研究[D]. 南京:东南大学,2001.

Research and Development of Micro Inertial Instruments

DING Heng-gao, WANG Shou-rong, HUANG Qing-an,
QIU An-ping, WAN De-jun, ZHOU Bai-ling, SU Yan

(Southeast University, Nanjing 210096, China)

Abstract　In this paper, the research and development for the design and fabrication technique of micro inertial instruments are presented with emphasis on the fundamental theory. A viewpoint about the computer-integrated micro manufacture unit is proposed.

Keywords　Micro inertial instrument; Micromachined gyroscope; Computer Integrated Micro Manufacture Unit (CIMMU)

《微纳米加工技术及其应用》序

(2004年12月)

本书集崔铮博士多年来的实践经验与研究成果，并结合近年来国际上的最新进展，综合介绍了微纳米加工技术的基础，包括光学曝光技术，电子束曝光技术，聚焦离子束加工技术，X射线曝光技术，各种刻蚀技术和微纳米尺度的复制技术。内容的编排上遵循了微纳米加工的从光刻制图到刻蚀与图形转移的基本步骤。对各种加工技术的介绍着重讲清原理，列举基本的工艺步骤，说明了各种工艺条件的由来，并注意给出典型工艺参数。充分分析了各种技术的优缺点，以及在应用过程中的注意事项。强调实用，避免烦恼的数学计算。并用专门一章介绍了微纳米加工技术在现代高新技术领域的应用，包括超大规模集成电路技术，纳米电子技术，光电子技术，高密度磁存储技术，微机电系统技术，生物芯片技术和纳米技术。通过实例说明了现代高新技术与微纳米加工技术的不可分割的关系，并演示了如何灵活应用微纳米加工技术来推动这些领域的技术进步。本书与国内外同类出版物相比，显著特点是将用于超大规模集成电路生产与用于微机电系统、微传感器系统制造的微纳米加工技术综合介绍，并加以比较。

本书首次将纳米加工归纳为平面工艺、探针工艺和模型工艺三种主要类型。突出了微纳米加工技术与传统加工技术的不同之处。一般微米以下，100纳米以上仍习惯上称为微细加工或微加工。制作100纳米以下的结构才是真正意义上的纳米加工。而纳米加工除了以上所说的平面工艺、探针工艺和纳米模型工艺之外，还包括分子工程实现的纳米结构。这就是所谓自上而下（Top – Down）和自下而上（Bottom – Up）的加工技术。平面工艺是典型的自上而下的加工技术。自下而上则依赖于分子自组装过程，更多地涉及生物与化学反应，而不是传统意义上的加工技术。所以分子自组装工程一般不归类于微纳米加工技术。但在大多数情况下，分子自组装也要依赖自上而下的微细加工技术来构筑自组装的平台。纳米加工技术不可能孤立存在。纳米尺度的物理化学现象通常需要通过微米结构的器件或系统过渡到宏观世界。除了扫描微探针加工技术之外，大多数纳米加工技术是在微米加工技术基础上发展起来的。因此微米与纳米加工实际上不可分割。

全书既注重基础知识又兼顾微纳米加工领域近年来的最新进展。并列举了大量参考文献供进一步深入研究。因此不论对初次涉足这一领域的高等院校的本科生或研究生，还是已经有一定工作经验的专业科技人员，都具有很好的参考价值。特此郑重推荐给广大读者。

丁衡高

2004 年 12 月 1 日

微型化陀螺研究进展和展望

(2005 年 9 月)

摘　要　本文分析了 MEMS 陀螺和集成光学陀螺研究现状，以陀螺灵敏度与噪声之比作为一个度量，对微型化陀螺性能进行了初步的比较和分析，指出了今后这两种陀螺可能达到的技术指标和研究中需要克服的主要技术难点。

1　引言

MEMS 陀螺和集成光学陀螺是实现陀螺微小型化的两条重要技术途径。就目前情况来看，MEMS 陀螺已接近实用化，特别是国际上，已有报道可提供精度为 1（°）/h 的 MEMS 陀螺，但国内要达到这一指标仍然存在一些困难。集成光学陀螺的研究在国内外也已进行了相当长的时间，取得了一些初步结果，但离实用化要求仍然有相当一段距离。对于这两种陀螺可能达到的技术指标以及发展前景是人们普遍关心的问题。本文以陀螺灵敏度与噪声之比作为一个度量值，简要回顾微型陀螺目前的研究水平，并从理论上对今后可能达到的技术指标进行预测，通过分析指出了今后微型化陀螺所需要克服的关键技术，从而为这两种微型化陀螺的研究提供若干参考。

本文发表于《2005 年惯性技术科技工作者研讨会论文集》。合作作者：金仲和，丁衡高，马慧莲，丁纯，周柯江，王跃林，袁祖武。

2　MEMS 振动式陀螺

微机械陀螺由于具有体积小、重量轻、成本低等特点，越来越受到重视。但是目前，特别是国内，大部分微机械陀螺的精度尚不是很高，使其在许多高精度应用场合的应用受到限制。

对于振动式 MEMS 陀螺来说，由单位角速度引起的陀螺敏感方向上的振动幅度 B_y 为

$$B_y = \frac{2\omega B_x}{\omega_y^2 \cdot \sqrt{\left[1 - \frac{\omega^2}{\omega_y^2}\right]^2 + \frac{1}{Q^2}\left[\frac{\omega}{\omega_y}\right]^2}}$$

上式即为用振动幅度表示的陀螺的灵敏度。式中，B_x 为微机械陀螺驱动方向上的振幅；Q 为检测方向上的品质因素；ω 为驱动方向上的振动频率；ω_y 为检测方向上的本征频率。由上式可以看到，灵敏度随着振动幅度而线性增加，而当驱动频率与检测方向上的频率接近时，可得到 Q 的放大。

陀螺灵敏度与陀螺噪声之比也称为陀螺分辨率，是陀螺可能分辨的最小角速度值，它是陀螺性能指标的重要组成部分。要提高分辨率，在提高灵敏度的同时必须降低噪声。对于 MEMS 陀螺来说，陀螺灵敏度与驱动方向上的振幅、检测方向上的品质因素成正比。对于实际的 MEMS 陀螺来说，振动幅度、Q 值等均是有限的，典型的 MEMS 陀螺灵敏度用电容变化来表示的话一般为 $10^{-19} \sim 10^{-20}$ F/[(°)/h]，即 $0.1 \sim 0.01$ aF/[(°)/h]。因此陀螺分辨率的提高很大程度上依赖于对噪声的抑制，以下主要就 MEMS 陀螺中存在的几种噪声来源分析各自对陀螺分辨率的影响。

(1) 陀螺 Brownian 噪声

就空气下工作的陀螺传感器本身来说，一般文献认为 Brownian 噪声是主要的噪声来源。该噪声与陀螺的结构参数、工作带宽、检测方向 Q 值等有关。通过优化陀螺结构设计，可以降低这两种噪声。D. J. Seter 等人早在 1996 年就研究了一种静电驱动、光检测的陀螺，理论分析表明陀螺机械噪声等效分辨率可达 1 (°)/h；Yong Chen 等人在 2003 年研究了一种基于滑膜阻尼的音叉式陀螺，理论分析表明该陀螺 Brownian 噪声为 0.112 (°)/h/Hz$^{1/2}$。但是，到目前为止，大部分报道的陀螺系统实际测试得到的分辨率基本在 50~100 (°)/h/Hz$^{1/2}$。例如 Y. Chen 等人研制的陀螺实际测试噪声为 100 (°)/h/Hz$^{1/2}$ 的水平，这与 Brownian 噪声的等效分辨率差几乎 3 个数量级。这些结果表明尚有其他噪声机制在起主要的制约作用。只有分析清楚这些噪声机制，才能预测哪些噪声是可以通过努力降低或消除的，哪些噪声是无法消除的，从而制约 MEMS 陀螺的分辨率。也只有分析清楚这些噪声的产生机制，才有可能提出相应的抑制措施，从而有效提高陀螺的分辨率。

(2) C-V 变换电路噪声

MEMS 陀螺一般采用电容敏感结构，其灵敏度换算为电容变化量的话，大致在 0.01~0.1 aF/(°)/h 量级。一般陀螺静态电容为几 pF，要使系统分辨率达到 1 (°)/h，C-V 转换电路必须能在几 pF 静态电容下检测出 0.01~0.1 aF 的电容变化量，这对 C-V 转换电路提出了很高要求。但目前国内 C-V 转换电路分辨率在 1~10 aF 的水平，这等效于 10~100 (°)/h 的噪声水平。因此要提高陀螺灵敏度，C-V 转换电路的分辨率尚需有较大幅度的提高。

在实际应用中，目前主要采用图 1 两种结构形式的 C-V 转换电路。这两个电路均采用固定频率、幅度的载波（正弦波或方波）去激励可变电容，用电容的变化去调制该载波的幅度，然后采用相干解调将电容调制在载波上的幅度变化检测出来，从而得到电容的变化值。在这一过程中，噪声将主要来自几个方面，一是电路中运算放大器、电阻的热噪声；二是相关电路，包括混频器（乘法器）、滤波器等带来的噪声；三是载波自身、衬底偏压、电源电压、寄生电容带来的噪声。

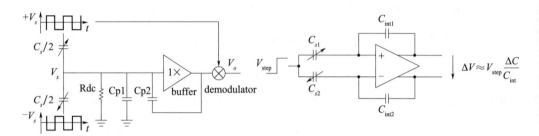

图 1 两种 C‑V 转换电路示意图

理想情况下这两种电路分辨率受限于热噪声，经过适当选择器件，理论上比较容易达到 0.1 aF 甚至 0.01 aF 量级。但实际应用中，C‑V 转换电路分辨率远远没有达到这个水平，从目前国内情况来看，要得到 10（°）/h 或更高精度的 MEMS 陀螺，C‑V 转换电路显然需要下大力气进行研究和提高。国外有报道表明 0.01 aF 分辨率的 C‑V 转换电路已经实用化，相信这是研制成功 1（°）/h 陀螺的重要技术支撑。

（3）陀螺驱动信号噪声

MEMS 陀螺的工作是以驱动方向上的振动为基础，陀螺输出信号与驱动方向上的振动幅度成正比。当驱动信号带有噪声时，该噪声会直接进入陀螺输出信号中，特别是该噪声的频谱位于驱动信号中心频率附近时，影响会更大。而驱动信号的漂移，不管是幅度的还是频率的，均将直接影响陀螺的稳定性。

MEMS 陀螺一般工作在自激振荡状态下，使振动频率等于本征频率。由于自激振荡是陀螺与外围电路构成闭合环路，环路内任一部分引入的噪声与稳定性均将影响系统的噪声和稳定性。另外由于闭环分析的复杂性，对这方面的分析报道比较少。但可以预计这方面的影响会是比较大的，有必要进行详细分析。

（4）直接耦合噪声

由于器件制作工艺不理想，可造成器件结构质量分布不均匀、弹性梁不平衡，这些问题均会导致驱动方向到检测方向的机械耦合，即使在没有角速度输入情况下，检测方向也会有振动；而驱动力沿陀螺检测方向有小的分量，也会引起陀螺在检测方向振动。这两种寄生振动在频率上与陀螺效应导致的检测方向振动完全相同，在实际检测电路中难以区分。上述两种误差一方面会使陀螺零点绝对值增加，同时也会引入额外的噪声和漂移。对于这些噪声，一方面需要从工艺及结构设计入手，需求降低直接机械耦合，另一方面也需要对这种耦合进行建模，以寻求抑制的方法。

从上面的初步分析可以看到，MEMS 陀螺的噪声除了陀螺器件自身有关噪声外，还有外围处理电路带来的噪声。上面仅就几种简单的噪声进行了定性分析，事实上，还有其他多种噪声可能影响陀螺的输出，例如机械结构的热弹噪声等。单就数值结果来看，目前国内 MEMS 陀螺所报道的技术指标基本与 C－V 转换电路所能达到的指标在同一数量级，可见 C－V 转换电路很可能是制约陀螺性能的主要因素。但是由于上述分析仅就部分报道而得到的结论，不能完全反映实际情况。比较严谨的做法应该是建立 MEMS 陀螺与外围电路的数学模型，并在模型中引入各模块的噪声，分析各部分噪声的传递过程，以及对最终输出信号的影响。在这一噪声模型的基础上，对 MEMS 陀螺及外围电路进行噪声的测试、分析，使理论和实际测试结果相互印证，从而建立起可信的噪声理论。进一步通过对模型的详细分析，可以发现哪些噪声对系统起制约作用，以及是否有可能通过优化结构、调整参数来有效抑制该噪声及其影响，最终提高陀螺整体性能。同时也可预测 MEMS 陀螺最终可能达到的性能指标，为进一步的研究工作指明道路。

3 集成光学陀螺

集成光学陀螺在上世纪 90 年代前期曾兴起一股研究的热潮，由于 MEMS 陀螺以及光纤陀螺的快速发展，导致人们对这一领域的研究在一段时间内失去了兴趣。但在 2004 年前后，这一领域的研究又开始引起人们重视。主要原因在于，一方面光纤陀螺微小型化比较困难，另一方面 MEMS 陀螺要达到较高精度存在困难，而集成光学陀螺在实现微小型化的同时可保证比较高的精度。因此可能在光纤陀螺、MEMS 陀螺之间找到合适的应用场合。

集成光学陀螺与光纤陀螺一样可以采用两种方式，无源谐振腔式和干涉式。以下主要就这两种方式在目前工艺和器件条件下可能达到的指标及制约因素进行理论分析和对比。

对于干涉式陀螺（图 2），正反两路光的相位差与输入角速度成正比

$$\Delta \phi_s = \frac{2\pi LD}{\lambda c}\Omega$$

反映到探测器上即为光强的变化

$$P_I = P_0 \cdot \frac{1}{8} \cdot 10^{-(4\alpha_C + \alpha_L \cdot L)/10} \cdot [1 + \cos(\Delta \phi_s - \pi/2)]$$

上式中已经考虑到为获得最大灵敏度所需要引入 $\pi/2$ 的相位偏置，

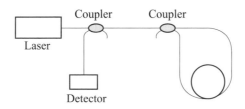

图 2 干涉式光学陀螺系统结构简图

从而使干涉仪工作在余弦响应曲线的拐点上。

对于图 3 所示的谐振式光学陀螺，正反两路光频率与谐振腔锁定后，在有角速度输入情况下，两路光的频率偏差为

$$\Delta f_s = \frac{\Delta L}{L} f_0 = \frac{LD}{c} \Omega \cdot \frac{1}{L} \cdot \frac{c}{\lambda n} = \frac{D}{\lambda n} \Omega$$

在谐振式光学陀螺与干涉式光学陀螺中影响精度的误差因素种类基本相同，包括散粒噪声、瑞利背向散射、光克尔效应、偏振波动、法拉第效应，以及温度波动等等。但由于光路结构和原理上的不同特点，这些误差因素的影响程度和所起作用有所差别。以下就散粒噪声、瑞利背向散射、光克尔效应三种噪声进行比较分析，所有比较分析的对象均为目前技术条件下可能达到的集成光学波导环，而不是光纤环。在分析中，波导环的直径在 2～4 cm 范围，损耗在 0.1 dB/cm 以下，光功率为 2 mW。

（1）散粒噪声限制的极限分辨率

一般来说，散粒噪声是无法克服的，因此对光学陀螺来说，当克服了其他噪声影响后，其精度最终将受限于散粒噪声。所以经常称散粒噪声限制的分辨率为极限分辨率（有时也称为极限灵敏度）。

谐振式光学陀螺由光探测器散粒噪声限制的分辨率可表示为

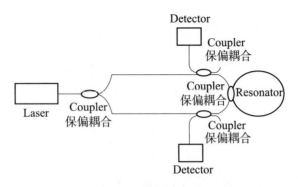

图 3 谐振式光学陀螺系统结构示意图

$$\delta\Omega_R = \frac{\sqrt{2}\lambda c}{DLF}\sqrt{\frac{2eB}{P_D R_D}}$$

而在干涉式光学陀螺中，由散粒噪声限制的检测精度可表示为

$$\delta\Omega_I = \frac{\lambda c}{2DL}\sqrt{\frac{2eB}{P_D R_D}}$$

上两式中，λ 为工作波长；c 为真空中的光速；e 为电子电量；B 为检测带宽；P_D 为达到光电探测器的平均光强；R_D 为探测器的相应灵敏度；D 为环直径；L 为环总长度；F 为清晰度。

图 4 给出了典型参数下谐振式光学陀螺极限灵敏度与谐振腔长度的关系，图中光波导传输损耗 α_L 分别为 0.01 dB/cm 和 0.1 dB/cm，谐振腔耦合器损耗 α_C 分别为 0.1 dB 和 0.2 dB，其中，传输损耗为 0.01 dB/cm 波导环的极限灵敏度数据已乘以 10，以便能在图中看上去更加清楚。由图可见，随着波导圈数增加，分辨率逐渐线性下降。这主要是由于长度增加后，损耗增加导致清晰度下降，而在谐振式光学陀螺中清晰度的下降将严重制约分辨率。另外一个值得注意的问题是，0.1 dB/cm 波导损耗下，陀螺分辨率不能达到 50（°）/h/Hz$^{1/2}$，这就意味着只有损耗很小的光波导材料才适合于较高精度谐振式光学陀螺。

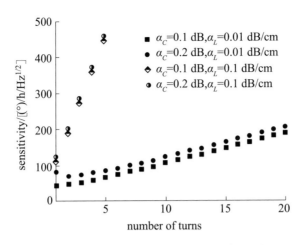

图 4 光波导环形腔直径为 4 cm，谐振腔耦合器耦合系数为 0.01 时，谐振式光学陀螺极限灵敏度与谐振腔长度的关系

注：为清楚起见，下方两条曲线为实际值的 10 倍

当然图 4 是尚未对耦合器耦合系数进行优化的结果，图 5 给出了在 0.01 dB/cm 波导损耗条件下，谐振式光学陀螺最佳极限灵敏度、最佳耦合系数与谐振腔长度之间的关系。由图可见，采用多圈结构后，4 cm 直径的环可以达到（2～3）（°）/h/Hz$^{1/2}$ 的极限分辨率。如果进一步提高损耗指标，可望实现更高精度的集成光学陀螺。例如目前国际上有报道的集成光波导最低损耗已能达到 0.001 dB/cm 量级。

图 6 给出了干涉式陀螺的极限灵敏度与长度之间的关系，由图可见，干涉式陀螺的精度对损耗并没有像谐振式那么敏感。但是在长度小于 60 cm 时，其极限灵敏度很难达到 50（°）/h/Hz$^{1/2}$ 以内。从图 6(b) 中可以看到，在光波导传输损耗为 0.01 dB/cm 条件下，只有当环形腔长度达到 1.3 m 以上才能达到这个水平，这相当于在 4 cm 直径下，约需 10 圈以上的光波导。

从上面比较可以看出，对于集成光学陀螺来说，如果仅比较散粒

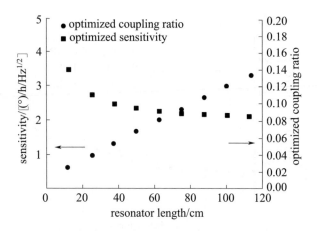

图 5　谐振式光学陀螺最佳极限灵敏度、最佳耦合系数和谐振腔长度的关系

噪声限制下的极限分辨率，采用谐振式方案比干涉式方案要优越一些。而干涉式方案只有在多圈结构下，才可能达到一定的精度。

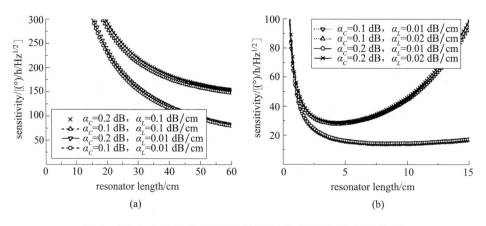

图 6　干涉式光学陀螺极限灵敏度和环形腔长度的关系

（2）背向散射噪声

背向散射主要是由于光纤内部介质或散射体的不均匀性、光纤通路中的焊接点以及器件的耦合点引起的。背向散射最终会以噪声或漂移的形式在陀螺输出中体现出来，从而限制陀螺的分辨率（图7）。

对于谐振式光学陀螺，由背向散射限制的陀螺分辨率表示为

$$\Omega_R = \frac{\lambda c}{\pi DL} \ln \frac{1}{1-a_0} \frac{\theta_0^2}{4}$$

式中，λ 表示真空光波长；c 表示真空中的光速；a_0 表示光波导总传输损耗系数；θ_0 表示光波导的数值孔径，$\theta_0 = n_1 \sqrt{2\Delta}$，已知 $n_1 = 1.45$，$\Delta = 0.25\%$。

对于干涉式光学陀螺，由背向散射引起的陀螺分辨率表示为

$$\Omega'_R = \frac{\lambda c}{DL} \ln \frac{1}{1-a_0} \frac{\theta_0^2}{4}$$

从上面两式的比较中可以知道：在结构参数相同的情况下，由背向散射引起的误差，干涉式光学陀螺是谐振式光学陀螺的 π 倍。

对于干涉式陀螺来说，背向散射可以直接采用非相干光源（如超辐射二极管）来抑制。但对于谐振式陀螺来说，需要对强相干光源采用抑制载波的调制方法，以实现对背向散射噪声的抑制。一般来说如果能将两路光的载波抑制到千分之一，由背向散射引起的伪旋转将下降到原来的百万分之一，从而达到 0.1（°）/h 的水平。

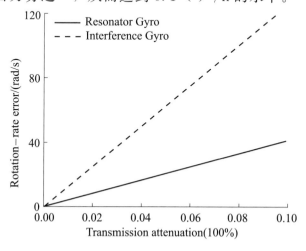

图 7　谐振环传输损耗与背向散射所引起的陀螺漂移的关系

(3) 克尔效应

光波导的折射率 n 随着光强而变化的非线性现象称为光波导的克尔效应。光学克尔效应是一种三阶非线性光学效应。表达式为

$$n = n_0 + a\xi P / A_{eff}$$

式中，n_0 为光波导的正常折射率；ξ 为光波导材料的非线性折射率系数（即克尔系数）；P 为光功率；$a(2/3 \leqslant a \leqslant 1)$ 为偏振因子；A_{eff} 为光波导芯区有效截面积。从上式不难看出，克尔效应与光波导的有效折射率和传输的光强有关。在光波导环中，克尔效应是由 CW 或 CCW 光强不平衡所引起光波导折射率系数微扰造成的。它所产生的输出偏离正比于 CW、CCW 传输光波之间的光功率差；只有当两光束的输入功率相等时，所产生的输出偏离才为零。

对于谐振式光学陀螺，当陀螺静止时，克尔效应引起的漂移可表示为

$$\Omega_K = \frac{\lambda c \xi}{4\pi D} \cdot \frac{F}{\pi} \cdot \frac{\Delta P_i}{A_{eff}}$$

式中，ξ 表示正比于光学克尔效应系数的一个量，对于熔融石英，$\xi = 7.4 \times 10^{-11}$ cm/W；F 表示光波导环清晰度；A_{eff} 表示光波导芯区有效截面积（即光束传输的横截面积）；ΔP_i 表示光波导环相反两路输入功率差。从上式中可以知道克尔效应与光波导环两路的入射功率差、光波导环精细度成正比，与谐振腔直径、光波导芯区有效截面积成反比。

在干涉式光学陀螺中，由克尔效应引起的漂移可表示为

$$\Omega'_K = \frac{\lambda c \xi}{4\pi D} \cdot \frac{\Delta P_i}{A_{eff}}$$

从上面两个公式可以看到，在光波导环尺寸等条件一致的情况下，谐振式光学陀螺是干涉式光学陀螺的 F/π 倍；这是可以理解的，因为

在谐振式光学陀螺中，由于谐振作用光波导内的强度被增强了大约 F 倍。图 8 给出了在集成光学陀螺典型参数下，采用谐振式和干涉方案受克尔效应限制的检测精度。由图可见，要将克尔效应的影响抑制到 10°以内，分光比精度要达到 0.01％的水平，在实际光路中，很难采用固定分光比的光功分器达到如此高精度的分光比，一般可采用可调光功分器实现自动控制，当两路光的输入光功率的相对差异减小为 0.001％，对于谐振式光学陀螺由克尔效应引起的陀螺漂移降为 0.78（°）/h，对于干涉式光学陀螺由克尔效应引起的漂移为 0.04（°）/h。

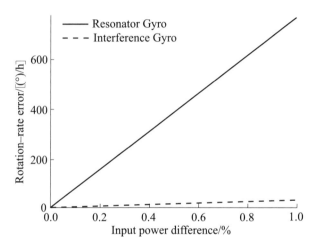

图 8 谐振式陀螺、干涉式陀螺中输入光功率差与所引起的伪旋转的关系

由上面三种噪声的分析可以看到，对于集成光学陀螺来说，在现有的技术条件下，要实现 1（°）/h 的陀螺是很有希望的，当然要达到这个目前尚有许多困难需要克服。另外，在集成光学陀螺中采用谐振式和干涉式方案各有优缺点。由于散粒噪声将最终限制陀螺可能达到的极限精度，因此在微小型化陀螺中，采用谐振式方案相对占优。但由于本文的分析仅限于少数几种噪声，并且分析仍然显得比较粗略，因此，本文只能是抛砖引玉，有待于更多的理论和实际研究工作的开展。

4 总结

对于 MEMS 陀螺，由于体积、成本等方面的优势，只要能达到 10（°）/h 的水平，就会存在巨大的市场，应该讲这一目标在不久的将来国内就能实现。但总体来说，国内对 MEMS 陀螺的噪声理论和测试方面的工作开展得不够深入，许多噪声的分析仍然停留在定性分析上，缺乏必要的定量分析手段，也就谈不上针对噪声进行优化设计。这一状况已经开始制约研究人员在陀螺精度方面的可能突破。因此有必要在噪声理论和测试方面加强研究工作，只有将基础研究工作做得扎实些，才有可能走得更远、更好。

对于集成光学陀螺来说，其长远目标仍应定位在 1（°）/h 甚至更高水平，以避免与 MEMS 陀螺正面争夺市场。单就这一目标来说，谐振式集成光学陀螺相对比较有希望。但是所需要克服的困难也是很多的，在理论上除了本文已经做了初步分析的散粒噪声、背向散射、克尔效应外，尚有其他多种噪声需要进一步分析。另外本文的分析也表明，采用多圈结构在研制较高精度的集成光学陀螺中仍然是必要的。

《微纳系统》译丛总序

(2012 年 6 月)

微机电系统(MEMS)出现于 20 世纪 80 年代中后期，是指可以批量制造的，集微结构、微传感器、微执行器以及信号处理和控制电路等于一体的器件或系统。其特征尺寸一般在 0.1~100 μm 范围。目前国际上通常将 MEMS 冠以 Inertial -，Optical -，Chemical -，Bio -，RF -，Power -等前缀以表示其不同的应用领域。MEMS 集约了当今科学技术的许多尖端成果，更重要的是它将敏感与信息处理及执行机构相结合，改变了人们感知和控制外部世界的方式。

MEMS 技术经过 20 多年的发展，诸如喷墨打印机中的喷嘴阵列、手机中的振荡器、陀螺、加速度传感器、磁场传感器、投影显示器中的微镜阵列等消费类电子产品，以及汽车防撞气囊中的加速度传感器、胎压检测系统中的压力传感器等等已经进入大规模生产。近年来我国出现了不少 MEMS 高新技术企业，很多大学也开设了 MEMS 课程，因此，无论是 MEMS 教学或科学研究，还是工业化产品开发，都迫切需要 MEMS 技术方面的信息资料，而 MEMS 是一个快速发展的前沿技术领域，信息资料分散于期刊论文、专利以及会议文集中，缺乏系统的归纳、分析与整理，不成体系。因此，应对我国 MEMS 发展需求，特别需要这方面的著作，虽然我国一些出版社已经购买版权并翻译出版了部分国外书籍，对我国 MEMS 技术的发展起到了积极推动作用，但这些书籍是零散的、缺乏整体规划，而发达国家的出版社在 MEMS 书籍方面，进行了有效的组织和规划，例如：Springer 出版社

2005 年出版了 *MEMS/NEMS Handbook*：*Techniques and Applications*（共 5 卷）、2007 年开始出版 *MEMS Reference Shelf*（目前已经出版 9 本），Wiley – VCH 出版社 2004 年开始出版 *Advanced Micro and Nanosystems Series*（目前已经出版 7 本）。面对国外 MEMS 快速发展的形势和我国对 MEMS 书籍的迫切需求，及时系统地规划、遴选、组织并翻译出版国外 MEMS 书籍很有必要，东南大学出版社 2005 年开始出版的《微纳系统》系列译丛就是这方面的尝试。

MEMS 设计、制造、封装、可靠性以及测试等共性技术推动了 MEMS 的发展，市场与应用需求则牵引了 MEMS 技术的进步，而微米与纳米技术的结合给 MEMS 带来了许多新的机遇。

1 设计

MEMS 工作过程涉及机械能、电能、磁能、热能和化学能等及其之间的耦合，工作原理复杂，因此，理解其工作过程，提高或优化其性能，需要有效地设计工具。另一方面，MEMS 制造工艺的模型与模拟，可降低试制成本，优化工艺流程。总而言之，MEMS 设计技术与工具的发展能够优化产品性能，降低产品研发成本，缩短产品研发周期。

MEMS 设计通常包括：

（1）器件级设计

器件级设计是根据器件结构，建立器件工作的微分方程，利用有限元或边界元等数值方法，采用合适的边界条件，进行偏微分方程的求解，从而给出器件的性能，这是 MEMS 设计最早发展的技术，目前已经有商用软件 Coventor 和 Intellisense 等可以使用。另一方面，由于 MEMS 器件特征尺寸通常在微米量级，宏观的物理规律仍可应用，因此如 ANSYS、ABAQUS、CFDX 等传统的偏微分方程求解器也在 MEMS 器件设计中广泛使用。器件级模拟计算量大、设计周期长，但是精度高。

(2) 系统级设计

器件往往不能单独使用，必须与驱动、检测或控制电路一起工作。系统级设计的前提条件是建立能够与电路分析工具实现无缝联接的 MEMS 器件的宏模型。宏模型是根据器件结构，采用合适的近似或算法，将器件工作的偏微分方程降阶为常微分方程，进而对常微分方程求解，给出器件的终端特性。这种方法通常速度快，但精度低。系统级模型容易对器件进行优化设计，且可与电路一起进行分析和优化，为激励-响应-控制（反馈）的闭环系统设计提供有效手段。系统级模拟运行的平台包括 SABER、SPICE、SIMULINK 等。

(3) 工艺级设计

根据器件级或系统级设计所确定的几何结构，就可选择合适的工艺进行制造，工艺级设计包括工艺流程设计和工艺模拟。工艺模拟是通过建立每一步制造工艺的物理模型，采用合适的数值算法，结合掩模版图和工艺流程文件，模拟出 MEMS 器件的拓扑结构。目前 Coventor、Intellisense 等设计工具的工艺模拟模块只能完成部分单步工艺模拟，尚不能提供由不同工艺次序所完成的器件结构及其分析，因此不能分析工艺偏差、材料参数偏差对 MEMS 结构或器件性能的影响。

目前，限于 MEMS 设计工具的能力，MEMS 设计主要依靠使用者的知识与专业水平。由于不同层次的设计过程存在着相互脱节的问题，还没有形成有机集成的设计环境，不能够完整地实现自上而下 (Top-down) 的 MEMS 设计过程，设计效率比较低。MEMS 设计者的目的是希望制造出性能符合要求的器件或系统。而制造过程中几何尺寸偏差、材料参数偏差等会使其偏离设计者的允许范围，因此，"试差"的设计方法仍然占据主流。在高性能 MEMS 研制方面，设计者需要与制造工艺紧密结合。同时，由于目前的 MEMS 设计工具是从现有微电子设计工具或机械设计工具衍生而来，无论从市场角度看还是从工具性能看，都需要与大型设计工具集成。

（4）NEMS 设计

在 20 世纪 90 年代后期开始发展的纳机电系统（Nano Electro Mechanical System，NEMS）是纳米科学技术的重要分支之一，NEMS 是指以纳米材料或结构所产生的量子效应、界面效应、局域效应和纳米尺度效应为工作特征的器件和系统，可实现超高灵敏度或选择性的敏感、探测与执行。

描述 MEMS 工作的模型是基于连续介质的理论，在物理上的连续意味着在数学上可采用微积分，因此 MEMS 模型、模拟及其设计方法主要是以有限元为代表的数值方法。固体、液体和气体都被分解为分子（或原子）的聚集体，而原子又被分解为原子核和电子，表面/界面上的分子（或原子）不同于内部，纳米材料独特的性质和优异的性能由其尺寸、表面结构及其粒子间的相互作用决定。例如，长、宽、厚分别为 100 nm、10 nm、10 nm 的硅纳米线，约有 10% 的原子在物体表面或者靠近物体表面，表征材料力学特性的杨氏模量、热学特性的热导率等与表面性质相关并出现尺寸依赖现象，而电学特性、磁学特性、光电特性等出现量子限制。

连续介质的描述忽略了粒子的个性，而微观粒子具有波粒二象性，就计算模型而言，经典粒子的运动由玻耳兹曼方程描述，而粒子的波动性由薛定谔方程描述。以纳米线为例，长度方向尺度远大于原子间距，可认为是连续介质，而截面尺度在纳米尺度范围，原子特性显现。在描述力学特性时，我们原则上可以用分子动力学方法计算每个原子与其他原子的相互作用行为，进而了解其力学性质，但是，这种情况下的原子数太多以至于无法实现。因此，在处理这类问题时，需要原子模拟方法和连续介质的有限元方法相结合。在描述其电学特性时，在长度方向电子是自由的，因而能量是连续的，而在截面方向，由于尺度限制，能量是量子化的。对于纳米线结构，局部尺度是纳米，可以用原子模拟方法，而纳米线的两端与微米尺度结构相连，这种微米尺度区域需要连续介质理论描述。因此，在 NEMS 器件中，几何空间的多尺度导致使用不同的物理描述方法，如密度泛函、分子动力学、Monte Carlo 方法等。

2 制造

MEMS 所用材料主要有半导体硅、玻璃、聚合物、金属和陶瓷等。由于所用材料不同，习惯上，将 MEMS 制造分为 IC（集成电路）兼容的制造技术和非 IC 兼容的制造技术。IC 所完成的功能主要利用了硅单晶的电学特性，而硅单晶也有良好的机械特性，例如，硅单晶的屈服强度比不锈钢的高、努氏硬度比不锈钢的强、弹性模量与不锈钢的接近，同时，硅单晶几乎不存在疲劳失效。硅单晶良好的机械特性以及微电子已经建立起来的强大工业基础设施，使其成为 MEMS 的主流材料。

（1）IC 兼容的微制造技术

由于微电子制造技术基本上是一种平面制造工艺，为在芯片上制造可动部件，需要微机械加工技术。

硅微机械加工技术主要包括表面微机械加工技术、体微机械加工技术、硅片直接键合技术以及这些技术的相互融合。

1965 年美国 Westinghouse 电气公司的 H. C. Nathanson 等人提出硅表面微机械加工技术，在 20 世纪 80 年代中后期得到发展，20 世纪 90 年代出现的气相 HF 牺牲层释放技术，大大提高了表面微机械加工技术的生产成品率和效率，利用表面微机械加工技术制造的典型产品有 ADI 公司的加速度传感器、TI 公司的微镜阵列投影显示器等等。

硅体微机械加工技术包括湿法刻蚀和干法刻蚀，KOH 湿法各向异性刻蚀于 1967 年由美国 Bell Lab 的 H. A. Waggener 等人提出，在 20 世纪 80 年代中后期得到发展，20 世纪 90 年代由日本京都大学 O. Tabata 等人发明的 TMAH 湿法各向异性刻蚀与 IC 工艺线兼容，进入工业化应用；反应离子刻蚀（RIE）是 IC 工艺，1994 年，德国 Bosch 公司采用电感耦合等离子（ICP）方法发明了 DRIE（深反应离子刻蚀）技术，它是硅体微机械加工的基本技术之一，利用体微机械加工技术制造的典型产品有 Freescale 公司的压力传感器、ST-microelectronics 公司的加速度传感器、Akustica 公司的麦克风、SiTime 公司和 Discera 公司的振荡器、HP 公司的喷墨打印机微喷嘴阵列等等。

1986 年，美国 IBM 公司的 J. B. Lasky 等人和日本东芝公司的 M. Shimbo 等人分别独立开发出硅片直接键合技术，它是硅三维结构制造的主要技术之一，利用硅片直接键合技术制造的典型产品有 NovaSensor 公司的压力传感器等。

（2）单片集成化制造技术

MEMS 微传感器需要信号放大、信号处理和校准，MEMS 微执行器需要驱动和控制。因此，在应用中，MEMS 器件需要和微电子专用电路（ASIC）集成，这种集成可以是单片集成也可以是多片集成，至于采用哪种方式集成，取决于系统要求和成本。单芯片集成是将传感器及执行器与处理电路及控制电路同时集成在一块芯片上，多片集成实际上涉及了封装技术。

  CMOS MEMS 技术是一种单芯片集成技术，它利用集成电路的主流 CMOS 工艺制造 MEMS。MEMS 器件与电路单片集成的主要优点有：

  1) 可以实现高信噪比。一般而言，随着传感器的面积减小，其输出的信号也变小，对于输出信号变化在 nA（电流输出）、μV（电压输出）或 fF（电容输出）量级的传感器，敏感位置与外部仪器引线的寄生效应会严重影响测量，而单片集成可降低寄生效应和交叉影响。

  2) 可以制备大阵列的敏感单元。大阵列的单元信号连接到片外仪器时，互连线制备及可靠性是主要问题。对于较小阵列，引线键合等技术就可以满足要求，但对于较大阵列，互连问题会影响生产成本和器件成品率甚至不可能实现大的阵列。因此，采用片上多路转换器串行读出，不仅降低了信号调理电路的复杂性，而且大大减少了键合引线的数量，提高了可靠性和成品率。

  3) 可以实现智能化。除信号处理功能外，诸如校准、控制以及自测试等功能也可以在芯片上实现。单片集成方式已经促成了多种 MEMS 产品商业化，如加速度传感器、数字光处理器以及喷墨头。

  但是，使用 CMOS MEMS 技术，可用材料被限制到 CMOS 材料以及和 CMOS 工艺兼容的材料，其制备与封装工艺也有较多限制。

  硅基 MEMS 的发展基本上是借鉴了 IC 工业的成功之处，即集中化批量制造，提供高性能价格比产品。但 MEMS 又与 IC 有较大的差别，IC 有一个基本单元，即晶体管，利用这个基本单元的组合并通过合适的连接，就可以形成功能齐全的 IC 产品；在 MEMS 中，不存在通用的 MEMS 单元，而产品种类繁多，因此，MEMS 加工不可能采用 IC 产业集中化制造的模式，而适合于分类集中制造。

（3）非 IC 兼容的微制造技术

由于硅材料耐磨性差以及特殊环境的使用问题，非 IC 兼容加工技术的发展可满足 MEMS 不同材料和结构的需要以及特定应用（如生物化学环境和高温环境等）的需要，1985 年德国 W. Ehrfeld 小组开发出的 LIGA（光刻电铸成型）技术以及后来发展起来的 UV－LIGA（紫外光-光刻电铸成型）技术是非 IC 兼容的主要加工技术，此外还有激光三维加工技术、微细电火花加工技术、热压/注射成型加工技术、微纳米压印技术等。

（4）纳米制造

纳米制造为 NEMS 发展提供支撑。目前，纳米制造的方式可分为由上而下和由下而上两大类。由上而下的技术路线是传统微制造工艺向纳米尺度延伸的必然产物；由下而上的方法则另辟蹊径，利用原子、分子组装构筑出复杂的结构。

光学光刻技术不但是微制造的主要技术，也是纳制造的主要手段。虽然光学光刻技术一度成为微纳结构加工的主要限制因素，但是随着短光波长技术的应用以及移相掩模、光学临近校正、浸没式光刻、多重曝光等新技术的发展，光学光刻在大批量生产中已经达到 22 nm 的工艺水平。22 nm 工艺仍采用深紫外浸没式光刻技术，该技术结合多重曝光有望延伸到 16 nm 甚至 11 nm 技术节点，但是能否支撑足够高成本代价仍有待观察。此外，极紫外光刻、离子束光刻、电子束光刻、纳米压印等下一代光刻或光刻替代技术有望取代目前的光学光刻成为 10 nm 以下大批量加工的关键技术。其中，极紫外光刻技术甚至有可能在 16 nm 工艺中率先被采用。

电子束、离子束不但可以用来光刻,也可以直接将固体表面的原子溅射剥离,被更广泛地作为一种直写式加工工具;此外,与化学气体配合还可以在衬底材料表面直接沉积出相应结构,成为一种用途广泛的纳制造工具。无论是溅射剥离还是辅助沉积,加工精度与束斑尺寸直接相关。目前的电子束系统、离子束系统分别能够轻易获取 5 nm 的电子束、离子束,透射电子显微镜系统中甚至可以得到 0.5 nm 的电子束斑。尽管电子束加工已经能够普遍实现 2~6 nm 线宽的雕刻加工,10~20 nm 线宽的结构沉积,最小 2~3 nm 的量子点沉积;聚焦离子束加工能够实现 3~5 nm 的雕刻,但是该类加工方法受限于加工效率,仅适用于单个器件的加工。短期内还难以应用于大批量生产,电子束、离子束加工技术为更小尺寸结构加工提供了一种可能的发展方向。

由下而上的技术思想经历了近 20 年的发展,形成了一系列以分子自组装为基础的加工技术。当然,目前自组装技术作为纳加工手段还相当原始,大多数情况下还是与纳米光刻等传统技术相结合,以此进入主流纳加工技术领域。纳米球光刻就是由上而下和由下而上技术相结合的典型技术,它利用自组装技术形成的纳米球阵列作为掩模进而加工高密度点阵图形。目前,通过纳米球光刻制作的阵列点最小可以达到 10 nm 左右。另外,蘸笔纳米光刻是两种加工思想相结合发展起来的又一种纳加工技术。该技术利用蘸有特殊液体的扫描探针直接书写出光刻图形,经历了 10 余年的发展,已经能够制作出最小线宽在 10~15 nm 的图形。尽管还不完善,但随着加工尺度的进一步缩小,由下而上的加工技术越来越显示出它的优越性,重新受到纳米加工界的关注。

此外,只有将纳米结构与微米结构互连后,才能与宏观世界联系起来,因此,同时实现纳米尺度制造和微米尺度制造的跨尺度制造方法也是值得关注的方向。

3 封装

MEMS 封装的目的是为其提供物理支撑和散热，保护其不受环境的干扰与破坏，同时实现与外界信号、能源及接地的电气互连。MEMS 含有可动结构或与外界环境直接接触，因此 MEMS 封装比 IC 封装更复杂。一般来说，IC 制造中采用的低成本封装技术只能适用于一部分 MEMS，而大多数 MEMS 器件中含有可活动部件，往往需要采用特殊的技术和材料才能实现其电信号与非电信号的相互作用，而且器件种类繁多，大大增加了封装的难度和成本。MEMS 封装包括单芯片封装、多芯片封装、圆片级封装和系统封装（SiP）等封装技术，可实现非气密、气密和真空封装，封装过程需要考虑电性能、电磁性能、热性能（等物理场）、可靠性等问题。MEMS/NEMS 封装设计与模型、封装材料选择、封装工艺集成以及封装成本都是在开发新型 MEMS/NEMS 封装技术时需要考虑的问题。

随着 MEMS/NEMS 技术在消费类电子、医疗以及无线传感网等中的广泛应用，为了实现低功耗和小体积，要求将完整的电子系统或子系统高密度地集成在只有封装尺寸的体积内，即 SiP 技术。封装内包含各种有源器件，如数字集成电路、射频集成电路、光电器件、传感器、执行器等，还包含各种无源器件，如电阻、电容、电感、无源滤波器、耦合器、天线等。未来电子产品将所有的功能集成在一个很小的体积内，因此，非常窄节距的倒装芯片凸点、穿透硅片的互连技术、薄膜互连技术、三维芯片堆叠技术、封装堆叠技术、高性能的高密度有机基板技术以及芯片、封装和基板协同设计与测试技术等成为 SiP 的关键技术。

4 可靠性

MEMS 可靠性是指 MEMS 器件在实际环境中无故障工作的能力。MEMS 可靠性一般分为制造过程中的可靠性（包括制造过程、划片、超声键合引线、封装等）、工作过程中的可靠性以及环境影响可靠性。为了保证 MEMS 的可靠性，还需要对材料、工艺、器件、系统等的可靠性进行测试、表征和预测。MEMS 在工作过程中的可靠性可以分为四类：

1) 没有可动的部件（例如压力传感器、微喷嘴等）；

2) 有可动但没有摩擦或表面相互作用的部件（例如谐振器、陀螺等）；

3) 有可动和表面相互作用的部件（例如继电器、泵等）；

4) 有可动并有摩擦和表面相互作用的部件（例如光开关、光栅等）。

在 MEMS 器件的设计过程中，为了避免失效从而提高器件的可靠性，往往根据器件的某些失效机理来改进设计方案。常见的可靠性设计包括：为了避免粘附引入凸点和防粘附层；为了避免断裂设计平滑过渡的变截面；为了避免介电层电荷注入而取消介电层或改变介电层的位置；为了避免可动结构的粘附和断裂而引入止挡结构等。这些设计在很大程度上改善了相应的失效，从而提高了器件的可靠性。

MEMS 器件在制造过程中也会引入各种失效因素，进而影响器件的成品率以及使用中的隐患。这些因素主要包括：制造工艺中的各种残留污染、材料沉积或刻蚀中形成的各种缺陷、不同材料构成的 MEMS 结构中的残余应力、热失配引入的热应力、圆片切割和处理造成的碎屑污染和划痕、封装、微互连中的热机械效应以及气密性等引起的环境条件变化和污染。

机械装置的运动包括弹性运动和刚体运动（或整体运动），弹性装置借助柔性结构（如弹簧和扭转杆等）运动；而刚体装置借助铰链和轴承运动。刚体装置允许部件积累位移，而弹性装置将部件限制在固定点或固定轴附近运动。由于 MEMS 器件表面接触、滑动和摩擦引起的诸多问题还没有解决，因此目前 MEMS 产品均使用了柔性连接方式。

5 应用技术的发展

MEMS 具有微型化的特征以及可高精度批量制造，与其他科学技术的结合，会产生新的应用领域，例如：

1970 年，美国 Kulite 公司研制出硅加速度传感器原型，1991 年，美国 Draper 实验室 P. Greiff 等人发明硅微机械陀螺。陀螺传感器与加速度传感器构成了惯性传感器及其系统，目前在电子类消费品、汽车、航空航天以及军事等领域有广泛应用。

1987 年，美国 UC Berkeley 的 R. S. Muller 小组和 Bell Lab 的 W. N. S. Trimmer 小组利用多晶硅表面微机械加工技术，研制出自由移动的微机械结构（微马达、微齿轮）；1991 年，美国 UC Berkeley 的 K. J. Pister 小组研制出多晶硅铰链结构，自此，微机械操作、微组装、微机器人成为新的研究分支。

1989 年，美国 UC Berkeley 的 R. T. Howe 小组研制出横向驱动梳状谐振器，它是目前微机械振荡器、微机械滤波器、加速度传感器、角速度传感器（陀螺）、电容式传感器等的基本结构。

1980 年，美国 IBM 公司的 K. E. Petersen 发明硅扭转扫描微镜，它是光学扫描仪、数字微镜器件、光学开关等的基本结构；1992 年，美国 Stanford 大学的 O. Solgaard 等人发明 MEMS 光栅光调制器，实现了微机械对光的操作，自此，Optical MEMS（光微机电系统）分支出现。光 MEMS 在光通信技术、显示技术、光谱分析技术等领域有广泛应用。

1990年，美国Hughes实验室的L. E. Larson等人研制出微机械微波开关。自此，RF MEMS（射频/微波微机电系统）分支出现，用微机械加工技术制造芯片上无源元件（电容、电感、开关等）、组件（滤波器、移相器）以及单芯片微波系统研究进入热潮。RF MEMS在雷达、通信等领域有广阔的应用前景。

1990年，瑞士Ciba-Geigy制药公司的A. Manz等人研制出微全分析系统（μTAS）或称为芯片上实验室（Laboratory on a chip），这是目前微流控分析芯片的原型。自此开始了微型泵、微型阀门、微型混合器、微型通道等对微尺度下的流体操作器件研究。微流控在生物领域的应用是近年来MEMS最活跃的方向之一，具有降低分析成本、缩短反应时间、提高精度、多功能集成等优点，在分析化学、医疗、药物筛选等领域有广阔应用前景。

1995年，美国MIT的J. H. Lang, A. H. Epstein和M. A. Schmidt等人开始了微型气动涡轮发动机研究；2000年，美国Minnesota大学Kelley小组研制出基于MEMS技术的微型直接甲醇燃料电池原型。另外，诸如压电振动能量收集、热电能量收集、电磁能量收集等技术的发展，促进了Power-MEMS（动力微机电系统）分支出现。动力微机电系统在无线传感网、医疗、土木工程结构健康监测等领域有广阔的应用前景。

6 微米/纳米技术的结合

试验已经证实，硅基NEMS器件能够提供高达10^9 Hz频率、10^5的品质因数、10^{-24} N的力感应灵敏度、低于10^{-24} cal的热容、小到10^{-15} g的质量以及10^{-17} W的功耗。由于纳米尺度材料或结构的量子效应、局域效应以及表面/界面效应所呈现的奇特性质，可以大幅度提高 MEMS/NEMS 的性能，也可能使以前不可能实现的器件或系统成为可能。例如，2004 年英国 University of Manchester 的 K. S. Novoselov 和 A. K. Geim 成功制备出可在外界环境中稳定存在的单层石墨烯（Graphene），其特异的性质如量子霍尔效应、超高迁移率、超高热导率和超高机械强度已经引起人们的广泛重视，是目前材料和凝聚态物理领域的研究热点之一，而当气体分子吸附在石墨烯表面时，吸附的分子会改变石墨烯中的载流子浓度，引起电阻突变，可实现单分子检测。但实际上，只有将纳米结构与微米结构互连后，才能与宏观世界联系起来，通过微米技术进行集成，可将基于纳效应的功能和特性转变成新的器件和系统，因此，MEMS 技术可作为纳米科学走向纳米技术的桥梁。例如，20 世纪 80 年代出现的隧道扫描显微镜、原子力显微镜以及近场显微镜等，它的探针最前面的部分是"纳"，后面就是"微"和"电"，三者集成在一起，协调工作。因此，微米纳米技术相互融合已成为趋势和发展主流。

7 市场

据有关咨询机构（例如 Yole，iSuppli，SPC，MANCEF，NEXUS，ITRS）的统计与预测分析，MEMS 产业在 2000 年全球销售总额约为 40 亿美元，2005 年约为 68 亿美元，2010 约为 100 亿美元。目前的主要产品包括微型压力传感器、惯性测量器件、微流量系统、读写头、光学系统、打印机喷嘴等，其中汽车工业和信息产业的产品居主导地位，占总销售额 80% 左右。

值得关注的是，2012 年 5 月由美国 MEPTEC（微电子封装与测试工程协会）在 San Jose 举办的第 10 届国家 MEMS 技术讨论会中，研讨的主题是"Sensors: A Foundation for Accelerated MEMS Market

Growth to $1 Trillion"。这次研讨会有来自学术界、工业界、咨询公司以及设备供应商的代表，他们认为以物联网为主要代表的市场快速增长正对传感器提出巨大需求，估计在 2020 年左右其产业链达到 1 万亿美元，而且计算、通信和感知技术的融合有可能成为第三次工业革命。而目前 MEMS 制造、封装和测试缺乏工业标准，产品研发周期较长，是通向 1 万亿美元产业的瓶颈。

一方面，MEMS 前期开发的技术已经开始进入产业化，另一方面，MEMS 与纳米技术等其他新技术的交叉研究方兴未艾。面临这种发展趋势，无论是高等学校教学或科学研究，还是工业部门产品开发，都需要及时系统地学习并总结前人的知识和经验。

东南大学黄庆安教授长期从事 MEMS 教学和科研工作，经常关注国际微米/纳米技术的最新进展及有关 MEMS 技术信息，他带领的团队在 MEMS CAD、RF MEMS、CMOS MEMS、MEMS 可靠性、NEMS 以及微传感器等方面进行了长期研究，此次东南大学 MEMS 教育部重点实验室与东南大学出版社合作，组织翻译出版《微纳系统》系列译丛，将会促进我国 MEMS 教学、科研以及产业化的发展。《微纳系统》系列译丛涉及面广，从选题、翻译、校对到出版等工作量巨大，为此，向为翻译该书付出辛勤劳动的师生们表示敬意。

希望《微纳系统》系列译丛的出版对有志从事微米/纳米技术及 MEMS 研发的广大师生和科研人员有所帮助。

微纳电极阵等离子体微系统

（2024 年 5 月）

摘　要　发现微纳尺度结构效应能调控电荷分布，在空间中激励形成极化带电场，极化带电场中原子、分子间相互作用的特异性产生了极化带效应。设计产生极化带效应的微纳电极阵结构，通过非硅微/纳加工技术集成形成微纳电极阵等离子体微系统（NPMEMS），NPMEMS 可集中或分布式地调控物质状态、产生极化带等离子体，利用其特殊的理化性能可大幅提高应用系统效率，或解决多个领域中应用技术的机理性难题。

关键词　微机电系统；微纳电极阵；等离子体

1　物理原理和技术特征

等离子体的特征包括两个方面[1]：一是从物质角度讲，含有与其他物态不同的特征组分：自由电子、离子和受激中性粒子。二是从能量角度讲，它与固态、液态和气态并称为物质的一种存在状态，在宏观的统计特性上，其组成粒子部分地或整体上具有比其他物态更高的能量；在微观的激发弛豫过程上，其特征组分在个体上寿命短、不稳定，一直处于快速产生、快速消失的过程中，需要外部能量的持续供

本文发表于《传感技术学报》2024 年第 37 卷第 5 期。合作作者：丁衡高，侯中宇。

给才能作为一类组分存在或发挥效用，而能量供给的物理机制则决定于电磁力的微观作用过程和粒子间碰撞的介观弛豫过程。

人工等离子体一般需要从其他物质状态演化而来，其核心过程在于电离源不断提供能量使得中性的被电离物质发生电荷分离。科学界和工业界在近百年中主要发展了三类等离子体产生方法[2-4]：1) 主要靠加速自由电子碰撞电离的产生方法，例如辉光放电、射频微波放电、高能电子束放电等；2) 主要靠加热气体分子的产生方法，例如碱金属燃烧和电焦耳热等；3) 主要通过电离辐射产生光电离的方法，例如紫外光电离、高能射线电离等。由于电离射线的高效可控产生难度依然很高，人工等离子体源目前仍以前两类方法为主，在物理过程上均依赖于热效应：其微观本质是自由电子和分子平动动能向分子内部轨道电子能量的转化，而宏观上则表现为局部生热、传热和热功转化，系统整体显著不均匀熵增。系统的混乱程度增加与其状态的可设计、可控和产生效率提升之间存在根本性矛盾，这一规律导致为提高等离子体产生范围、提高电子密度，普遍存在功耗高、温度高、应用条件苛刻等局限性，成为应用技术上的机理性障碍。

微纳电极阵等离子体产生方法的基本思想源于上海交通大学的研究工作，具有原创性[5-6]，但其较为完整的动理学和热力学理论形式和微纳技术体系，则是在2008年后针对高效可控新型等离子体源的特殊应用需求，逐步发展起来的。其物理原理在于：利用微纳集成工艺实现由特殊微纳结构组成的电极阵，激励产生与传统电场有根本差别的"极化带电场"（传统电场在原子尺度上三个维度都具有均匀性，极化带电场在一个维度上原子尺度级别具有极高的电场梯度，在另外两个维度上具有微观宏观均匀性。形成极化带电场不必是产生强电场，而是通过设计电磁场的空间分布控制电磁力对原子、分子的作用模式，这种模式不必是一种高强度作用力驱动产生，而是由于分布特性相关的电磁作用力驱动产生），这种电场能够使普通放电介质激发形成"超极化介质"，产生"极化带效应"，在低功耗和低平均场强条件下，产生和维持比传统电场更高的等离子体密度，并具有更大的扩散范围，使小体积、低功耗和长时稳定工作的等离子体发生器成为可能。

微纳电极阵极化带效应是一种微纳结构效应，微纳电极阵结构的电通量收敛一方面形成了极化带电场，另一方面强化了电极系统表面电场。在等离子体演化过程中，前者导致了气体中的极化带过程，我们称之为"极化带电场效应"；后者导致了电极上的极化带过程，我们称之为"极化带电极效应"。我们将两者统称为微纳电极阵极化带效应，理论和实验研究表明，它们共同决定了极化带等离子体的状态特性和演化特性。

在极化带电场作用下，强极化气体分子受激产生亚稳态的过程主要有两种：其一是强极化分子在电场中可较长时间保持取向性排列，直接相互作用发生场致激发形成亚稳态；其二是强极化分子被极化带电场捕获，并与微纳电极发生直接相互作用形成亚稳态粒子。亚稳态粒子自发退激概率极低，因此寿命长而扩散迁移距离远[7]，在其越渡路径上，通过与自由电子等其他粒子群的非弹性碰撞，实现高通量密度的低温软电离。因而，"极化带电场效应"发挥着向轨道电子内能高效转移与传递极化带电场能量的作用：第一，由于强极化分子受激跃迁量子效率高[8]，因而在微观上导致极化带电场向分子内部结构自由度转移和储存能量的高效性；第二，由于亚稳态粒子电离碰撞截面（概率）显著高于基态，因而在介观上导致优先获得能量的组分向低能组分能量弛豫过程的高效性；第三，由于亚稳态粒子所获得的极化带电场能量"储藏"在分子内部，活跃的电离过程对自由电子等粒子群的温度要求显著降低，以无规碰撞焦耳热形式耗散掉的一部分电荷迁移运动能量显著下降，构成物质状态演化中的负熵化环节，因而在宏观上导致等离子体产生和维持过程的高效性。理论计算表明，不考虑具体实现上的任何约束而对极化带电场分布特性进行优化，相对于平板电极，微纳电极阵极化带等离子体达到 $10^{12}/cm^3$ 相同量级的自由电子密度，功耗可降低 6 个量级，平均电子温度降低至 0.1 倍～0.5 倍，并可通过分区、分级调节电场分布，将局部区域平均电子温度提升至 1～10 eV。在当前工艺水平和具体应用场景下，实测功耗仍可降低 2～4 个量级。在功耗低于 1 W、气体温升 1～10 K 的条件下，自由电子密度可以达到 10^{14}～$10^{15}/cm^3$ 量级。由此可见，极化带电场效应决定了物质状态及其演化的基本模式，是极化带等离子体过程中起到决定性作用的物理效应。而"极化带电极效应"主要与极化带等离子体在空间和能谱上的具体分布模式有关，我们目前主要通过微纳电极材料的选择，优化等离子体热力学稳定性，强化其化学活性物质组分及浓度分布的可控性。

极化带电场实质上激发形成了一种自由空间电磁能向分子内部电磁能转化的特殊模式，我们提出构建"极化带激子"这一准粒子模型描述极化带效应影响等离子化过程的基本规律，而极化带等离子体是一种能够以极化带激子传递能量的物质状态。极化带激子因极化带电场与分子间相互作用而被激发生成，并通过与分子、自由电子等各类粒子间的碰撞弛豫过程传递极化带电场能量。这一"准粒子"使得极化带等离子体演化过程存在自由电子以外的第二种能量转化模式，形成了物理动理学意义上的能量转化、转换特殊通道（这一通道会使得包括自由电子在内的各类粒子间通过碰撞向分子内部转移能量的过程，更接近服从经典连续的碰撞动力学统计特性，换言之，平动动能向结构势能转化概率会大幅提升，亦即在等离子化必然的"加热"熵增过程中形成了负熵化的"降温"环节）。极化带激子调控和利用物质内能的模式对高效控制物质状态具有普遍的物理意义，相对于通过热效应改变物态，它无须首先加热物质，再通过热功或传热过程激发和电离分子以实现物态转变，而是直接通过结构化电磁力调控分子结构内能，相对于通过脉冲激光等强场效应改变物态[9-10]，它能够实现更宽频谱范围电磁场在更灵活的时域、空间范围和温度条件下对分子内部结构做功。在物质内部介观尺度相互作用所决定的非平衡热力学性质上，与热效应正熵化正比于有用功相反，分子间相互作用内能在极化带电场中的激发、累积与分子排列的有序性水平正相关，使得负熵化产生有用功作用于物态演化过程成为内禀属性。产生和控制极化带激子是一种开发和利用分子结构内能的普遍方法，它不但可用于等离子体物态的产生，也能够用于形成例如仅含有大量亚稳态和激发态粒子而非自由电子和离子的其他特殊物态。

因此，与传统意义上基于自由电子或分子热效应的等离子体产生方法相比，NPMEMS技术本质上就是要构建足以激发、控制和利用高通量极化带激子的微纳系统平台，在粒子间碰撞过程的介观尺度上发挥控制能量转化模式的效用，进而在等离子体或其他物态转化生成过程中构建负熵化环节，使相变过程收敛到可高效维持的低熵热力学状态。负熵化环节构成了极化带等离子体产生过程的核心特异性，在等离子体状态形成以后，其对等离子体组成粒子在各自由度间的热力学弛豫模式也会产生根本性影响，进而会影响等离子体在电磁（等离子体的电磁特性源于自由电子与离子所组成的偶极子在外场电磁力作用下的受激辐射场与入射场的耦合，其对电磁波传输各类效应的有无和强弱，在物理上可归结为其复介电常数实部与虚部时空分布的特征，其实部主要决定于电子密度,而虚部主要决定于电子与气体分子相互作用中的碰撞弛豫过程,因此决定于电子温度和气体分子的能量状态)、化学(等离子体的化学性质决定于等离子体中气体分子在各个自由度的能量状态,从物理动理学和热力学的角度讲,进而决定于转化外部能量最活跃、与各类粒子碰撞频率最高的成分的能量弛豫过程)和气体动力学意义上诸多重要的理化性能(根据粒子物理标准模型,电磁力相对强度比引力高 36 个量级,仅比强力低 2 个量级[11]。飞行器流场的能量源于流体自身惯性,而黏性等流体性质则源于气体分子间的电磁相互作用,通过在小尺度上巧妙设计电磁场分布,以影响电磁力对微观系统的作用模式,可对大尺度宏观过程产生根本性影响。如果能结合飞行器的结构设计巧加利用,还能用以改造高超流场的边界层结构,实现高效减阻)。总之,NPMEMS 器件通过微纳结构效应设计与控制电磁场的空间分布,在链接微观与宏观的介观尺度水平上,调控形成负熵化环节以提高产生和维持等离子体状态的效率和可控性,为"四两拨千斤地"利用电磁相互作用[12]的特异性,创造了新途径。

2 发展历程

NPMEMS 技术经过近二十年时间,从观察微纳尺度内小于电离能的低电势条件下电荷分离和电子聚集现象开始,经理论与实验相结合的探索,在理论上对等离子体高效产生的微纳尺度结构效应("极化带效应")物理机制进行了数理描述。在对关键效应的理论条件实现了实验验证的基础上,创新微纳制造技术实现了这一机理的器件化,并通过联动激励技术布控成阵产生状态可控的等离子体,实现了在特定应用领域的功能化。

在基础研究中,对极化带电场的产生方式、在等离子体中的基本场致效应和物理过程，系统开展了理论描述和实验表征。基于微纳加工技术国家重点实验室在厚胶工艺上的积累，边发展基础工艺制备样品开展实验研究，边发展数理模型开展理论研究。发现了等离子体径向均匀化发展、超长时延介质阻挡正反馈累积、亚电离能低电压宏观尺度电离和低功耗跨态相变、极化带效应所导致的场致特异性现象等。提出了极化带等离子体产生过程的基本理论模型，对极化带效应基本物理机制实现了数理描述和实验验证。

在应用基础研究中，为模拟仿真极化带等离子体演化过程，以实现 NPMEMS 器件微纳结构和布控激励模式的优化设计，研究发展了极化带准粒子动理学模型的计算方法及其基础软件；提出"惰性背景真空模型"，开发了极化带等离子体演化过程基础软件。针对飞行器流场和电磁特性调控应用方向，发展了极化带等离子体电磁散射特性计算软件，并正在开发高速流场中极化带等离子体演化过程和热力学特性的计算软件。

在应用研究中，针对发展空天应用等离子体源芯片化集成制造工艺难题，2009 年开展了专用自动化工艺设备的研制工作，2013 年起将机器学习技术、自动化技术和工艺过程物理场计算技术相结合，研制了智能化工艺优化装置，突破了 NPMEMS 芯片集成化制造核心工艺，形成了较完整的核心功能结构微纳制备工艺体系。针对在表面产生和控制等离子体包覆层的技术需求，发展了片内和片间极化带效应的联动激励方法，初步掌握了激励系统的小型化技术。面向复杂极端空天条件下的应用研究工作，同时带动了在其他应用方向上的技术发展。

3　应用方向

NPMEMS 的应用主要分为三种类型：一是器件型应用，二是系统型应用，三是复合型应用。这种划分方法是从系统的角度和过程的角度区分 NPMEMS 器件应用方式，当 NPMEMS 器件在应用系统中所发挥的功能与其他环节或部件构成物理过程上的耦合关系，就是一种复合型应用；如果其应用对其他环节或部件构成约束，或者其他环节或部件对其应用构成约束，就是一种系统型应用；如果既没有物理过程上的耦合关系，也没有分系统间的约束关系，就是一种器件型应用。在电磁特性调制、减阻、生物安全、能源安全和粮食（食品）安全领域，通过研制原理样机、样件，其关键技术效果通过了相关科研院所、上海计量院等机构第三方测试验证，受到相关应用单位的重视。

NPMEMS 应用研究确立了四个近期重点突破方向：减阻降热、电磁特性调控、核生化侦防消和空间电荷探测。在减阻降热和电磁特性调控主攻方向上，多年来与总体单位协作开展应用基础研究工作，针对应用的紧迫性和现实性，明确需求牵引和技术特征具体要求，主要关注各类移动平台流场的力、热、电磁特性高效调控。

在核生化侦防消方向上，一是为实现空气和物表的生化战剂绿色、无耗材等离子体洗消净化，亟需解决高场强、气体温度高和能量小范围集中倾向所导致的高热和不稳定问题，研究基于极化带效应调控激励分子内部能量的 NPMEMS 等离子体洗消技术，改变电离频率对自由电子高能量状态的依赖关系，形成绿色高效等离子体洗消防护系统。二是为实现对成分、浓度复杂突变的化学战剂在线侦、检、探测，提升侦防消一体化水平，亟需解决广谱分析式气体检测器微小型化难题，研究利用极化带效应选择性调控分子能量状态的 NPMEMS 侦检技术，可形成微纳气体探测器、色谱分离器和高效软电离源。

在空间电荷探测方向上，基于空间电荷电动力学过程极化带效应远距离感知机理，通过 NPMEMS 感应电荷分布模式的时域信号读出电子学映射关系,构建辐射电磁场空间分布模式与感应电荷分布模式的空间映射关系,最终还原空间电荷分布模式与辐射电磁场分布模式的映射关系,从而实现对暗弱空间目标的探测识别。其系统构型低剖面、轻载荷、高集成,可形成散射抑制与辐射探测共生的高效智能系统。

为实现 NPMEMS 在上述领域的应用技术研究,整合关键工艺、形成芯片集成制造以及面向新应用的器件级和系统级研制能力,是我们当前攻关工作的重点。

4 主要体会

NPMEMS 应用研究工作的核心目标是解决一些传统技术路线难以解决的物理机理性难题,提高 NPMEMS 各型器件和系统在各应用方向上的技术成熟度水平,为解决领域技术瓶颈问题提供新方法、新器件和新系统。做好这一工作我们有三点主要体会:第一是要高度重视基础研究、应用基础研究和各阶段充分的实验验证工作。这些工作做好了,应用研究、技术开发就有了扎实的基础,良好的基础工作也给我们攻坚克难以巨大力量和坚定信心。对于那些从科学上讲 NPMEMS 用上去能解决瓶颈问题的重要需求方向,技术上相对难实现、有风险也勇于创新突破。第二是要有专用的微纳加工工艺平台作为器件研制的保障。NPMEMS 器件是极化带效应的载体,是实现各类应用的物质基础,由于其功能材料、微纳电极阵结构固有的复杂性,要由多种工艺通过多个自主非标装备的复合来实现。建立专用线平台,有利于从全流程各个工艺环节把控质量,提高合格率,有一定的批量才能保证应用研究的需求。工艺上从核心结构的单点突破过渡到工艺整合的集成性突破,固化为核心工艺段的工艺线,是 NPMEMS 从基础研究和应用基础研究为主,过渡到以应用研究为主所要跨越的主要门槛。要下大力以"磨刀不误砍柴工"的精神提升工艺水平,研制和产出符合实际使用要求的高质量器件。第三是要在科研和技术研发的实践中成长一支"攻坚不畏难"的队伍。在有限甚至匮乏的客观条件下,仍能集中每个人的智慧有组织地攻关,高效迭代突破科研和技术研发中间的科学技术难题。科技创新和创新成果的应用不可能一帆风顺,必然是机遇与挑战并存。在这样的条件下坚持和成长起来的人们,通常对自己承担的业务有高度责任感,能够在困难的环境中依然保持紧迫感和使命感,始终科学客观地理论联系实际努力学习,不断提高自己的工作能力,是实现 NPMEMS 工程应用的宝贵力量。

5 几点启示

NPMEMS 从理论假想起步经过近二十年科学和应用技术研究,在实践中深化了我们对一些规律性的认识,重点有以下四点启示:

第一,发现和利用好微纳尺度内结构效应的特异性,构建负熵化环节,是研制高效等离子体发生器成功的关键,也是发展高效能微纳系统的重要途径。从热力学的观点看,微纳电极阵激励产生的结构效应,提升原子、分子内部结构获取外部能量受激跃迁的效率,降低了产生和维持等离子体状态所需的能流强度,等效于一种内生的有效能量,显现为 NPMEMS 运行中的负熵化环节。推而广之,任何一个微纳系统的目标性能都必然伴随着一个目标热力学状态,实现高性能的本质特征,是高效利用外部能量产生和维持系统运行底层所需的热力学状态。因此,构建结构效应与负熵化过程之间的普遍理论关系,通过微纳结构设计产生负熵化环节,自觉地思考微纳系统性能提升的热力学本质特征,建立效率与性能的内在联系,对创新发展微纳系统技术具有普遍意义。熵是广延量,一个系统的熵是由各分系统的熵构成的。根据目标需求整合各类叠加关系、耦合关系，对创新发展复合型高效微纳系统具有重要现实意义。

第二，极化带效应是一种微-纳尺度功能结构相结合才可能激发形成的物理效应，其电场分布通过微纳人工结构跨尺度叠加产生，单纯在纳米尺度工作的物理效应不能实现，单纯在微米尺度有效的物理效应也不能实现。之所以如此，是由于等离子体这种物质状态的性质决定于组成粒子的微观量子化行为，并同时决定于大量此类微观系统在物理动理学特征尺度上的相互作用及其宏观统计行为，极化带电场对气体分子的作用模式必须严格服从上述物理过程的规律，才可能构建负熵化的能量转移转化机制。为此，需要从系统运行的具体规律出发，形成严格的理论设计方法和微纳集成工艺。

第三，微纳结构效应的器件化技术，面临微纳结构显微形态、材料及工艺流程极端复杂性的挑战，并集中体现在微纳功能结构加工工艺上面，主要难点有两个：一是跨尺度特性强导致工艺质量评价难；二是工艺系统敏感性高导致过程控制难。为此，需要构建能够体现工艺过程和结构特征完整性的工艺系统辨识方法体系，找到工艺设备与工艺物理场多模态参数的内在关联，融合计算物理技术、人工智能技术和自动化技术，形成一条"机理感知型"微纳工艺研发新途径。这一物理学方法和数据驱动方法有机结合的工艺研发路线主要特征在于：一是可自主归纳在工艺过程中物理化学特征量空间分布和时变特性与工艺效果之间的规律性联系；二是能够基于小样本多通道自动化工艺实验技术，提升辨识工艺参数体系、优化工艺流程的效度和信度水平；三是能够应用机器学习技术跨尺度地评估工艺效果，同时输出工艺过程物理场、化学场的优化结果。这一创新微纳工艺技术途径，大幅提升了 NPMEMS 的研制、生产水平，也为发展新型微纳器件打下了工艺基础。

第四，极化带等离子体并非将任意结构特征的纳米材料作为电极使用就可以产生。在我们的实验中发现，简单地引入纳米结构作为电极有时不但不会增加效率，甚至可能显著强化等离子体的热效应和热不稳定性[13]。因此，必须发现和利用微观和介观尺度能量调控过程产生特殊宏观热力学效应和状态的物理条件，微纳结合地通过严格的理论设计和加工工艺制造趋近，才能形成由极化带物理机制激发和维持的高效等离子化过程。极化带电场对束缚态电子在微观尺度量子化的特异性作用，激发了粒子间在介观尺度能量弛豫过程中的动理学特异性，这种介观尺度的特异性在特定条件下与外部电磁力场耦合，能够产生宏观影响并显现为热力学状态或其演化过程的特异性。实践表明，不能绕开内在规律而通过外力"强迫"一个系统在生成或维持某种目标热力学状态的过程中产生负熵化环节，而必须找到和利用这个系统在微观和介观尺度上的特殊内禀属性，形成物质内能累积和宏观有序化的内在调谐机制，使得负熵化成为这个系统热力学状态转变与维持中的自发环节，而理论与实验相结合地利用微纳结构效应是找到这种特殊条件的行之有效的途径。微纳技术应考虑如何既能调控量子化特征显著的微观系统，又能把这种调控效应过渡到宏观。仅仅想到或者仅仅依靠有限个微观体系自身的改造，往往是不够的。在很多情况下，还需要能够同时改造足够数量微观体系之间的相互作用机制，才能真正形成热力学意义上的自发稳态以实现高效可控和技术应用，这就是我们面临的任务与挑战。

参 考 文 献

[1] Fridman A, Kennedy L A. Plasma Physics and Engineering [M]. 2nd Edition. Boca

Raton, FL, USA: CRC Press, 2011.

[2] Fridman A. Plasma Chemistry [M]. New York, NY, USA: Cambridge University Press; 2008.

[3] Raizer Y P. Gas Discharge Physics [M]. Heidelberg, Germany: Springer – Verlag; 1991.

[4] Chen F F. Introduction to Plasma Physics and Controlled Fusion [M]. Cham, Switzerland: Springer International Publishing, 2016.

[5] Hou Z Y, Cai B C, Liu H. Mechanism of Gas Breakdown Near Paschen's Minimum in Electrodes with One – Dimen – sional Nanostructures [J]. Applied Physics Letters, 2009, 94 (16) : 163506.

[6] Hou Z Y, Zhou W M, Wang Y Y, et al. Direct Current Dielectric Barrier Discharges under Voltages below the Ionization Potential of Neutrals in Electrode Systems with One – Dimensional Nanostructures [J]. Applied Physics Letters, 2011: 98 (6): 063104.

[7] Chupp T E, Fierlinger P, Ramsey – Musolf M J, et al. Electric Dipole Moments of Atoms, Molecules, Nuclei, and Particles [J]. Reviews Moderm Physics, 2019, 91 (1): 015001.

[8] Martin R M. Electronic Structure, Basic Theory and Practical Methods [M]. Cambridge, UK: Cambridge University Press, 2004.

[9] Harilal S S, Phillips M C, Froula D H, et al. Optical Diagnostics of Laser – Produced Plasmas [J]. Reviews Modern Physics, 2022, 94 (3): 035002.

[10] Gonoskov A, Blackburn T G, Marklund M, et al. Charged Particle Motion and Radiation in Strong Electromagnetic Fields [J]. Reviews Modern Physics, 2022, 94 (4): 045001.

[11] Tully C G. Elementary Particle Physics in a Nutshell [M]. Princeto, NJ, USA: Princeton University Press, 2011.

[12] Colonna G, D'Angola A. Plasma Modeling: Methods and applications, Bristol, UK: IOP Publishing, 2022.

[13] Deng X X, Ding H G, Hou Z Y. Thermal Characteristics of Stabilization Effects Induced by Nanostructures in Plasma Heat Source Interacting with Ice Blocks [J]. International Journal of Heat and Mass Transfer, 2023, 202: 123695.

Micro/Nano Electrode Array Plasma MEMS

DING Heng-gao[1], HOU Zhong-yu[2]

(1. Former General Armament Department, Beijing 100034, China;
2. National Key Laboratory of Advanced Micro and Nano Manufacture Technology, Shanghai Jiao Tong University, Shanghai 200240, China)

Abstract We find that effects resulted from micro/nano scale structures can regulate space charges which excite and lead to the electric field distribution featuring the flux convergence band structure. It is here referred to as the polarization band effect, which stems from the specific field induced interactions among atoms and molecules. The micro/nano electrode array structures are designed and fabricated by using the non-silicon micro/nano processing technology, forming nanoelectrode arrays-based plasma microelectromechanical systems (NPMEMS). The integrated NPMEMS device can be used to regulate inner energy states of matters and generate plasma based on the polarization band effect, all within a single chip-size limited local area or extending into a large volume space with the deployment of a distributed array of multiple devices. Its special physical and chemical properties can be utilized to greatly improve the efficiency of potential application systems or solve mechanism-level challenges in plasma-related applications of multiple fields.

Keywords MEMS; Micro/ nano Electrode array; Plasma

科学技术研究

叉指式硅微加速度计的结构设计

(1998 年 11 月)

摘　要　为了满足对中等精度的微机械加速度计的需求，讨论了一种硅片溶解法制造的平面叉指式加速度计，其敏感轴平行于检测质量平面。为了分析该种加速度计的分辨率和固有频率，分析了其机械模型和数值解，并且用有限元方法对几种设计尺寸进行了仿真。着重研究了几个关键尺寸，如检测质量厚度和梁的宽度的改变对分辨率和固有频率的影响，并且给出了此种结构在当前条件下所能得到的测量分辨率极限。最后，给出了两种尺寸的设计，其大小约为 1.5 mm×1.5 mm，检测质量厚 10 μm 和 15 μm，相对于当地重力加速度，分辨率分别为 $5×10^{-3}$ 和 $3×10^{-3}$。

关键词　加速度计；分辨率；有限元分析（FEA）

加速度计是一种重要的惯性仪表。在过去的十多年中，建立在微电子工艺基础上的微结构工艺日趋成熟，各种形式的微机械加速度计纷纷诞生。为了适应军民两方面的广泛的应用需求，以及与微机械硅陀螺的工艺兼容，以将来胜任短时间的导航任务，我们实验室从 1994 年开始进行微机械惯性仪表的研究，已经研制出薄片溶解法加工的偏心扭摆结构的加速度计样机。它敏感垂直于基片的加速度。

结构的厚度靠浓硼扩散的厚度和抛光工艺来控制，厚度范围可达

本文发表于《清华大学学报（自然科学版）》1998 年第 38 卷第 11 期。合作作者：袁光，丁衡高，高钟毓，董景新。

亚微米至几百微米的范围，通常为几十微米。检测质量厚度的增加提高了灵敏度，此时 Brownain 噪声可以忽略不计。SOI 工艺避免了双面光刻和双面高精度对准接合，因此工艺简单，成品率较高，成本较低。美国 Draper Lab、Ford 公司、U. C. Berkley 大学和日本 Toyohashi 工业和 Yamatake-Honeywell 公司在这方面均有研究。为了在同一芯片中完成三轴加速度测量，最近，我们设计了另一种 SOI 工艺基础上的平面叉指式加速度计。

1 微结构原理

图 1 为其结构示意图，敏感元件是一个微机械的双侧梳齿结构。这种加速度计敏感轴与基片平行，检测质量为"H"形。"H"形的四根细梁将检测质量固定于基片上，检测质量可以自由地沿垂直于细梁的方向运动。叉指由中央质量杆向外侧伸出，每个叉指为可变电容的一个活动电极；固定电极与活动电极交错配置，因检测及加力电路需要，在每两个相邻的动指间有两个定指。固定电极形状为"T"和"L"形状，以增大接合面积。设有 $2n_1$ 对检测指、n_2 对加力指。摆片总长度 L_0，宽 W，厚 h。指长 l_0，交叉部分长 l。动指和静指形成一对差动检测电极 C_{t1}、C_{t2}，一对加力电极 C_{f1}、C_{f2}。

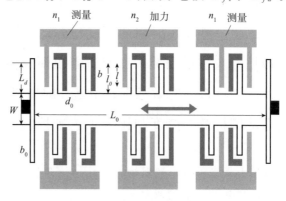

图 1　叉指结构示意图

该结构检测、加力电极分开并且完全对称，比较文献［1］中不完全对称的结构，从力学角度，本结构受力对称、均匀；另外，检测电极分别位于两端，很容易通过改变接口电路的引线方式来改变差动方式，从而便于后续检测电路的改变，而无须改变版图设计、重新制版。同时，采用薄片溶解法，比较文献［2］中硅表面工艺，结构厚度增加，分辨率提高；比较文献［3］中的 SFB（silicon fusion bonding）方法，薄片溶解法分布电容小，并且可以和已有的扭摆加速度计及轮式陀螺工艺兼容，以便今后在同一芯片上完成三轴加速度计及惯性测量单元。图 2 为该加速度计的伺服系统原理方框图。

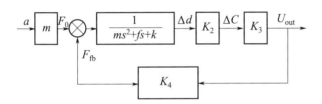

图 2　加速度计伺服控制系统原理方框图

2　性能参数分析

2.1　检测

设没有加速度输入时，摆片处于平衡状态，动、静指间距为 d_0，则

$$C_0 = C_{t10} = C_{t20} = \frac{\varepsilon S}{d_0} = 4n_1 \frac{\varepsilon h l}{d_0} \tag{1}$$

加速度 a 作用时，摆片将移动一个小位移 Δd，检测电容随之改变

$$C_{t1} = 4n_1 \frac{\varepsilon h l}{d_0 - \Delta d}, C_{t2} = 4n_1 \frac{\varepsilon h l}{d_0 + \Delta d} \tag{2}$$

$$\Delta C = C_{t1} - C_{t2} = C_0 \frac{2\Delta d}{d_0} \tag{3}$$

开环时，有

$$F_{\text{out}} = ma = K\Delta d \tag{4}$$

$$m = \rho h [L_0 w + (4n_1 + 2n_2)b] \tag{5}$$

叉指结构灵敏度公式为

$$S = \frac{\Delta C}{a} = \frac{2m C_0}{d_0 K} \tag{6}$$

其中，a 为输入加速度，m 为摆片总质量，K 为四悬梁的刚度。K 由梁的长度和截面尺寸确定。通常，为降低刚度，同时不加大结构平面尺寸，多采用折叠梁结构，无法通过简单的材料力学计算求得。不过可以根据简单的悬臂梁加以定性的分析，将单个细梁简化为一当量长度为 L_d 的悬臂梁（图3）。

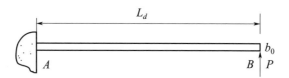

图3 单梁受力示意图

在力 P 作用下，B 点位移为

$$W_B = \frac{P L_d^3}{6EJ} \tag{7}$$

其中，$P = \frac{1}{4}F = \frac{1}{4}ma$，$J = \frac{1}{12}hb_0^3$，$E$ 为硅的弹性模量。

$$W_B = \frac{PL_d^3}{6EJ} = \frac{FL_d^3}{24EJ} = \frac{maL_d^3}{24EJ} = \frac{maL_d^3}{2Ehb_0^3} = \Delta d \tag{8}$$

所以四个细梁总的刚度为

$$K = \frac{F}{\Delta d} = \frac{24FJ}{L_d^3} \qquad (9)$$

将式（8）代入式（3），可得

$$\Delta C = C_0 \frac{2\Delta d}{d_0} = \frac{C_0 m a L_d^3}{E d_0 h b_0^3} \qquad (10)$$

设最小可测电容值为 $\Delta C_{\min}$，则由上式可得最小可测加速度 $a_{\min}$ 为

$$a_{\min} = \frac{E d_0 h b_0^3}{C_0 m L_d^3} \Delta C_{\min} \qquad (11)$$

敏感元件的自然频率为

$$\overline{\omega} = \sqrt{\frac{K}{m}} = \sqrt{\frac{24EJ}{mL_d^3}} = \sqrt{\frac{2Ehb_0^3}{\rho h S L_d^3}} = \sqrt{\frac{2Eb_0^3}{\rho S L_d^3}} \qquad (12)$$

2.2 力反馈

电容极板间作用力为

$$F = \frac{\varepsilon A U^2}{2d^2} \qquad (13)$$

其中，A 为极板面积。所以两个差动反馈力为

$$F_1 = \frac{n_2 \varepsilon h l}{2(d_0 - \Delta d)^2} (V_{\text{ref}} - V_{\text{fb}})^2 \qquad (14)$$

$$F_2 = \frac{n_2 \varepsilon h l}{2(d_0 + \Delta d)^2} (V_{\text{ref}} + V_{\text{fb}})^2$$

$$F_b = F_1 - F_2 \approx \frac{n_2 \varepsilon h l}{d_0^2} \cdot \left[4(V_{\text{ref}}^2 + V_{\text{fb}}^2) \frac{\Delta d}{d_0} - V_{\text{ref}} V_{\text{fb}} \right] \approx -\frac{4 n_2 \varepsilon h l}{d_0^2} V_{\text{ref}} V_{\text{fb}}$$

$$(15)$$

又

$$C_{F0} = \frac{n_2 \varepsilon h l}{d_0} \qquad (16)$$

$$F_b \approx -\frac{4n_2\varepsilon hl}{d_0^2}V_{ref}V_{fb} = -\frac{4C_{F0}}{d_0}V_{ref}V_{fb} \qquad (17)$$

闭环时

$$F_b = F_{out} = ma \qquad (18)$$

满量程时

$$F_{bmax} = -\frac{4C_{F0}}{d_0}V_{ref}V_{fb} = ma_{max} \qquad (19)$$

从而闭环满量程

$$a_{max} = -\frac{4C_{F0}}{md_0}V_{ref}V_{fb-max} \qquad (20)$$

2.3 结构设计

（1）设计依据

技术指标为：测量范围 $a = \pm 10$ g（g 为重力加速度，以下同），分辨率 $= 1 \times 10^{-3}$ g，冲击 $a = 100$ g。工艺参数为：结构厚度 $10 \sim 15$ μm，电容（指）间隙 2 μm，固定指电极间隙 >30 μm，电极引线宽 10 μm，平面尺寸约 1.5 mm$\times$1.5 mm。最小可检测电容：0.5 fF。

（2）尺寸确定

受限于平面尺寸、加工条件及接合强度、面积要求，取 $l_0 = 140$ μm，$L_0 \approx 1.4$ mm，$b = 5$ μm，$d_0 = 2$ μm，$l = 120$ μm，$W = 200$ μm。

为在有限平面尺寸内增大 4 个细梁的长度，并且在垂直于摆片所在平面内对结构进行保护，在梳指轴两端各增加一个小矩形，尺寸为 50 μm$\times$80 μm。

因为相邻的动指间有两个定指，所以总共可有动指数

$$\frac{L_0}{3(b+d_0)} = \frac{1\,400}{3\times(5+2)+2t} \approx 63$$

取整十数，共 60 个动指。由式（18）取 $n_2 = 12$，$2n_1 = 48$。

将以上数据代入前面推导的公式中，$h = 15~\mu\mathrm{m}$ 时，$C_{t10} = C_{t20} = 0.382~\mathrm{pF}$，$m = 12.302\,4~\mu\mathrm{g}$；$h = 10~\mu\mathrm{m}$ 时，$C_{t10} = 0.255~\mathrm{pF}$，$m = 8.201\,6~\mu\mathrm{g}$。

3 有限元分析

有限元模态分析得出的前 4 个模态的变形见图 4～图 7，频率值在表 1 中给出。其中，第一模态为位于平面内的平行移动，是检测中需要的运动方式；第二模态为垂直所在平面上下运动；第三模态为结构绕平面中一轴上下摆动；第四模态为位于平面内的扭动。第一模态以外的其他模态均为所不希望的干扰运动，必须尽量抑制它们，即拉开它们与第一模态频率之间的差距，从而降低结构的交叉耦合。

图 4 第一模态变形

1) 由有限元仿真数据可得出

$$\frac{f_{1\mathrm{T}1}}{f_{1\mathrm{T}1\text{-}10}} = \frac{2.005\,1}{1.989} = 1.008$$

即 T1（$h = 15~\mu\mathrm{m}$，$b = 3~\mu\mathrm{m}$）和 T1 - 10（$h = 10~\mu\mathrm{m}$，$b = 3~\mu\mathrm{m}$）的第一模态频率相差 0.80%。

图 5　第二模态变形

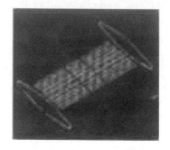

图 6　第三模态变形

图 7　第四模态变形

同理可得 T2（$h=15\ \mu m$，$b=2\ \mu m$）和 T2-10（$h=10\ \mu m$，$b=2\ \mu m$）的第一模态频率相差 0.96%，说明第一模态频率与结构厚度无关。

$$\frac{f_{1T1}}{f_{1T2}}=\frac{2.005\,1}{1.095}=1.831\approx\sqrt{\left(\frac{3}{2}\right)^3}=1.837$$

这和理论分析式（12）$\overline{\omega}=\sqrt{\dfrac{2Eb_0^3}{\rho SL_d^3}}$，即与厚度无关是吻合的。

2) 由前四个模态的频率和变形知，第二、三模态的频率 f_2、f_3 是第一模态频率 f_1 的几倍，即交叉干扰小，在灵敏度分析时，可不予考虑。

4 几种不同尺寸的性能比较

用 Super SAP[4] 有限元软件对几个不同尺寸的结构进行仿真，表 1 给出仿真结果。其中，文件名最后的字母表示输入加速度 a 的方向；f_1，f_2，f_3 为结构前三个模态的频率；$\sigma_{\max}$ 为最大应力；$a_{\max}$ 为开环时强度上允许承受的最大加速度。除了梁宽 b_0 和厚度 h 变化，其余尺寸均相同。

各结构所能承受的最大加速度、止挡保护间隙、最小敏感加速度如表 2 所示。其中，$a_{\min}$ 为加速度计分辨率，S 为灵敏度，$a_{\max}$ 为该方向开环时强度允许最大加速度值，dx 为 x 方向止挡应留的间隙。

结果分析：

1) 因为目前来讲，最小可检测电容为 0.5 fF，由上面仿真结果可发现，当细梁宽 2 μm、结构厚 15 μm 时（这个尺寸已是当前所能达到的最好的），此时，加速度计分辨率约为 $3\times 10^{-3} g$；厚 10 μm 时，分辨率约为 $5\times 10^{-3} g$。

2) 若想进一步提高分辨率，应增加细梁的长度、增加结构度或增加叉指数量等。就目前工艺条件，是可能达到灵敏度 $=10^{-3}\ g$ 的。

3) 增加结构厚度虽然不影响第一模态的频率，但很有效地抑制了第三模态，同时，也有助于提高第二模态的频率。所以，增加结构厚度可以大大降低交叉耦合。

表1　叉指加速度计有限元仿真结果（受力分析均为一个 g 的加速度）

文件名	$h/\mu m$	$b/\mu m$	f_1/kHz	f_2/kHz	f_3/kHz	σ_{max}/kPa	位移 $dx/\mu m$	位移 $dy/\mu m$	位移 $dz/\mu m$
T1X	15	3	2.005	6.849	12.788	19.0	7.858×10^{-4}	7.949×10^{-7}	1.2×10^{-10}
T1Y	15	3	2.005	6.849	12.788	316.8	7.812×10^{-8}	6.205×10^{-2}	7.64×10^{-10}
T1Z	15	3	2.005	6.849	12.788	148.3	8.139×10^{-7}	2.796×10^{-6}	5.453×10^{-3}
T1-10X	10	3	1.989	5.137	9.6745	19.4	7.922×10^{-4}	5.173×10^{-7}	9.198×10^{-8}
T1-10Y	10	3	1.989	5.137	9.6745	313.5	7.619×10^{-8}	6.307×10^{-2}	1.51×10^{-12}
T1-10Z	10	3	1.989	5.137	9.6745	196.6	1.257×10^{-6}	2.482×10^{-6}	9.808×10^{-3}
T2X	15	2	1.095	4.831	8.595	44.7	2.664×10^{-3}	4.59×10^{-7}	1.61×10^{-10}
T2Y	15	2	1.095	4.831	8.595	725.3	1.157×10^{-7}	2.078×10^{-1}	7.74×10^{-10}
T2Z	15	2	1.095	4.831	8.595	241.2	8.351×10^{-6}	1.974×10^{-6}	1.082×10^{-2}
T2-10X	10	2	1.048	3.701	6.8816	43.8	2.699×10^{-3}	4.558×10^{-7}	7.204×10^{-11}
T2-10Y	10	2	1.084	3.701	6.8816	714.0	5.395×10^{-8}	2.123×10^{-1}	9.069×10^{-9}
T2-10Z	10	2	1.084	3.701	6.8816	353.0	1.224×10^{-6}	4.211×10^{-6}	1.849×10^{-2}

4) 以上仿真，许用应力采用了最保守的值，因此上面的结构设计符合强度要求。

研究平面叉指式加速度计，对于今后三轴加速度计或加速度计单元的研制都是很有必要的。

表 2 上述几种尺寸的加速度计的分辨率、抗冲击能力

结构名	$a_{min}/$ ($\times 10^{-3} g$)	a_{max}/g			$dx/\mu m$	$dz/\mu m$
		x	y	z		
T1	10.6	3 552	220.935	471.853	3	3
T1-10	15.8	3 613.4	223.274	355.898	3	3
T2	3.15	1 566.99	96.51	290.2	4	3
T2-10	4.62	1 598.6	98	197.96	4	4

参 考 文 献

[1] Kuehnel Wolfgang. Modelling of the mechanical behavior of a differential capacitor acceleration sensor. Sensors & Actuators, 1995, A48: 101-108.

[2] Kuehnel Wolfgang, Sherman Steven. A surface mircomachined silicon accelerometer with on-chip detection circuity. Sensors & Actuators, 1994, A 45: 7-16.

[3] Van Drieenhuizen B P, Maluf N I, Opris I E, et al. Force-balanced accelerometer with mG resolution, fabricated using silicon fusion bonding and deep reactive ion etching. In: Transducers '97, 1997 International Conference on Solid-State Sensors and Actuators, 1229-1230.

[4] Super Sap 说明书.

Design a Multi-finger Silicon Micromechanical Accelerometer

YUAN Guang, DING Heng-gao, GAO Zhong-yu, DONG Jing-xin

(Department of Precision Instruments and Mechanology,
Tsinghua University, Beijing 100084, China)

Abstract In order to fulfill the need for a moderate accurate accelerometer, a multi-finger parallel accelerometer based on silicon-on-insulator (SOI) process has been designed. Its sensitive direction is parallel to the proof mass. A mechanical model and mathematical solution are presented, which have been developed in order to calculate the sensitivity and frequency behavior. The finite element analysis (FEA) method is also be used. This paper emphases on the discussion of the changes of the sensitivity, frequency, the acceleration measurements range, caused by the changes of the main dimensions such as the thick of the proof mass, the width of the beam of the acceleration sensor. In the end, two designs are described. The size of the sensor is about 1.5 mm×1.5 mm, the proof masses are 10 μm and 15 μm thick and the sensitivities relative to gravitational acceleration are 5×10^{-3} and 3×10^{-3}, respectively.

Keywords Accelerometer; Resolution; Finite Element Analysis (FEA)

微型光学陀螺仪中声表面波声光移频器的研究

(1999年2月)

摘 要 微型光学陀螺仪是基于Sagnac效应,采用先进的集成光学技术研制的新型光学陀螺仪。研制微型光学陀螺仪的关键在于用频率调制来实现频率伺服,对主要误差进行有效抑制,实现高精度Sagnac频差测量。声表面波声光移频器正是解决上述技术关键的一个集成光学核心器件。针对微型光学陀螺仪对声表面波声光移频器所提出的衍射效率为3%~5%,频移范围±300 kHz,短期频率稳定度优于10^{-8}这些特殊要求,依据声光理论进行了分析与特殊设计。采用平面集成光学工艺,在K8玻璃基片上研制出声表面波声光移频器。实验结果表明,该器件已初步满足设计要求。

关键词 微型光学陀螺仪;声表面波;声光移频器;频率稳定度

微型光学陀螺仪(MOG)是一种基于无源环形谐振腔的光学陀螺仪,结构如图1所示。图中粗箭头与双向箭头为光连接,其余为电连接。其基本原理是Sagnac效应[1]

$$\Delta f = \frac{4A}{n\lambda p}\Omega \tag{1}$$

本文发表于《清华大学学报(自然科学版)》1999年第39卷第2期。合作作者:张斌,潘珍吾,丁衡高,里·尼·阿斯尼斯。

式中，Δf 为顺时针和逆时针方向两束光的谐振频率 f_1 与 f_2 之差；A 为谐振腔所包围的面积；n 为谐振腔光波导的折射率；λ 为光波波长；p 为谐振腔周长；Ω 为谐振腔所敏感的角速率。

MOG 通过测量在波导谐振腔中相向传播的两束光的相位差来确定旋转物体的转动角速率，由于每束光的总相移可达 10^{10} rad，但需要测量的相位差一般却只有 10^{-7} rad，因此，在目前的技术条件下，要直接测量出光频变化的瞬时相位极其困难。为此提出频率调制的思想，即，如果能使入射到波导谐振腔中的两束光发生频移，而频移量可以精确控制，以使得在任意时刻由两光束的频差所引起的相位差与 Sagnac 相移大小相等方向相反，则两光束的频差就正比于待测的旋转角速率，通过测量频差就可以测得角速率[1]。这一方案的核心是对每束光实现精确的频移控制。在谐振式光学陀螺仪、布里渊激光陀螺仪以及 MOG 中广泛采用声光移频技术来实现这一控制[2-5]。因此，目前在提高光学陀螺仪闭环系统精度的研究中，声光移频技术成为热点之一。MOG 是一全固态、全集成的微型化系统，它对声表面波声光移频器提出了十分特殊的要求。即，声光移频器不仅要有很高的工作频率和频率稳定度，而且还要有较大的频移和一定的衍射效率。本文的主要目的在于提出一种有效方案来解决上述技术关键（图1）。

1 声表面波声光移频器的设计

声表面波声光移频器主要由声光介质、叉指换能器及驱动电源等组成。声光移频器的工作原理是一种非线性光学效应，即声光效应[6]。声光移频器的设计包括材料、工艺选择，叉指换能器及控制电路的设计。该设计要保证声光移频器满足以下技术指标：衍射效率 3%～5%，工作频率 200 MHz，移频范围 ±300 kHz，短期频率稳定度优于 10^{-8}。

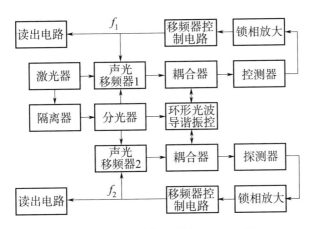

图 1　MOG 频率伺服控制系统图

1.1　结构设计

叉指换能器图形复杂，设计与加工是研制声光移频器的基础，其周期应与声波波长 Λ 相当

$$\Lambda = V_R / f = 12.5 \ \mu m \tag{2}$$

式中，V_R 为声波的波速；f 为声波的频率。

叉指换能器每个电极的宽度 w 为

$$w = \Lambda / 4 \tag{3}$$

为了得到更高的频率转换效率，必须满足以下布拉格衍射条件

$$2\pi\lambda L / (n\Lambda^2) \gg 1 \tag{4}$$

式中，n 为基片的折射率系数；λ 为自由空间中的光波长；L 为声孔径。

而布拉格衍射角 θ_B 为

$$\theta_B = \arcsin \frac{\lambda}{2n\Lambda} = 1.98° \tag{5}$$

最大频率偏移量设计为

$$\Delta f = \frac{1.8 V_R n\cos\theta_B}{\lambda}\Delta\theta = 1\ \text{MHz} \tag{6}$$

式中，$\Delta\theta$ 为声波发散角，$\Delta\theta = \Lambda/L$。由式（2）及式（6）可推出，声孔径应满足

$$L \leqslant L_{\max} = \Lambda/\Delta\theta = 6.5\ \text{cm} \tag{7}$$

选 $L = 0.5\ \text{cm}$，可验证满足布拉格条件 $2\pi\lambda L/n\Lambda^2 = 170 \gg 1$。此时由式（6）得 $\Delta f = 13\ \text{MHz}$，于是叉指换能器的齿对数可取为 20。光波的衍射效率为

$$\eta = \frac{I_1}{I_0} \approx \sin^2\left(\frac{\pi}{\lambda\theta_B}\sqrt{\frac{M_2}{2}P_a c^2 \frac{L}{H}}\right) \tag{8}$$

式中，I_1 为一级衍射光的光强；I_0 为零级光的光强；P_a 为声功；c 为声场与光场的覆盖系数；M_2 为与衍射效率有关的声光介质材料系数；H 是楔形体的高度（图2）。

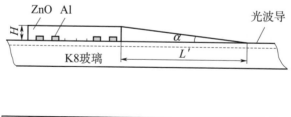

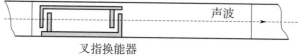

图 2 声光移频器的结构设计

图 2 给出了所设计的声光移频器结构。考虑由 ZnO 薄膜到基片过渡斜面处的损耗、声波传播损耗及相应的机电耦合系数，可知输入电功率 <1 W 时，$\eta < 5\%$。根据 MOG 系统的要求，设计的最大衍射效率 η 为 3%～5%。图中 L' 是端部长度，α 是 ZnO 薄膜楔形体的倾角，

小于 3°。ZnO 薄膜楔形体是一项很有意义的新设计，工艺和实验表明，它的斜度均匀性以及薄膜本身的压电特性是保证衍射效率的关键。在声光移频器的研究中，这是一次新的成功的尝试。

1.2 材料与工艺

考虑到材料的均匀性、温度特性、老化特性，以及与环形腔的可集成性，选用 K8 玻璃为基片，ZnO 薄膜为声光介质材料，叉指电极的材料选用 Al 来抑制高频时因电极质量而引起的二次负载效应。薄膜工艺选用真空溅射 ZnO 技术，叉指电极的制备采用光刻工艺。

1.3 控制电路设计

图 1 已给出了 MOG 频率伺服控制系统的原理示意，它的难点主要在于敏感低角速率时需要很高的频率稳定度。当陀螺敏感 10（°）/h 的角速率时，移频器必须保证最小频移量为几个 Hz；当陀螺仪工作在动态时，又要求移频器的最大频移量为 ±300 kHz，而移频器的工作频率为 200 MHz，因此要求频率稳定度优于 10^{-8}。如果采用一般的双晶振双路压控振荡器分别激励方案，那么在技术上存在相当大的难度。为此提出一种新的方案，如图 3 所示。这一控制电路的主要特点是双路共用同一晶振，这样晶振的频率不稳定性对双路所产生的误差为同源误差，在有用的频差信号 Δf 中该误差就可以被成功地抑制，从而简化了控制电路，并且将频率稳定度的要求降低了一个数量级。通过选用合适的压控振荡器完全可以实现。

2 实验结果

在玻璃基片上首先采用钾钠离子交换工艺制成平面光波导，再按图 2 所设计的掩膜研制出质量较高的叉指电极，然后在温度 260 ℃、

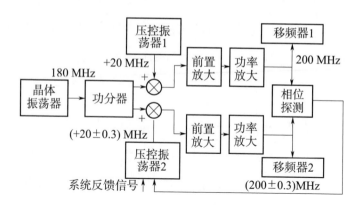

图 3　高稳定度声光移频电路框图

电流 0.4 A、溅射时间 1.5 h 的工况条件下在该基片及电极表面溅射 ZnO 薄膜，并研制出满足设计要求的薄膜楔形体。测得的薄膜驻波系数为 2.1。

楔形体端部特性如图 4 所示，图中 $\alpha_1=0.95°$，$\alpha_2=0.8°$，$\alpha_3=0.56°$，$\alpha_4=0.24°$。实验测得声光衍射效率为 4%（输入功率为 1 W）。移频器在布拉格衍射状态下的工作情况如图 5 所示。

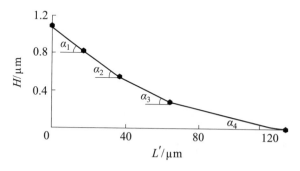

图 4　ZnO 薄膜楔形体端部特性

图 5　声光布拉格衍射

3　结束语

在 MOG 中，要求采用声表面波声光移频器作为其闭环控制的核心器件。本文针对相应技术指标进行了分析与设计，实验表明，在 K8 玻璃基片上研制出的声表面波声光移频器已初步满足设计要求，其中 ZnO 薄膜楔形体结构这一新设计较好地保证了所要求的衍射效率。为有效地提高声光移频器的频率稳定度，提出采用单晶振双路压控电路来代替一般的双晶振双路压控振荡器激励方案，以保证 MOG 对分辨率的特殊要求。在下一步的工作中，这些研究结果几乎完全可以移植到硅基片上，以便在硅基片实现光路与电路的全集成。这是本文为将声表面波声光移频器有效应用于微型光学陀螺仪所做的有益的探索与尝试。

这项工作得到了清华大学精密仪器与机械学系章燕申和汤全安教授的悉心帮助与指导。该系研究生张斌等同学也给予了热情的帮助。在此向他们以及给予大力协助的俄罗斯国家光学研究院和圣·彼得堡光学及精密机械学院表示诚挚的谢意。

参 考 文 献

[1] Zhang Yanshen, Zhang Bin, Ma Xinyu. Techniques for developing a miniature resonant optic rotation sensor. In: The Institute of Navigation, ed. Proceedings of the 52nd Annual Meeting "Navigational Technology For The 3rd Millennium". USA: The Institute of Navigation, 1996, 719 – 723.

[2] Lawrence A W. The Micro – Optic Gyro. Symposium on Gyro Technology, 1983: 161 – 166.

[3] Takiguchi K. Method to reduce the optical kerr – effect induced bias in an optical passive ring resonator gyro. IEEE Photonics Technology Letters, 1992, 4 (2): 203 – 206.

[4] Fischer S, Schroder W, Meyrueis P, et al. Brillouin ring laser gyroscope with acousto – optical modulators. In: Symposium Gyro Technology. Germany: Stuttgart University, 1995, 11.0 – 11.18.

[5] Байбородин Ю В. Интегрально – оптический кольцевой пассивный резонатор для оптических гироскопов. Квантовая Электроника, 1992, 19 (2): 191 – 193.

[6] 徐介平. 声光器件的原理、设计和应用 [M]. 北京: 科学出版社, 1982.

Research on Surface Acoustic Wave Acoustooptic Frequency Shifter for a Micro Optic Gyro

ZHANG Bin, PAN Zhen – wu, DING Heng – gao,

L. N. ASNIS

(Department of Precision Instruments and Mechanology,

Tsinghua University, Beijing 100084, China;

S. I. Vavilov State Optical Institute, Russia)

Abstract The micro optic gyro (MOG) is a new optic gyro based on Sagnac effect using advanced integrated optic technology. The key technologies in development of MOG are the frequency modulation technology for frequency servo and elimination of main gyro error, and the measurement of Sagnac frequency difference with high accuracy. The key to solve these problems is adopting the acoustooptic (AO) frequency shifter. Theoretical analysis and design based on AO theory are performed for special requirements of MOG. The requirements are that acoustooptic diffraction efficiency is 3%~5%, frequency shift is ±300 kHz, and frequency stability in shorttime is less than 10^{-8}. The AO frequency shifter was made on K8 glass by means of planar integrated optical technique. Experiment results show that its main characteristics can satisfy the design requirements.

Keywords Micro optic gyroes; Surface acoustic wave; Acoustooptic frequency shifters; Frequency stability

一种微型隧道效应磁强计的设计

(2000年8月)

摘　要　隧道效应传感器是一种灵敏度很高的位移传感器，其原理已经被用于微加速度计和红外传感器的设计。文中介绍了基于这一原理的微型磁强计，对传感器的固有频率和驱动电压进行了分析和计算。

关键词　MEMS；隧道效应；磁强计

1　引言

磁强计是用来测量磁感应强度的传感器。磁场测量技术的应用已深入到工业、农业、国防以及生物、医学、宇航等各个部门。例如，卫星上使用磁强计作为姿态测量的主要传感器，同时具有测量空间磁场的功能。测量磁场的主要方法有电磁感应法、霍尔效应法、磁阻效应法及磁通门法等。

由于微机电系统技术的发展，目前出现了一些新原理的传感器，隧道效应式传感器就是近年来发展起来的高灵敏度的传感器，其原理已经被应用于加速度计和红外传感器的设计。它具有灵敏度高、噪声低、温度系数小及动态响应性能好的特点。依此原理设计制作了一种磁强计，

本文发表于《仪表技术与传感器》2000年第8期。合作作者：朱俊华，丁衡高，叶雄英。

体积为 10 mm×10 mm×0.6 mm，分辨率达到了 10^{-6} T/Hz$^{1/2}$，带宽 10 kHz[1]，计划将其用于直径为 100 mm 的微小卫星上，测量空间磁场。

2 隧道效应式磁强计的原理

隧道效应传感器的基本原理与隧道扫描显微镜相同。隧道电流的大小与针尖和电极的间距呈指数关系[2]

$$I_{\text{Tunneling}} = I_0 e^{-\alpha\sqrt{\Phi}s} \tag{1}$$

式中，α 为 10.25 eV$^{-1/2}$ (nm)$^{-1}$；Φ 为针尖和电极表面间的势垒，对于空气中的金电极，该值在 0.05～0.5 eV 之间；s 为针尖与电极之间的距离。

图 1 所示为微型隧道效应磁强计的结构图，它由上层的玻璃衬底和下层的硅片组成。在驱动电极和偏置电极之间加上一定电压，静电力使微梁变形，当针尖和电极之间的间距约为 1 nm 时，就会产生隧道电流，在梁背面的平面线圈内通上交变电流，梁在 Lorentz 力的作用下上下振动，测量隧道电流的大小，就能得到梁的变形量和磁感应强度的大小。薄膜的上表面和下表面都有一层 0.2 μm 厚的 SiO$_2$ 作绝缘层。

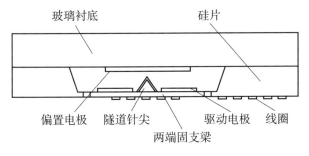

图 1 微型磁强计结构图

隧道效应的噪声是典型的 $1/f$ 噪声[1]。在低频时噪声较大，随着频率升高，噪声逐渐变小。因此，隧道效应加速度计一般只能作动态测量，而磁强计在线圈中通以交变电流，可以将工作点移至噪声较小的高频段，因此可以测量静态磁场，有利于提高分辨率。

3 传感器的主要性能参数计算

薄膜梁是传感器中唯一的可动部件，它决定着传感器的主要性能，因此对梁的参数进行优化是提高性能的关键。一般来说，MEMS 加工的薄膜都有一定的内应力，应力对薄膜的机械特性有较大影响，在计算中必须考虑应力的影响。

3.1 固有频率

微梁的固有频率是决定传感器动态性能的一项重要参数。对内应力为 σ 的两端固支梁，其一阶固有频率 ω_R 为[3]

$$\omega_R^2 = \lambda_1^2 \lambda_2^2 \cdot \frac{h^2}{12\rho l^4} \cdot \frac{E}{1-\nu^2} \qquad (2)$$

$$\lambda_1^2 - \lambda_2^2 = 12 \frac{l^2}{h^2} \cdot \frac{\sigma(1-\nu^2)}{E} \qquad (3)$$

$$2\lambda_1 \lambda_2 (\cos\lambda_2 \operatorname{ch}\lambda_1 - 1) + (\lambda_2^2 - \lambda_1^1)\sin\lambda_2 \operatorname{sh}\lambda_1 = 0 \qquad (4)$$

式中，h 和 l 分别为梁的厚度和长度；ν 为泊松比；E 为杨氏模量。

将式（3）代入式（4），利用牛顿迭代法，可以解得 λ_1 和 λ_2，再代入式（2）即可解得 ω_R。

对复合梁的应力和杨氏模量可以通过下式得到[4]

$$\sigma_c = \left(\sum_i \sigma_i h_i\right)/h_c \qquad (5)$$

$$E_c = \left(\sum_i E_i h_i\right)/h_c \qquad (6)$$

式中，c 为复合梁；i 为复合梁的各层。

在这里，取重掺杂硅和二氧化硅的残余应力分别为 24 MPa 和 -230 MPa，杨氏模量为 150 GPa 和 50 GPa。图 2 所示为计算得到的固有频率和梁的尺寸关系曲线，梁的厚度：1 为 4 μm；2 为 5 μm；3 为 6 μm。

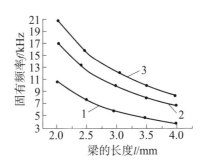

图 2　薄膜的固有频率与梁的尺寸关系

3.2　驱动电压

考虑到传感器的耐冲击性和成品率，加工时在隧道针尖和电极之间有 1 μm 的间距，需要施加一个电压使薄膜变形以便产生隧道电流，驱动电压的大小是传感器的一项重要的性能指标。

设薄膜梁变形挠度曲线函数为

$$w(x) = w_0 \cos^2(\pi x/l) \tag{7}$$

由纵向应力引起的变形能为

$$U_a = \frac{(\sigma_0 + \Delta\sigma)^2}{2E} \cdot hbl \tag{8}$$

式中，σ_0 是薄膜的初始张应力。

$$\Delta\sigma = E\frac{\Delta l}{l} = \frac{1}{2l}\int_{-l/2}^{l/2}\left(\frac{dw}{dx}\right)^2 dx \tag{9}$$

当电压为 V 时，两电极之间的静电能为

$$U_e = -\frac{1}{2}\int_{-l/2}^{l/2}\rho V dx = -\frac{1}{2}\varepsilon_0 b V^2 \int_{-l/2}^{l/2}\frac{1}{d-w(x)}dx \tag{10}$$

式中，d 为两电极间的间距，$d = 8 \ \mu m$。

对于薄膜变形，可以忽略弯曲引起的变形能，因此总势能

$$I = U_a + U_e$$

由总势能驻值定理 $\partial I/\partial W_0 = 0$ 和式（7）～式（10）可求得驱动电压

$$V^2 = \left(1 - \frac{w_0}{d}\right)^{\frac{3}{2}} \cdot \frac{d^2}{\varepsilon_0} \cdot \left(\frac{2\pi^2 \sigma_0 h \omega_0}{l^2} + \frac{\pi^4 h E \omega_0^3}{2l^4}\right) \quad (11)$$

图 3 为驱动电压与梁的长度和厚度的关系曲线。梁的厚度：1 为 $4 \ \mu m$；2 为 $5 \ \mu m$；3 为 $6 \ \mu m$。

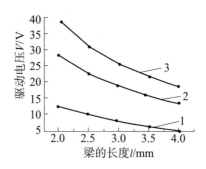

图 3　传感器的驱动电压和梁的长度、厚度的关系

4　结论

由式（2）或式（11）可以看出，微梁的固有频率、驱动电压和梁的宽度都没有关系，而梁长度增加、厚度减小时，驱动电压和固有频率都会减小。当长度增加时，梁的刚度变小，会使驱动电压降低，灵敏度增加。但增加长度，一方面受工艺条件的限制，另一方面会使固有频率降低。当传感器工作时，必然会受到建筑物、工作台振动的干扰，这些振动的频率一般在 1 kHz 以下，为抑制噪声，必须使薄膜的固有频率远远高于噪声的频率。因此，选择薄膜的尺寸为 3 mm ×

1 mm，厚度为 5 μm，此时薄膜的谐振频率为 10 kHz，驱动电压为 18 V。

参 考 文 献

[1] L M Miller, J A Podosek, et al. A μ - magnetometer based on electron tunneling. Proceedings of the 9th Annual International Workshop on Micro Electro Mechanical Systems, 1996, Feb (11～15): 467 - 472.

[2] T W Kenny, S B Waltman, et al. Micromachined silicon tunnel sensor for motion detection. Appl. Ohsy, Lett, 1992: 58 (1).

[3] X Y Ye, Z Y Zhou, Y Yang. Detemination of the mechanical properties of microstructures. Sensors and Actuators, 1996, A (54): 750 - 754.

[4] O Tabata, K Kawahata, S Sugiyama. Mechanical property measurements of thin films using load - deflection of composite rectangular membrance, Proc. IEEE Workshop on Microelectromechanical Sysemts, 1989, (Feb): 152 - 156.

[5] S T Cho, K Najafi, K D Wise. Scaling and dielectric stress compensation of ultrasensitive boron - doped silicon microstructures.

The Design of a Micro Tunnel Magnetometer

ZHU Jun - hua, DING Heng - gao, YE Xiong - ying

(Department of Precision Instruments and Mechanology,

Tsinghua University, Beijing 100084)

Abstract The sensor based on tunnel effect is a kind of highly sensitive displacement and its inducer, the principle has been applied in the design of accelecrometer and infrared sensors. This paper introduces the micro magnetometer and analyses the natural frequency and driving voltage of the sensor as well.

Keywords　MEMS; Tunnel Effect; Magnetometer

Investigation of the System Configuration for Micro Optic Gyros

(2005 年)

The re-entrant system configuration of optic gyro was investigated at Tsinghua University for developing the micro optic gyro (MOG). In the system, the bi-directional light beams are circulating in the Sagnac sensing ring (SSR) for many round trips. The feasibility of the proposed system was proved by the test results on the fiber analog experimental setup.

1 Introduction

Since 1980, the micro-nano technology has been applied extensively in the field of micro sensors. One of the outstanding examples is the micro integrated vibration gyro on silicon substrate (MSG). The MSG is developed as a micro electro-mechanic system (MEMS) with very low price.

Obviously, optic gyros are facing the challenge in the market of navigation products. Since 1983, people tried to use the integrated optic

---

本文发表于 Вестник МГТУ им. НЭ. Баумана. Сер. "Приборостроение"（莫斯科国立鲍曼技术大学论文集《精密仪器制造》）2005 年第 4 期。合作作者：Y. S. Zhang, H. G. Ding。

(IO) ring cavity to replace the mirror system resonator in the conventional ring laser gyro. Northrop company in USA had developed the first prototype of micro optic gyro (MOG), in which a passive IO resonator was used. Unfortunately, the MOG program of Northrop company stopped on the half way and the first MOG had not been developed as a gyro product.

In 1997, the LETI of CEA in France has developed the solid-state optic gyro (SSOG), in which the eight-turn optic waveguide coil with the diameter of 30 mm and the length of 800 mm was used for replacing the sensing coil in the conventional interferometer fiber optic gyro (IFOG). The advanced IO technology on silicon substrate (IOS2) insured the quality of the sensing coil with the optic loss coefficient < 0.025 dB/cm. By using IOS2 technology, LETI developed many other devices successfully. But due to the low accuracy caused by the short sensing coil, the SSOG did not find application in navigation system.

Since 1995 the authors at Tsinghua University started to investigate the MOG with IO sensing ring or coil. The goal of the investigation is to develop the MOG with the performance better than the existing miniature optic gyros, such as the GG1308 of Honeywell company and the μFORS of LITEF company. The required performance of the MOG to be developed is shown in Table 1.

Table 1 Characteristics of the MOG to be developed

Range of measurement	500(°)/s
Resolution	0.1~1(°)/h

续表

Bias stability	1~10(°)/h
Scale factor errors	100~500 ppm
Bandwidth	200 Hz

For investigating the proposed re – entrant interferometer MOG, a fiber anlogue experimental setup was developed at Tsinghua University, in which the elements of conventional IFOG are used, such as the high quality super luminescent diode (SLD), the multi – functional IO phase modulator chip (MIOC), and the all digital closed loop electronics (ADCL).

2 Proposed interferometer system configuration

The setup of the proposed MOG system and the main elements are shown in Fig. 1 and Table 2, respectively. In the system configuration, an input / output coupler A is designed for transmitting the bi – directional light beams into the Sagnac sensing ring (SSR) and then back to the MIOC.

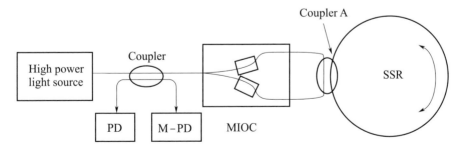

Fig. 1   The system configuration of re – entrant interferometer optic gyro

Table 2　Parameters of the system in comparison with IFOG

Element	Conventional IFOG	The setup
1.3 μm SLD	Output power > 200 μW	Output power > 600 μW
SSR (or coil)	Length 500~1 000 m	Length < 200 m
Beam splitter (50 : 50)	Extinction ratio > 20 dB The same Insertion loss < 0.2 dB	
MIOC Splitting ratio (49.6/50.4)	Insertion loss < 3 dB Extinction ratio > 40 dB Bandwidth 500 MHz	The same
Coupler A (90 : 10)	No need	Extinction ratio > 20 dB Insertion loss < 0.2 dB

In case of open loop operation, the optical intensity detected by the photodiode (PD) is the sum of the optical responses of light beams with different number of round-trips. The resolution of the MOG system can be determined as

$$\Delta \Omega_{\min} = \frac{\lambda c}{2\pi MLD} \frac{\sqrt{4h\upsilon \cdot T(\phi_s)}}{\sqrt{\eta_D t_i} \cdot S(\phi_s) \sqrt{P_{in}}}$$

where M is the number of round-trips circulated; λ, c are the wavelength and the light speed, respectively; L, D are the length and the diameter of SSR, respectively; h is Plank's constant; υ is the light angular frequency; η_D, P_{in} are the efficiency of PD and the power of the SLD, respectively; $T(\phi_s) = 4p + 2q \sum_{m=1}^{M} \eta^{2m} [1 + \cos(2m\phi_s)]$ is the light transmission coefficient inside the SSR; $S(\phi_s) = \left| \frac{dT(\phi_s)}{d\phi_s} \right|$ is the scale factor.

In the system configuration with passive SSR, the resolution of the system depends significantly on the power coupling ratio of coupler A.

In order to ensure the maximum SF and the best resolution, the coupling ratio (90 : 10) should be selected.

3 Development of the super luminescent diode

For the proposed MOG system, a superluminescent diode (SLD) with the output power of 0.6~1.0 mW is required. Suppose the coupling efficiency between the SLD chip and the PM fiber pigtail is around 30%, the output power of the SLD chip should be higher than 3 mW at the injection current of 100 mA and the temperature of 20 ℃.

To realize the SLD, low optical loss, high internal quantum efficiency, and high gain are the three key features for the epitaxial layers. The epitaxial structures for the emission wavelength at 1 300 nm were grown by metal organic chemical vapor deposition (MOCVD) on S doped n - type (100) - eriented InP substrate. All the layers were lattice - matched to the InP wafer except for the QWs, which were grown with a compressive strain.

We have designed and fabricated three epitaxial layers with different structures, doping profile in p - cladding layers, and the number of QWs. Finally, we selected the structure, which contains eight QWs. In the structure, a graded - refractive - index separate confinement heterostructure (GRIN - SCH) is used for realizing four energy steps from 1.1 to 1.3 eV with thicknesses of 80, 60, 40, and 20 nm, respectively. Within each step the energy is further linearly graded. The etch stop layer in the p - cap layer in 20 nm thick 150 nm above the SCH region. The p - InP cap layer is linearly graded doped from a concentration of 2×10^{17} cm^{-3} to 9×10^{17} cm^{-3} within 500 nm. The characteristics of the developed SLD are shown in Table 3.

Table 3 Characteristics of the developed SLD

Parameter	Measured result
Internal quantum efficiency	$\eta_i = 93\%$
Internal optical absorption at threshold	$\alpha_i = 11$ cm^{-1}
Characteristic temperature (20~60 ℃)	$T_0 = 58.2$ K

In order to suppress the spectrum modulation, facet reflectance must be minimized. We have designed the SLD, which is composed of a J – shape waveguide with 2 μm ridge and a tapered rear absorption region. The curvature of the waveguide suppresses the Fabre – Perot (FP) mode so that the device has only a single – pass amplified spontaneous emission (ASE). The tapered part is used to enhance the ASE, which has an open angle of 5°. The whole cavity is tilted against the facet with an angle of 8°.

The structure is etched to a depth of 1.4 μm. The side wall of the ridge and the surface of the absorption area are coated with 200 nm SiO$_2$ deposited by electron – cyclotron plasma – enhanced chemical vaper deposition (ECP – PECVD). The ridge part of the SLD is 1.4 mm long and the whole absorption section is 1.5 mm long. After cleavage, the front facet of the SLD is coated with one layer of SiN$_X$ as the anti – reflection film deposited by ECP – PECVD.

Characteristics of the developed SLD chip are shown in Table 4.

Table 4 Characteristics of the developed SLD chip

Parameter	Test result
Output power at 400mA and 20 ℃	25 mW
Full – width at half – maximum (FWHM)	26~28 nm
Fluctuation on the spectrum (Ripple)	0.2 dB

4 Experimental investigation

The setup in closed loop operation is different from the conventional IFOG. In order to avoid the overlap of the signals with different numbers of round-trips, a pulsed phase modulation approach is used for separating and selecting the needed signal. The pulse width is selected to be equel to the group transition time τ through the SSR and the modulation frequency is selected to be $2/(N+1)$ of the SSR eigen frequency.

In the passive SSR, the CW and CCW light beams can only propagate 2~4 round trips, after that, they will be shadowed by the shot noise of the PD and can not be detected. Therefore, in the case of passive SSR, $N=4$ is chosen. In the setup, the length of SSR is 200 m, and the pulsed phase modulation frequency is 200 kHz. The time interval between the two pulses is N_τ.

The phase biasing is realized by the MIOC. The peak voltage is designed to be equel to the half-wave voltage of MIOC for alternately realizing the optimal phase bias of $\pm\pi/2$.

In order to improve the Signal – Noise – Ratio (SNR), the pulsed phase modulation has been modified by using a pair of positive and negative modulation pulses with the same amplitude and width τ. In this case, a subtrator should be used in the ADCL for processing the paired signals.

As mentioned before, the small coupling ratio α will lead to waste the SLD output light power and cause the saturation of the pre-amplifier circuits after the PD due to the increment of the DC component. In this case, the Sagnac effect signals of light beams after 3~4 round trips will be difficult to be detected. Therefore, in the setup the pre-amplifier has been re-designed by adding a special cascade for completely cutting off the DC component before amplification.

In order to investigate the performance of the proposed system for MOG, a specially designed ADCL electronics has been developed, in which the above mentioned phase modulation approach is realized. The setup with the ADCL electronics provides the digital readout as a gyro.

In the drift rate test, the samples of the Sagnac phase shift signals after 2 round-trips are selected with the sampling frequency equal to the modulation frequency (200 kHz). In every eight sampling periods, one digital data processing cycle has to be completed for providing the digital readout data. The data processing cycle includes signal sampling, demodulation, integration and digital ramp signal generation.

The random drift rate of the setup was tested with the integration time of 10 s and the sampling length of 0.5 h. The standard deviation of the drift rate is 1.25 (°)/h.

The same drift rate test has been carried out on the setup, but with the additional monitoring PD (M - PD in Fig. 1). As shown by the test results, with the same integration time and the same sampling length, the standard deviation of the drift rate has been reduced to 0.75 (°)/h.

The test results proved the effectiveness of the M - PD in the improvement of accuracy.

5 Comparison with the resonator system configuration

In 1997, the authors have developed a fiber analogue experimental setup of the resonator system configuration for MOG, in which a short PM fiber optical SSR was used. The system configuration and its elements are shown in Fig. 2 and Table 5, respectively.

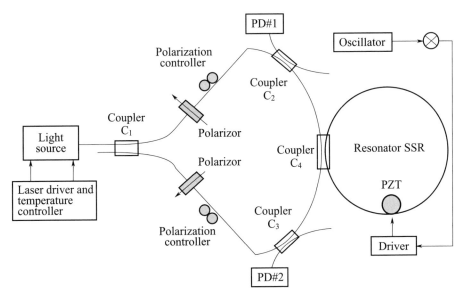

Fig. 2 The fiber analogue resonator system configuration for MOG

Table 5 Elements used in the setup

Element	Parameters
SSR	Length 10 m
PM couplers C_1, C_2, C_3 Coupling ratio 50 : 50	Extinction ratio $>$ 20 dB, Insertion loss $<$ 0.2 dB
PM coupler C_4 Coupling ratio 10 : 90	Extinction ratio $>$ 20 dB, Insertion loss $<$ 0.2 dB
Polarizor	Extinction ratio $>$ 20 dB
Polarization controller	Extinction ratio $>$ 40 dB

Considering only the coupling ratio and losses in SSR and C_4, the calculated finesse of the resonator is around 50.

In the resonator MOG, the SSR and the light source are two key elements, which should be developed carefully. The best approach is to combine the SSR and the light together in one element by doping the optical waveguide to form an active SSR.

In 1997, the task to develop an active SSR was difficult to be completed, we had to use separately the passive SSR and the light source with extremely narrow linewidth.

Firstly, a high power 25 mW Nd – YAG laser was used in the setup and the measured finesse of the passive SSR is around 35~40. According to the measured finesse, the linewidth of the Nd – YAG laser has been calculated as around 240 kHz. This value coincides with the real linewidth of the Nd – YAG laser.

Considering that the Nd – YAG laser is not suitable to be used in the MOG product, we have developed a fiber Bragg grating FBG – LD with narrow linewidth. As shown by the measured result, the linewidth is around 1.5 MHz.

The FBG-LD was used in the setup for replacing the Nd-YAG laser. As shown in the resonant waveform in the SSR (Fig. 3), the actual finesse of SSR is about 20~25. Obviously, the linewidth of the light source reduced the finesse of SSR significantly.

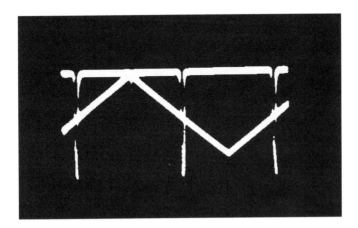

Fig. 3 The resonant waveform in the fiber analogue setup

In comparison with the resonant system configuration, the proposed interferometer system configuration is easier to be realized on the base of the matured elements, used in IFOG, including the light source SLD, the MIOC, and the closed loop control electronics.

For developing the MOG with higher accuracy, in both system configurations, the active waveguide SSR is necessary to be used, because in the proposed interferometer system configuration, the number of round trips of light beams in the passive SSR is < 5.

6 Conclusions

1) In the near future, the proposed re-entrant interferometer

system configuration with passive SSR is feasible for the development of the MOG with moderate accuracy. This kind of MOG is realistic to be developed, because the elements of the MOG are compatible with those used in IFOG, which are matured industrial products.

2) For developing the MOG as a navigation grade gyro, the active SSR is necessary to be developed in either the interferometer or the resonator system configuration.

References

[1] A W Lawrence. The Micro - Optic Gyro, Symposium Gyro Technology, Stuttgart, Germany (1983).

[2] A Yu, A S Siddiqui. Novel Fiber Optic Gyroscope with a Configuration Combining Sagnac Interferometer with Fiber Ring Resonator, Electronics Lett. Vol. 28, No. 19, pp. 1778 - 1779 (1992).

[3] P Motteir. Integrated Optics at the LETI, International Journal of Optoelectronics, 9 (2), pp. 125 - 134 (1994).

[4] Y S Zhang, B Zhang, X Y Ma. Techniques for Developing a Miniature Resonant Optic Rotation Sensor, Proceedings of the 52nd Annual Meeting, June 19 - 21, Cambridge, MA, pp. 719 - 723 (1996).

[5] H G Ding, Y S Zhang, et al. Key Technologies of Micro Inertial Measurement Unit, Chinese Journal of Scientific Instrument, Vol. 17, No. 1, pp. 31 - 35 (1996).

[6] P Motteir, P Pouteau. Solid State Optical Gyrometer Integrated on Silicon, Electronic Letters, Vol. 33, No. 23, Nov., pp. 1975 - 1977 (1997).

[7] Y S Zhang, X Y Ma, B Zhang, Q A Tang, Z W Pan, Q Tian, A Y Zhang, M Li, M Zhang. Investigation of the Elements for Integrated Optic Gyro, The Second International Symposium on Inertial Technology (BISIT), October, Beijing, pp. 173 - 178 (1998).

[8] L Z Wu, Y S Zhang, H Schweizer. The Feasibility of Compact Gyroscope Realized by Semiconductor Ring Laser, Symposium Gyro Technology, Stuttgart, Germany (1999).

[9] C Ford, R Ramberg, K Johnsen, W Berglund, B Ellerbusch, R Schermer, A Gopinath. Cavity Element for Resonant Micro Optical Gyroscope, IEEE AES Systems Magazine, December, pp. 33–36 (2000).

[10] Y Zhang, F Gao, X Wu, W Tian, Z Hu, Q Tian, Z Pan, Q Tang. Investigation of the Re-entrant Integrated Optical Rotation Sensor, Symposium Gyro Technology, Stuttgart, Germany (2000).

[11] Y Zhang, W Tian, L Fu, H Schweizer. Experimental Research on a Novel Interferometric Fiber Optical Gyro with Light Beams Circulating in the Sagnac Sensing Ring, Symposium Gyro Technology, Stuttgart, Germany (2002).

[12] L Fu, H Schweizer, Y Zhang, L Li, A M Baechle, S. Jochum, G C Bernatz, S Hansmann. Design and Realization of High-Power Ripple-Free Super luminescent Diodes at 1300 nm, IEEE J. of Quantum Electronics, Vol. 40, No. 9, pp. 1270–1274 (2004).

A MEMS Hybrid Inertial Sensor Based on Convection Heat Transfer

(2005年6月)

Abstract This paper presents a micro electromechanical hybrid inertial sensor based on convection heat transfer. The sensor configuration consists of a small silicon etched cavity, a suspended central heater and four suspended thermistors, all of which are packaged in a hermetic chamber. The sensor has similar configuration with known thermal acceleromelers, but is different because it can sense dual axis accelerations and single axis angular rate. The working principle of the sensor, the numerical simulations of the convection flow and a primary experiment results are presented.

Keywords Inertial sensor; Convection heat transfer; Micro electromechanical system (MEMS)

1 Introduction

Most micromachined inertial sensors usually used a mechanical

---

本文发表于 The 13th International Conference on Solid – State Sensors, Actuators and Microsystems。合作作者：Rong Zhu, Yan Su, Henggao Ding。

structure including a solid proof mass attached to springs[1]. For example, several micromachined vibrating gyroscopes have been demonstrated, including vibrating shells[2], tuning-forks[3], and vibrating beams[4].

This paper reports for the first time on the preliminary design and analysis of a novel micromachined angular rate sensor that measures temperature gradients induced by the Coriolis acceleration acting on a gas flow sealed in a small chamber. Novelties also consist that the sensor acts not only as an angular rate sensor, but also as a dual-axis accelerometer. The design incorporates the recent advances in micromechanical thermal accelerometer[5-7] with micromachined vibrating gyroscopes to realize a low-cost, thermo-fluidic hybrid inertial sensor system. Since it does not require a solid proof mass and the gas proof mass imposes much less force on the mechanics of the sensor when accelerating, this inertial system is less prone to malfunction after undergoing high shock and strong vibration and is prospective in applications in harsh environment.

2 Operation and Design

The hybrid sensor system under consideration is based on the interaction of inertial and thermal properties in a laminar, internal chamber flow. The sensor has a similar configuration with a thermal accelerometer, which is comprised of a small silicon etched cavity with the size of 2 mm×1 mm×0.3 mm ($X \times Y \times Z$) shown in Fig. 1. Where X and Y are the central axes of the cavity surface. A suspended central heater placed at $X-Y$ plane of the cavity (named as work plane) heats up and lowers the density of the surrounding gas. Four suspended thermistors oppositely placed at the same work plane measure the local gas flow. The sensor is packaged in a sealed chamber to prevent external air flow from disturbing the sensor operation. When an acceleration is applied along Z - axis and the heater heats up, the gas flow in the region of the hermetic chamber is represented schematically in Fig. 2 (a) and (b). Fig. 2 (b) shows the gas flow at work plane displays opposite symmetrical characteristic.

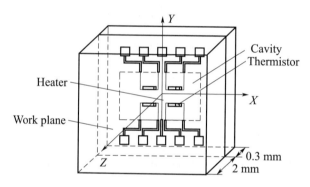

Fig. 1　Schematic view of the sensor configuration in a sealed chamber

The principle of detecting acceleration in X and Y directions based on convection heat transfer has been widely investigated[5-7]. In this paper, we only focus on the angular rate detection.

In work plane, the input angular rate rotation $\vec{\omega}_z$ of the sensor around Z - axis induces Coriolis acceleration $\vec{a}_y$ normal to the direction of flow $\vec{v}_x$

$$\vec{a}_y = 2\vec{\omega}_z \times \vec{v}_x \tag{1}$$

where $\vec{v}_x$ represents the flow velocity in X direction at work plane. This

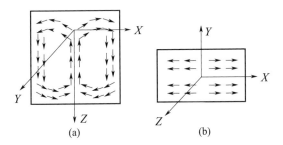

Fig. 2 Schematic view of the gas flow in region of hermetic chamber

Coriolis acceleration moves the opposite flows $\vec{v}_x$ to deflect in the opposite direction of Y as shown in Fig. 3 (a). The deflection causes opposite cooling effects between pair of opposite thermistors (i. e. R_{C1} and R_{C2} or R_{W1} and R_{W2}), and as a result, the resistances of R_{C1} and R_{C2} (also for R_{W1} and R_{W2}) change oppositely. Wheatstone bridge circuit as shown in Fig. 3 (b) converts the two pairs of resistance changes ($R_{C1} - R_{C2}$ and $R_{W1} - R_{W2}$) to the output voltage changes ΔR_C and ΔR_W, respectively. Since the opposite symmetric flows in X direction are deflected in the opposite direction of Y due to Coriolis acceleration. Therefore ΔR_C is proportional to the angular rate (ω_z) of the rotation with respect to the Z axis, whereas ΔR_W is proportional to the inverse angular rate ($-\omega_z$) of the rotation with respect to the Z-axis. The difference δv between ΔR_C and ΔR_W ($\Delta R_C - \Delta R_W = \delta v$) is thus proportional to the angular rate $2\omega_z$. This differential processing by using opposite symmetric flows in X direction can successfully eliminate side-effect of non-Coriolis accelerations in Y direction that move the opposite flows $\vec{v}_x$ to deflect in the same Y direction and are eliminated via $\Delta R_C - \Delta R_W$.

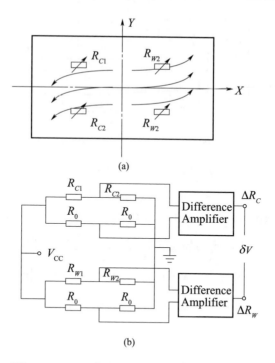

Fig. 3 The opposite flows in X direction are deflected in the opposite directions of Y due to the Coriolis effect, and Wheatstone bridge circuit outputs the angular rate

3 Numerical Simulation and Analysis

A CFD model is developed as following to describe the three-dimensional convection flow in the sensor chamber, which can be solved to obtain the transient velocity, temperature and pressure of gas in the chamber.

$$\frac{\partial \rho}{\partial t} + \nabla \cdot (\rho \vec{v}) = 0 \tag{2}$$

$$\frac{\partial (\rho \vec{v})}{\partial t} + \rho \vec{v} \cdot \nabla (\vec{v}) = -\nabla(p) + \rho \vec{a} + \nabla(\mu \cdot \nabla(\vec{v})) \tag{3}$$

$$\frac{\partial(\rho c T)}{\partial t} + \nabla \cdot (\rho c \vec{v} T) = Q - W + k \cdot \nabla^2(T) - \nabla(p\vec{v}) \quad (4)$$

$$\rho = \frac{p}{R \cdot T} \quad (5)$$

where $\vec{v}$, $\vec{a}$, ρ, p, μ, T, c, k, R, Q, and W refer to the velocity vector, acceleration vector, gas density, pressure, dynamic viscosity, temperature, gas specific heat, thermal conductivity of gas, gas constant, heat generation and wall heat transfer, respectively. The vector operator ∇ is defined as $\nabla \equiv i\frac{\partial}{\partial x} + j\frac{\partial}{\partial y} + k\frac{\partial}{\partial z}$. The wall heat transfer is given in terms of the boundary temperature difference by $W \approx h \cdot (T_S - T_B)$. Where h, T_S and T_B refer to heat transfer coefficient, temperature of gas and temperature of the wall, respectively.

The numerical simulation of the flow velocity was performed by using ANSYS software. Fig. 4 shows the flow velocity profile in X direction at the work plane when 1g (9.8 m/s^2) acceleration is applied along Z-axis of the sensor, where y represents the distance from X-axis. According to the above analysis, the convection flow is due to the local acceleration applied and the temperature difference between the gas and the wall. And the angular rate detection is based on the opposite symmetric flows along X-axis at work plane that are induced by the acceleration in Z direction. However the acceleration in the X direction will perturb the symmetry of gas flow. When 1g acceleration is applied in Z direction and additional acceleration A_x is applied in X direction, the asymmetric flow velocities in X direction is shown in

Fig. 5, which indicates that the flow in X direction is accelerated so as to break the opposite symmetry of the flows at the work plane induced by the acceleration applied in Z direction. Fortunately, we found when we differenced the X velocities between the two opposite flows in X direction, the difference of velocity was almost unchanged and independent of the accelerations applied in X direction. Fig. 6 shows the difference of X velocities between the two opposite flows in X direction at the work plane when the $1g$ acceleration and the additional acceleration A_x are applied along Z - axis and X - axis, respectively. The result reveals that the above differencing approach of detecting angular rate of the rotation using opposite symmetric flows along X - axis at work plane is still valid even when the accelerations in X direction are existent because the measurement of the angular rate is relied on the difference between two opposite symmetric flows at the work plane.

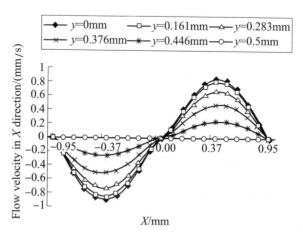

Fig. 4 Flow velocity profile in X direction at work plane

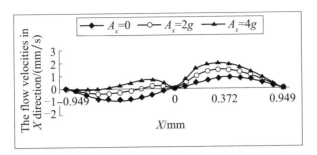

Fig. 5 The flow velocities in X direction at $y=0$ mm when the 1g acceleration is applied along Z - axis and the additional acceleration A_x is applied along X - axis

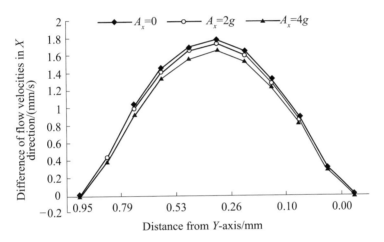

Fig. 6 Difference of X velocities between the two opposite flows in X direction at the work plane when 1g acceleration is applied along Z - axis and the additional acceleration A_x is applied along X - axis

The simulation results also show that the acceleration in Z direction correlates with the X velocity of the flow at the work plane so as to affect the scale factor of the sensor. As a result, the acceleration in Z direction must be known and employed as a parameter to determine the scale factor of the sensor for detecting the angular rate of the rotation around Z axis.

4 Primary Experimental Results

A prototype of the sensor was fabricated based on the same technology used for thermal accelerometers[5]. To test the sensing performance, the prototype was placed on a turntable and connected to a signal amplifier. The acceleration input ranged from －1 g up to ＋1 g was applied to X direction, then to Y direction. The output voltage over the applied acceleration was measured. Fig. 7 presents the results. Then the angular rate input ranges from －600 (°) /s up ＋600 (°) /s was applied around Z‐axis of the sensor, when the gravity acceleration (9.8 m/s²) is along Z‐axis. The output voltage over the applied angular rate was measured as Fig. 8.

5 Conclusion

The concept, the configuration design and flow simulations of a MEMS hybrid inertial sensor based on convection heat transfer are presented. The sensor with a simple configuration can detect dual axis accelerations and single axis angular rate. A primary experiment of prototype is performed to validate the effectiveness of the sensor. The absence of a vibrating proof mass should increase the shock resistance of the thermal sensor, and hence provide greater sensitivity as reported for thermal accelerometers.

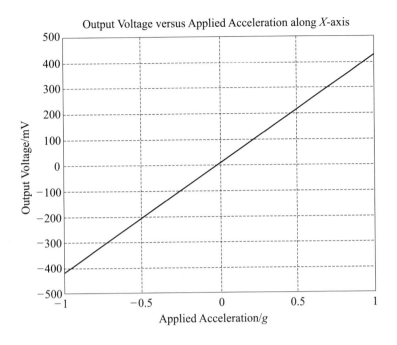

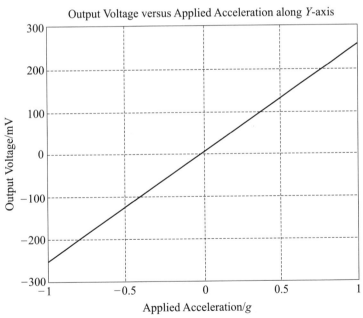

Fig. 7 Experimental results of the prototype for acceleration measurement

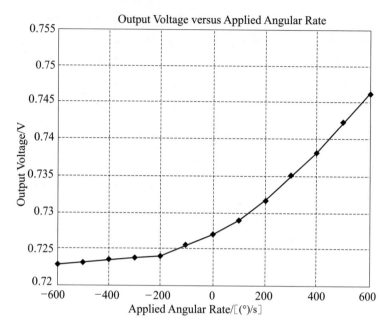

Fig. 8 Experimental results of the prototype for angular rate measurement

References

[1] Y Navid, F Ayazi, K Najafi. Micromachined Inertial Sensors, Proceedings of the IEEE 86 1640 – 1659 (1998).

[2] F Ayazi, K Najafi. A Harpss Polysilicon Vibrating Ring Gyroscope, J. MEMS, 10, 169 – 179 (2001).

[3] J Bernstein, et al. A micromachined comb-drive tuning fork rate gyroscope, Proceedings of the IEEE Micro Electro Mechanical Systems (MEMS), pp. 143 – 148, Fort Lauderdale, FL, USA, Feb. 7 – 10, 1993.

[4] K Maenaka, T Shiozawa. A study of silicon angular rate sensors using anisotropic etching technology, Sensors and Actuators A, 43, 72 – 77 (1994).

[5] A M Leung, J Jones, E Czyzewska, J Chen, and B Woods. Micromachined accelerometer based on convection heat transfer, Proceedings of the IEEE Micro Electro Mechanical

Systems (MEMS), pp. 627 - 630, Heidelberg, Germany, Jan. 25 - 29, 1998.

[6] F Maily, A Martinez. Design of a micromachined thermal accelerometer: Thermal simulation and experimental results, Microelectronics Journal, 344 (2003) .

[7] D Crescini, D Marioli. An inclinometer based on free convective motion of a heated air mass, Sensors for Industry Conference IEEE, pp. 11 - 15, New Orleans, Louisians, USA, Jan. 27 - 29, 2004.

Micromachined Gas Inertial Sensor Based on Convection Heat Transfer

(2006 年 1 月)

Abstract A micromachined gas inertial sensor based on the principle of convection heat transfer is presented in the paper. The configuration of the sensor consists of a small silicon etched cavity, a suspended central heater and four suspended thermistor wires, all of which are assembled and packaged in a hermetic chamber. The sensor has similar configuration with known thermal accelerometers, but is different and novel because it is not only as a dual – axis accelerometer but also as a single – axis gyroscope. Numerical simulations and primary experiments are performed to validate the effectiveness of the sensor.

Keywords　Inertial sensor; Convection heat transfer; Micro electromechanical system (MEMS)

1　Introduction

Most micromachined inertial sensors generally use a mechanical structure including a solid proof mass attached to springs[1]. For example, several micromachined vibrating gyroscopes including vibrating shells[2],

---

本文发表于 Sensors and Actuators A: Physical。合作作者: Rong Zhu, Henggao Ding, Yan Su, Zhaoying Zhou。

tuning – forks[3], and vibrating beams[4] have been demonstrated. Recently a new concept of a thermal accelerometer with no proof mass has been reported. The operating principle of this thermal accelerometer is based on free – convection heat transfer of a central heater in an enclosed chamber[5-9].

This paper reports on a preliminary design and analysis of a novel micromachined angular rate sensor that measures temperature gradients induced by the Coriolis acceleration acting on a gas flow sealed in a small chamber. Novelties also consist that the sensor acts not only as an angular rate sensor, but also as a dual – axis accelerometer. The design incorporates the recent advances in micromechanical thermal accelerometer with micromachined vibrating gyroscopes to realize a low – cost, thermo – fluidic hybrid inertial sensor system. Since it does not require a solid proof mass and the gas proof mass imposes much less force on the mechanics of the sensor when accelerating, this inertial system is less prone to malfunction after undergoing high shock and strong vibration and is prospective in applications in harsh environment.

2 Operating principle and design

The sensor under consideration is based on the interaction of inertial and thermal properties in a laminar, internal chamber flow. The sensor has a similar configuration with known thermal accelerometers, which is comprised of a small silicon etched cavity with a dimension of 2 mm×1 mm×0.3 mm ($X \times Y \times Z$) shown in Fig. 1. X – and Y – axes are central axes of the cavity surface, and Z – axis normal to the cavity surface constructs Cartesian coordinates with X - and Y - axes. A suspended central heater placed in the X - Y Cartesian plane (specified as working plane, abbreviated to WP) heats up and lowers the density of the surrounding gas. Four suspended thermistor wires placed symmetrically on both sides of the heater measure the local gas flow. The cavity, heater and thermistors are assembled and packaged in a sealed chamber with a height of 2.3 mm (Z) to prevent external airflow from disturbing the sensor operation. Note that the cavity needs to be placed on one side of the chamber and the cavity surface (i.e. WP) where the heater and thermistors are located is necessary to be situated at the position where the convection flow has the highest magnitude of the X velocity. This design can help to enhance the sensor's angular rate sensitivity.

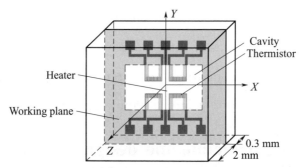

Fig. 1 Schematic view of the sensor configuration in a sealed chamber

The operating principle of detecting accelerations along X - and Y - axes based on asymmetric heat transfer in a sealed chamber has been widely investigated[5-9]. Here we only study the angular rate detection using this thermal sensor.

Inertial forces and temperature differences induce the gas (we used air in this study) sealed in the chamber regularly flowing. For instance, when an acceleration is applied on the reverse direction of the Z - axis and the central heater heats up, the gas flow in the region of the hermetic chamber is depicted in Fig. 2 (a) and (b). The gas flow in the WP is assumed to mainly move along X - axis and be inverse symmetric about Y - axis.

Considering in the working plane, applying an angular rate rotation $\vec{\omega}_z$ around the Z - axis will induce Coriolis acceleration $\vec{a}_c$ normal to the direction of flow $\vec{v}_x$. Where $\vec{v}_x$ represents X fluidic velocity vector in the WP and is inverse symmetric about Y - axis. Combining non - Coriolis accelerations $\vec{a}_y^0$, the acceleration vector $\vec{a}_y$ along the Y - axis is expressed by:

$$\vec{a}_y = \vec{a}_c + \vec{a}_y^0 = 2\vec{\omega}_z \times \vec{v}_x + \vec{a}_y^0 \tag{1}$$

i. e. $a_y = 2\omega_z v_x + a_y^0$ Opp. $a'_y = -2\omega_z v_x + a_y^0$ (2)

where a_y and a'_y respectively represent the accelerations along Y - axis on the two sides of the heater in the WP. It is obvious that the Coriolis accelerations $\vec{a}_c$ on the two sides of the heater point to opposite directions so as to move the flows $\vec{v}_x$ on the two sides of the heater to deflect in opposite directions of Y as shown in Fig. 3 (a). The flow deflection causes opposite cooling effects between the two opposite thermistors in pairs (specifically R_{W1} and R_{W2} forms one pair, R_{C1} and R_{C2} forms another pair), and as a result, the resistances of R_{W1} and R_{W2} (R_{C1} and R_{C2} also) change oppositely. Wheatstone bridge circuit as shown in Fig. 3 (b) converts the resistance changes of the two thermistor pairs $R_{W1} - R_{W2}$ and $R_{C1} - R_{C2}$ to the voltages $V_{\Delta R_W} > V_{\Delta R_C}$, and further gives an output voltage difference δV between them. If the Wheatstone bridge is supplied with a constant voltage, the voltage $V_{\Delta R_W}$ is proportional to $R_{W1} - R_{W2}$. Since the operating principle of this sensor is based on the heat convection it is reasonable to assume that the Grashof number ($Gr = \dfrac{a\rho^2 \beta T_h l^3}{\mu^2}$ with a, acceleration; ρ, gas density; β, gas coefficient of expansion; T_h, heater temperature; l, linear dimension and μ, gas viscosity), which governs the convection heat transfer process has an influence on the sensor. Based on this observation the voltage $V_{\Delta R_W}$ of the Wheatstone bridge is thus proportional to the acceleration a'_y. Similarly the voltage $V_{\Delta R_C}$ is proportional to the acceleration a'_y. The output voltage difference δV can be given by:

$$\delta V = V_{\Delta R_W} - V_{\Delta R_C} \propto a_y - a'_y = 4\omega_z v_x \tag{3}$$

Eq. (3) indicates that the voltage difference δV is proportional to the angular rate ω_z. This differential processing can successfully eliminate the side-effect of non-Coriolis accelerations $\vec{a}_y^0$, which are equal for the flows on both sides of the heater and are subtracted from the output δV via $V_{\Delta R_W} - V_{\Delta R_C}$.

3　Numerical simulation and analysis

A CFD model is developed as following to describe the three-dimensional convection flow in the sensor chamber, which can be solved to obtain the transient velocity, temperature and pressure of gas in the chamber.

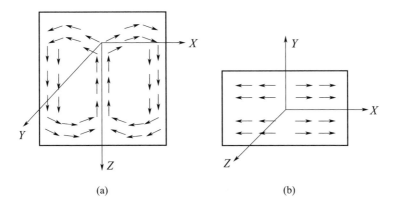

Fig. 2　Schematic view of the gas flow in region of the hermetic chamber

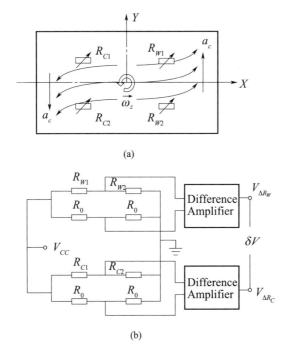

Fig. 3　The inverse symmetric flow in the WP are deflected in opposite directions of Y due to Coriolis effect, and Wheatstone bridge circuit outputs the angular rate

$$\frac{\partial \rho}{\partial t} + \nabla (\rho \vec{v}) = 0 \tag{4}$$

$$\frac{\partial (\rho \vec{v})}{\partial t} + \rho \vec{v} \nabla (\vec{v}) = -\nabla (p) + \rho \vec{a} + \nabla (\mu \nabla (\vec{v})) \tag{5}$$

$$\int_{vol} \left(\frac{\partial (\rho c T)}{\partial t} + \nabla (\rho c \vec{v} T) \right) d(vol) = \int_{s_1} Q d(s_1) - \int_{s_2} W d(s_2) +$$

$$\int_{vol} (k \nabla^2 (T) - \nabla (p \vec{v})) d(vol) \tag{6}$$

$$\rho = \frac{p}{RT} \tag{7}$$

where $\vec{v}$, $\vec{a}$, p, T, c, k, R, Q, and W refer to the velocity vector, acceleration vector, pressure, temperature, gas specific heat, thermal conductivity of gas, gas constant, heat generation and wall heat transfer, respectively. The vector operator ∇ is defined as $\nabla \equiv i \frac{\partial}{\partial x} + j \frac{\partial}{\partial y} + k \frac{\partial}{\partial z}$. Q and W refer to boundary conditions. Q is given in terms of the heat flux dependent on the applied heater power and the heater dimension. Generally thermistors in the sensor dissipate less heat power than the heater, for simplifying calculation we only consider the heater generation in the model. The wall heat transfer is given in terms of the boundary temperature difference by $W = h(T_S - T_B)$. Where h, T_S and T_B refer to film coefficient, temperature of the boundary gas in the chamber and temperature of the wall, respectively; vol refers to the volume of the element, s_1 refers to the surface of the element that specified heat flows acts over, s_2 refers to specified convection surface.

The behavior of the flow velocity in the chamber with dimension of

2 mm×1 mm×2.3 mm ($X \times Y \times Z$) is studied by using the numerical resolution of the CFD model with the commercial ANSYS FLOTRAN. The thermal properties of the gas in the chamber are temperature dependent but are assumed constant in the temperature range considered in order to limit the temperature dependence of the equations and ease up the calculations. The constant model parameters are assumed as $c = 716$ N·m/(kg·K), $k = 0.027$ W/(m·K), $\mu = 18 \times 10^{-6}$ N·s/m^2, $h = 10$ W/(m^2·K), $Q = 5 \times 10^5$ W/m^2. The chamber wall is assumed to be isothermal with $T_B = 293$ K. The gas pressure in the chamber is assumed to be 1 bar. Fig. 4 (a) shows the X fluid velocity profile at the WP for an acceleration of 1 g (9.8 m/s^3) applied on the reverse direction of Z. We can observe the inversion symmetry of the X velocity profile about Y - axis as assumed in the previous section. Compared with the X velocity, the Y velocity of the flow at the WP is weaker and symmetric about Y - axis as shown in Fig. 4 (b). For this simulation, a constant heat power is applied on the heater and it leads to a symmetric temperature profile at the WP presented in Fig. 4 (c), whose maximum temperature is about 680 K for a heater power of 14 mW. The rotational effect to the convection flow is further simulated. Fig. 5 shows the difference in the Y fluid velocity at the WP between the two flows with and without rotation of 600 (°)/s applied about Z - axis. It can be seen that the rotation load around Z moves the flows on the two sides of the heater to deflect in opposite directions of Y that is accordance with the assumption proposed in the previous section. Fig. 6 presents the simulated

quadratic differential temperature $\delta T = \delta T_W - \delta T_C$ among the four symmetrical positions about X- and Y-axes at the WP (the four symmetrical thermistors are assumed to be located at these corresponding positions), according to the X and Y distances for a rotation of 600 (°)/s and a heater power of 14 mW. The quadratic differential temperature δT in response to the rotation around the Z-axis is proportional to $(R_{W1} - R_{W2}) - (R_{C1} - R_{C2})$, and thus can be outputted by δV.

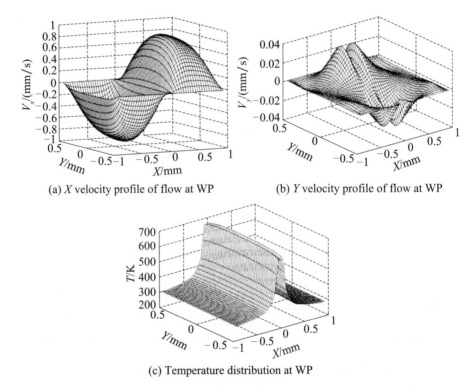

(a) X velocity profile of flow at WP

(b) Y velocity profile of flow at WP

(c) Temperature distribution at WP

Fig. 4 Convection flow velocity profile and temperature distribution at the WP

Note that we have suggested in the previous section that the angular rate detection via differential processing is based on the

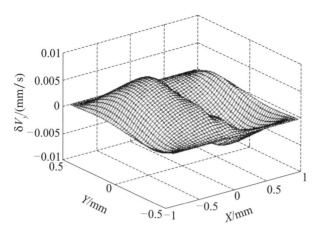

Fig. 5 Y velocity difference at the WP between the two flows with and without rotation of 600 (°)/s

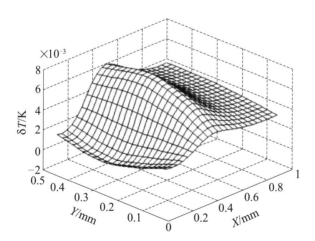

Fig. 6 Simulated quadratic differential temperature δT among the four symmetrical positions about X- and Y-axes at the WP versus the X and Y distances for a rotation of 600 (°)/s and a heater power of 14 mW

inversion symmetry of the flow moving along X-axis in the WP. However accelerations on the X-axis will disturb this inversion symmetry. We simulate the scenario when acceleration A_z is applied on the Z - axis and an additional acceleration A_x is applied on the X - axis. The X fluid velocity profile at the WP is shown in Fig. 7. It is found that the X velocity profile is asymmetric about Y - axis and the asymmetry becomes greater with the magnitude of the acceleration A_x increasing.

According to Eq. (3), the output of the Wheatstone bridge circuit δV is in fact proportional to the differential X fluid velocity between the two pairs of thermistors, i. e.

$$\delta V \propto a_y - a'_y = 2\omega_z (v_x - v'_x) \tag{8}$$

where v_x and v'_x refer to the X fluid velocities at two pairs of thermistors R_{C1}, R_{C2} and R_{W1}, R_{W2}, respectively. We then calculate the differential X velocity Δv_x between two opposite positions located symmetrically about Y - axis at the WP. The Δv_x distributions according to the distance to the heater with different accelerations A_x are shown in Fig. 8, which indicates that this differential X fluid velocity at the WP is independent on A_x, i. e. $v_x - v'_x$ in Eq. (8) is independent of A_x. As a result, the dependence of the sensor response to the rotation ω_z will not be affected by the acceleration of A_x. It reveals that the differential approach by using opposite flows along the X - axis to response the angular rate of the rotation keeps valid even the symmetry of the X fluid velocity is disturbed by the acceleration on the X - axis.

We alslo analyze the dependence of the acceleration applied on the Z - axis to X fluid velocity at the WP. Fig. 9 depicts the simulating results, illustrates that the magnitude of X flow velocity increases with

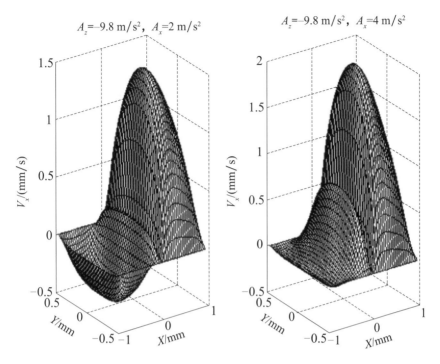

Fig. 7　X fluid velocity profile at the WP is asymmetric when the acceleration A_z is applied on the Z‑axis and the acceleration A_x is applied on the X‑axis

the magnitude of the acceleration A_z. Since the output voltage δV is proportional to the magnitude of the X fluid velocity according to Eq. (3), it reveals that the practical acceleration applied on the Z‑axis determines the sensitivity of the angular rate sensor, therefore must be known in prior or measured by using an accelerometer and regarded as a reference while detecting the rotation around Z‑axis. The sensitivity of the sensor is also found to increase with the heater power.

4　Primary experiment

A prototype of the sensor is fabricated using the same fabrication

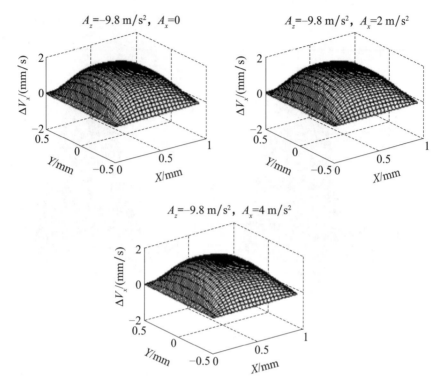

Fig. 8 Simulated differential X fluid velocity ΔV_x at the WP for an acceleration of A_z applied on Z and the additional acceleration of A_x applied on X

techniques of thermal accelerometers that have been already detailed in [8]. Platinum resistors are used as the heater and thermistors. The four thermistors have the resistances at room temperature of about 250 Ohm. The operating heater power is about 14 mW with the current of 8 mA. The supplied current to each thermistor in the Wheatstone bridge is 2 mA.

The sensitivity of dual – axis accelerometer is characterized by using earth's gravity as a reference. The acceleration in the range of $\pm g$ is applied on the X – and Y – axes of the sensor, respectively. Note

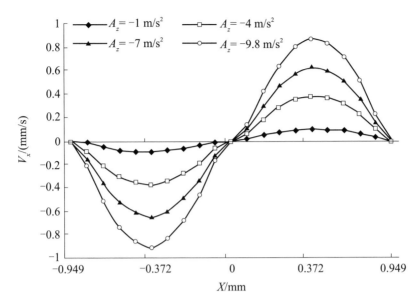

Fig. 9 X fluid velocity (at $y=0$) in the WP with the different acceleration of A_z

that the Wheatstone bridge circuit shown in Fig. 3 (b) needs to be modified for measuring the acelerations. When measuring the acceleration along the X - axis, the circuit is adjusted to get the output voltage δV proportional to $(R_{W1}+R_{W2})-(R_{C1}+R_{C2})$. While for measuring the acceleration along Y - axis, the circuit is adjusted to get the output voltage δV proportional to $(R_{W1}-R_{W2})+(R_{C1}-R_{C2})$. The output voltages of the sensor in response to the applied accelerations with an amplification of about 1 500 are shown in Fig. 10. Note that the offsets of the outputs have been balanced. As a result, the electric sensitivities for measuring X and Y acceleations are about 430 and 256 mV/g, respectively.

To test the performance of the angular rate detecting, the prototype is mounted on a turntable. The Z - axis of the sensor is aligned vertically so

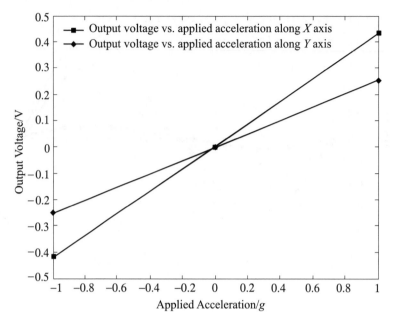

Fig. 10 Ouput voltage of the sensor vs. accelerations applied on the X - and Y - axes, respectively

that the earth's gravity acceleration is applied on the Z - axis of the sensor. The angular rate input ranges from −600 (°) /s up to +600 (°) /s is applied around Z - axis of the sensor. The output voltage over the applied angular rate with an amplification of about 36 000 is shown in Fig. 11. Note that the offset of the output signal has been balanced. It reveals that the sensitivity of this angular rate sensor is about 20 μV/ (°) /s for $A_z = 1$ g.

The dependency of the rotation sensitivity with the acceleration of A_z can be calibrated by using a dual - axis rotating table, as illustrated in Fig. 12. The table has two orthogonal rotating axes, i. e. X - and Z -axes, and the X - axis stands upright. The sensor is eccentrically

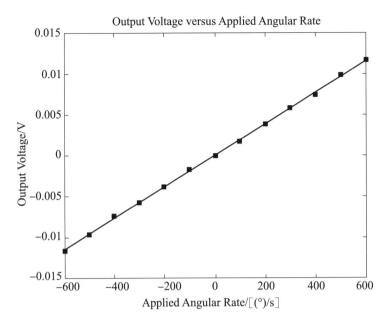

Fig. 11 Ouput voltage of the sensor vs. the angular rate applied around the Z-axis

placed on the Z-axis of the table and the Z-axis of the sensor is aligned with the Z-axis of the table. A rotation ω_1 around X-axis induces a centrifugal acceleration on the Z-axis of the sensor that is regarded as A_z. The rotation ω_2 around the Z-axis of the table acts as the input of the angular rate sensor. Therefore, by adjusting ω_1 and ω_2, we can obtain the dependence of the sensor output with the centrifugal acceleration A_z and the input angular rate that are both along the sensitive axis of the angular rate sensor.

By the above experiments, the sensor is proven to be tridimensional: the sensitive axes of the accelerometer are the X- and Y-axes and the sensitive axis of the angular rate sensor is the Z-axis. In virtue of the novel design, the three sensitive axes of the sensor possess a good orthogonality.

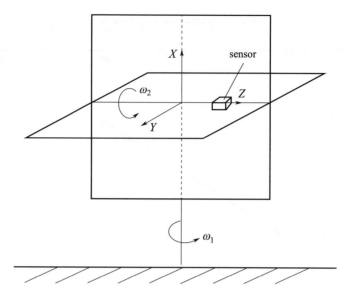

Fig. 12　Calibration system for the angular rate sensor

5　Conclusion

The concept, configuration design, flow simulation and primary experiments of a micromachined gas inertial sensor based on convection heat transfer are presented. The sensor with a similar configuration of known thermal accelerometers can detect applied accelerations in dual‐axis (i. e. X and Y) and can also detect an angular rate around another single‐axis (i. e. Z) if the acceleration applied on the Z‐axis is not null. The absence of a vibrating proof mass obviously increase the shock resistance of this thermal sensor, and hence provide greater reliability as reported for thermal accelerometers.

References

[1] Y Navid, F Ayazi, K Najafi. Micromachined inertial sensors, Proc. IEEE 86 (1998) 1640–1659.

[2] F Ayazi, K Najafi. A Harpss polysilicon vibrating ring gyroscope, J. MEMS 10 (2001) 169–179.

[3] J Bernstein. A micromachined comb-drive tuning fork rate gyroscope, in: Proceedings of the IEEE Micro Electro Mechanical Systems (MEMS), Fort Lauderdale, FL, USA, February 7–10, 1993, pp. 143–148.

[4] K Maenaka, T Shiozawa. A study of silicon angular rate sensors using anisotropic etching technology, Sens. Actuators A 43 (1994) 72–77.

[5] A M Leung, J Jones, E Czyzewska, J Chen, B Woods. Micromachined accelerometer based on convection heat transfer, in: Proceedings of the IEEE Micro Electro Mechanical Systems (MEMS), Heidelberg, Germany, January 25–29, 1998, pp. 627–630.

[6] F Mailly, A Martinez, A Giani, F Pascal-Delannoy, A Boyer. Design of a micromachined thermal accelerometer: thermal simulation and experimental results, Microelectr. J. 34 (4) (2003) 275–280.

[7] D Crescini, D Marioli. An inclinometer based on free convective motion of a heated air mass, in: Sensors for Industry Conference IEEE, New Orleans, Louisians, USA, January 27–29, 2004, pp. 11–15.

[8] F Mailly, A Giani, A Martinez, R Bonnot, P Temple-Boyer, A Boyer. Micromachined thermal accelerometer, Sens. Actuators A 103 (2003) 359–363.

[9] U A Dauderstädt, P J French, P M Sarro. Temperature dependence and drift of a thermal accelerometer, Sens. Actuators A 66 (1998) 244–249.

静电悬浮微机械加速度计设计

(2007年2月)

摘 要 设计了一种基于静电悬浮原理的微机械加速度计。它采用三明治结构的微敏感元件,电容式位移检测方案及静电力悬浮控制方案。着重对敏感元件的主要结构设计、参数的选取、支承静电力的计算以及量程极限、精度极限等问题进行分析,表明在目前技术条件下,可以实现量程约 $5\ g$,精度为 $10^{-5}g$ 的悬浮微加速度计。设计并加工出一种圆环形结构的敏感元件,已实现了 Z 轴方向的静电悬浮及加速度测量,证明设计方案是可行的,进一步研究可实现三轴悬浮加速度计。

关键词 导航设备;微机械;静电悬浮;加速度计;量程

静电悬浮的微机械加速度计结合了静电悬浮技术与 MEMS 技术的优势,具有灵敏度高、温度系数小、性能调整方便、三轴集成、可微小型化等特点。若对静电悬浮状态的惯性质量块进行加转,还可构成两轴转动陀螺,成为高度集成的加速度计/陀螺仪,使惯性组合更趋微型化。目前国外已有关于静电悬浮单轴加速度计和加速度计/陀螺仪的研究报道,研制出了具有一定性能的原理样机[1-5]。

本文将探讨静电悬浮三轴微机械加速度计的工作原理、性能计算和敏感结构参数设计。

本文发表于《清华大学学报(自然科学版)》2007年第47卷第2期。合作作者:刘云峰,丁衡高,董景新。

1 工作原理

静电悬浮微加速度计实现加速度检测的关键在于敏感质量的受控悬浮——通过闭环控制使悬浮质量块在静电力作用下稳定悬浮，始终保持其与固定电极的相对位置不变。通过控制电压的调节，悬浮质量块处于力平衡和力矩平衡状态。

实现静电悬浮须对质量块进行 5 自由度位置控制，分别是 X、Y、Z 方向平动，绕 X、Y 轴转动。静电悬浮的原理如图 1 所示，图中只示意了在 3 个自由度上的控制电极，即沿 Y、Z 方向平动及绕 X 轴转动，另两个自由度的情况类似。

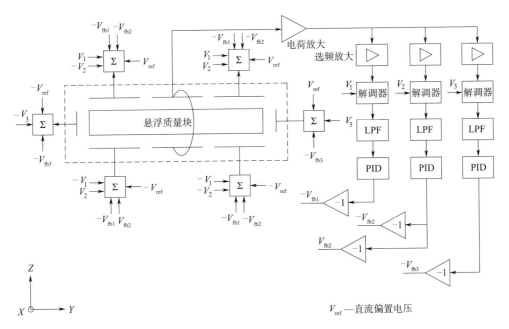

图 1 静电悬浮原理示意图

1.1 位移检测原理

如图 1 所示，由固定的控制电极与悬浮质量块构成若干对差动式可变电容器。依靠它们构成多个电容电桥，悬浮质量块在某个自由度上的位移将引起相应电容电桥的差动变化，检测这一微电容变化可以反映出在该自由度上的位移。为了同时检测多自由度上的位移，采用多路频分复用的方法：对不同自由度上的控制电极对加不同频率的高频载波信号，由公共电容极板输出多路混合的调制信号，对其进行选频放大、同频解调就可以获得悬浮质量块在相应自由度上的位移信息。

1.2 力反馈控制原理

为了对悬浮质量块施以差动的静电反馈力，须给控制电极预加一定的直流偏置电压，将检测到的位移信号送给控制器经校正后，得到的控制电压信号按负反馈原则作用到相应自由度的控制电极，与预载电压叠加，产生的静电反馈力（矩）使质量块回到检测零位，实现悬浮控制。控制电压的大小反映了悬浮质量块在该自由度上所受惯性力（矩）的大小，加速度得以测量。

1.3 悬浮质量块的虚地电位

实现稳定的静电悬浮的重要前提是保证悬浮质量块的电位恒定。这样才能通过改变控制电极的电位来调节对悬浮质量块所施静电合力（矩）的大小和方向。由于悬浮状态下质量块与外界没有连接，又处于由控制电极构成的复杂静电场中，要保持其电位恒定，需采取特殊措施。通过对控制电极的合理拆分，将质量块置于零电位是可行的方案[3,4]。

如图 2 所示，将图 2（a）中一对电极的每边拆分为图 2（b）所示的面积严格相等的两块或两组分电极，分别施加大小相等、极性相反的直流电压，可以保证处于电极之间的孤立导体始终保持零电位，不随位置移动而改变。图 3 示出了 V_1、V_2 取 6 V，悬浮质量块处于上、下控制极板间不同位置时，板内空间电势分布的有限元仿真结果，图中两条白线之间的区域为悬浮质量块，结果显示质量块保持在零电位。计算还表明 V_1、V_2 的取值不影响上述结论。由于悬浮质量块并无实质上的接地，所以视其为虚地电位。

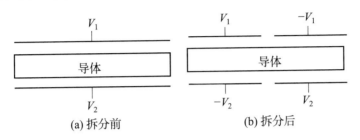

图 2　电极拆分示意图

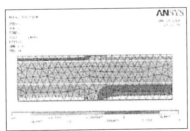

(a) 悬浮质量块向上偏移

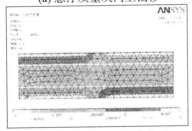

(b) 悬浮质量块处于平衡位置

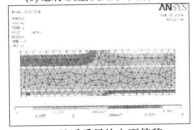

(c) 悬浮质量块向下偏移

图 3　极板间的电势分布仿真

2 性能分析

2.1 最大静电支承力与满量程

微敏感器件的结构形式按悬浮质量块的平面形状可分为圆盘形、圆环形、矩形、矩形环等，考虑到可转动性，则只有圆盘形和圆环形两种满足要求。以下性能估算只针对这两种结构。图 4 所示为敏感结构模型。设悬浮质量块质量为 m，厚度为 h，轴向（Z 向）上与控制电极的间隙为 d_A，径向上与控制电极的间隙为 d_r，悬浮质量块在 XY 平面上的投影面积为 A，轴向控制电极面积与 A 的比值为 n，单块径向控制电极所占弧角为 α。

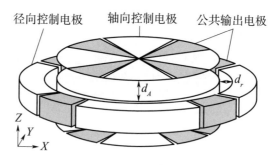

图 4 敏感结构模型

2.1.1 轴向（Z 轴）

在 Z 轴方向，设单侧控制电极与悬浮质量块构成的平板电容器的电容为 C，控制电极上施加的电压为 V，则单侧控制电极对质量块所产生的静电力为

$$F_1 = \frac{\partial W}{\partial d_A} = \frac{V^2}{2} \frac{\partial C}{\partial d_A} = -\frac{V^2}{2} \frac{\varepsilon \varepsilon_0 n A}{d_A^2} \tag{1}$$

其中，$W = V^2 C / 2$，为控制电极与质量块所构成平板电容的电场能；ε 和 ε_0 分别为空气的相对介电常数和真空介电常数。

采用两侧电极差动加力，闭环反馈控制，预载电压为 V_{ref}，反馈电压为 V_{fb}，当质量块有微小轴向位移 Δd 时，根据式（1）可以得总的静电力为

$$F = -\frac{\varepsilon\varepsilon_0 nA\,(V_{ref}-V_{fb})^2}{2\,(d_A-\Delta d)^2} + \frac{\varepsilon\varepsilon_o nA\,(V_{ref}+V_{fb})^2}{2\,(d_A+\Delta d)^2}$$

$$= \frac{2\varepsilon\varepsilon_0 nA(d_A^2+\Delta d^2)V_{ref}}{(d_A^2-\Delta d^2)^2}V_{fb} - \frac{2\varepsilon\varepsilon_0 nA(V_{ref}^2+V_{fb}^2)d_A}{(d_A^2-\Delta d^2)^2}\Delta d$$

(2)

闭环工作时，$\Delta d \approx 0$，且 $V_{fb} \leqslant V_{ref}$，故最大轴向静电力为

$$F_{max} = \frac{2\varepsilon\varepsilon_0 nA\,V_{ref}^2}{d_A^2} \tag{3}$$

最大可平衡 Z 轴加速度为

$$a_{z\text{-}max} = \frac{F_{max}}{m} = \frac{2\varepsilon\varepsilon_0 n}{\rho h}\left(\frac{V_{ref}}{d_A}\right)^2 < \frac{\varepsilon\varepsilon_0 n}{2\rho h}\left(\frac{V_b}{d_A}\right)^2 \tag{4}$$

其中，ρ 为硅质量块密度；V_b 为极板间静电击穿电压；预载电压须满足 $V_{ref} < V_b/2$。

由式（4）可知，Z 轴量程极限值与悬浮质量块的平面形状及大小无关，只取决于 h、n 和 V_b/d_A。

在目前的微加工工艺条件下，d_A 通常可以取 3~5 μm，h 的选取由深刻蚀工艺的刻蚀深宽比及径向电极间隙 d_r 取值来确定。深宽比一般小于 40，d_r 若取 3~7 μm，则 h 的取值范围在 200 μm 之内，V_b/d_A 的大小与极板间的真空度有关，由此估算出在不同结构参数、不同真空度下 Z 轴量程极限值如表 1 所示。表中斜体部分数据为现有技术条件下的推荐选值。

表1 不同条件下 Z 轴的量程极限 $a_{z\text{-max}}$

(单位:g)

$V_b \cdot d_A^{-1}/(\text{V} \cdot \mu\text{m}^{-1})$	$h/\mu\text{m}$			
	50	100	150	200
>3(1标准大气压)	17.4	8.7	5.8	4.4
>6(0.133 Pa)	69.8	34.9	23.3	17.4
>15(1.33 μPa)	436.2	218.1	145.4	109.1

注:n 取 0.5。

2.1.2 径向(X、Y 轴)

以 X 轴为例进行估算,Y 轴情况完全相同。如图 5 所示,取半径为 r 的圆盘形悬浮质量块外缘上一微弧段 $r\mathrm{d}\theta$,微弧段所对应的径向电极面积为 $\mathrm{d}A = hr\mathrm{d}\theta$,所产生的静电力在 X 轴上的分量为

$$\mathrm{d}F_1 = -\frac{\varepsilon\varepsilon_0 V^2 hr\cos\theta}{2d_r^2}\mathrm{d}\theta \tag{5}$$

已知 X 轴单侧控制电极所占弧角为 α,则可以得出单侧电极产生的静电力在 X 轴上的分量为

$$F_{X_1} = \int_{-\frac{\alpha}{2}}^{\frac{\alpha}{2}} \mathrm{d}F_1 = \frac{\varepsilon\varepsilon_0 V^2 hr\sin(\alpha/2)}{d_r^2} \tag{6}$$

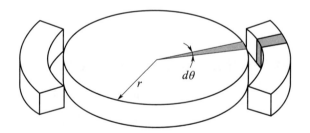

图 5 径向静电力计算模型

当质量块在 X 向发生微小位移 Δd 时,两侧差动控制电极的静电反馈力为

$$F \approx \frac{4\varepsilon\varepsilon_0 hr\, V_{ref} V_{fb} \sin(\alpha/2)}{d_r^2} - \frac{4\varepsilon\varepsilon_0 hr\,(V_{ref}^2 + V_{fb}^2)\sin(\alpha/2)}{d_r^3}\Delta d \quad (7)$$

闭环工作时，$\Delta d \approx 0$，且 $V_{fb} \leqslant V_{ref}$，故 X 轴的最大静电反馈力为

$$F_{max} = \frac{4\varepsilon\varepsilon_0 hr\, V_{ref}^2 \sin(\alpha/2)}{d_r^2} \quad (8)$$

最大可平衡 X 轴加速度为

$$a_{max} = \frac{F_{max}}{m} = \frac{4\varepsilon\varepsilon_0 \sin(\alpha/2)}{\rho\pi r}\left(\frac{V_{ref}}{d_r}\right)^2 < \frac{\varepsilon\varepsilon_0 \sin(\alpha/2)}{\rho\pi r}\left(\frac{V_b}{d_r}\right)^2 \quad (9)$$

对圆环形悬浮质量块，如果环的内、外侧均有控制电极，则 X 轴量程为

$$a_{max} = \frac{4\varepsilon\varepsilon_0 \sin(\alpha/2)}{\rho\pi(r-r_i)}\left(\frac{V_{ref}}{d_r}\right)^2 < \frac{\varepsilon\varepsilon_0 \sin(\alpha/2)}{\rho\pi(r-r_i)}\left(\frac{V_b}{d_r}\right)^2 \quad (10)$$

其中，r、r_i 分别为悬浮转子的内外半径。

由式（9）知，若质量块为圆盘形，则 X（或 Y）轴量程极限取决于结构参数 r、α 和 V_b/d_r。式（10）显示，对于圆环形结构，最大量程受环的宽度 $r-r_i$ 的影响，而与圆环的平面尺寸大小无关。

受深刻蚀工艺的限制，径向电极间隙 d_r 须取 2 μm 以上，并综合考虑深度 h 的取值及刻蚀深宽比能力进行确定。表 2 列出了不同结构参数，不同真空度条件下 X（或 Y）轴最大量程估算值。

表 2 不同条件下 X（或 Y）轴量程极限 a_{max}

（单位：g）

$V_b \cdot d_r^{-1}/(V \cdot \mu m^{-1})$	(r/mm, r_i/mm)					
	(1,0)	(1,0.7)	(1,0.9)	(2,0)	(2,1.7)	(2,1.9)
>3（1 标准大气压）	0.56	1.85	5.55	0.28	1.85	5.55
>6（0.133 Pa）	2.22	7.41	22.22	1.11	7.41	22.22
>15（1.33 μPa）	13.89	46.28	138.85	6.94	46.28	138.85

注：α 取 $\pi/3$。

由表1、表2可看出，在大气环境下，实现小量程（小于5g）的三轴静电悬浮微加速度计是可行的；若提高真空度，则可以通过提高预载电压来加大量程，但需较高压的直流电源，这对整体微小化集成不利，同时，真空封装也是一个需解决的关键问题。

基于以上理论计算的结果，针对小量程应用目标，设计了一种圆环形结构的悬浮加速度计结构（图6），主要参数见表3。

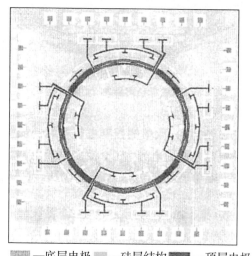

▨—底层电极 ▨—硅层结构 ▨—顶层电极

图6 悬浮加速度计敏感结构图

2.2 灵敏度与精度极限

由于微敏感元件的质量块没有机械支承结构，消除了机械弹力的影响，理想状况下敏感结构的微电容变化对输入加速度的灵敏度为无限高，即 $S = \Delta C/a = \infty$，这是静电悬浮加速度计的最大优点。

敏感元件的灵敏度高有利于提高加速度计的精度，但静电悬浮微加速度计的精度极限取决于敏感元件的热噪声和检测电路的电噪声水平[6]。

表3 悬浮加速度计敏感结构主要设计参数

r/mm	r_i/mm	h/μm	d_A/μm	d_r/μm	V_{ref}/V	n	α	各轴量程/g		
								X轴	Y轴	Z轴
1.750	1.600	200	5	7	7	0.5	$\pi/3$	1.64	1.64	5.06

对于静电悬浮的敏感元件，热噪声体现在悬浮质量块周围气体分子的热运动对其带来的扰动上，如果对敏感元件进行真空封装，减少气体分子数目，则热噪声水平将大为降低，不成为限制加速度计精度的主要因素。

电路噪声是通过影响位移（电容）检测电路的精度来制约传感器整体精度的。受电路噪声的影响，位移检测电路对电容变化的分辨率有限，决定了位移检测的分辨能力。悬浮时，质量块的位置误差最大为检测电路的位移分辨率值，此位置误差将造成悬浮质量块与两侧差动加力极板的间距有微小差异，影响了静电合力的大小，在其他外力不变情况下，质量块仍会偏移直至被检测到，再通过反馈回路调整控制电压将质量块拉回位移检测的零位附近，这种控制电压的改变并不是由输入加速度改变引起的，而是由各种扰动引起的微小位移未被位移检测电路敏感到所致。控制电压波动的范围可由式（2）计算出，它的大小决定了静电悬浮加速度计的精度。以表3中的敏感元件参数为例，目前电容检测电路的分辨能力为 2×10^{-17} F，则检测电路的位移分辨能力为 2.68×10^{-11} m，相应地引起 Z 轴输出波动量为 $3.21\times10^{-5}g$，故加速度计的精度低于 $3.21\times10^{-5}\ g$。要提高精度，必须通过降低电路噪声来提高位移（电容）检测电路的精度。如果电容检测电路的分辨能力达到 1 aF 级，则悬浮加速度计的分辨率可达到 $10^{-6}g$。

3 结语

基于前述工作原理及计算理论，设计了如图6所示的悬浮微加速度计敏感器件，并成功流片，目前已实现了 Z 轴悬浮和加速度检测，证明前述设计思路的可行性。进一步的研究工作，则是希望完成三轴静电悬浮微加速度计。

参 考 文 献

[1] Fukatsu K, Murakoshi T, Esashi M. Electrostatically levitated micro motor for inertia measurement system [C] // Proc Transducer'99. Sendai, 1999: 1558 - 1561.

[2] Murakoshi T, Endo Y, Fukatsu K, et al. Electrostatically levitated ring - shaped rotational - gyro/accelerometer [J]. Japanese Journal of Applied Physics, 2003, 42 (4B): 2468 - 2472.

[3] Houlihan R, Kraft M. Modelling of an accelerometer based on a levitated proof mass [J]. Journal of Micromechanics and Microengineering, 2002, 12 (4): 495 - 503.

[4] Toda R, Takeda N, Murakoshi T, et al. Electrostatically levitated spherical 3 - axis accelerometer [C] // Proceedings of the IEEE Micro Electro Mechanical Systems. Las Vegas: IEEE, 2002: 710 - 713.

[5] Kraft M, Farooqui M, Evans A. Modelling and design of an electrostatically levitated disc for inertial sensing applications [J]. Journal of Micromechanics and Microengineering, 2001, 11: 423 - 427.

[6] 高钟毓, 赵长德, 张嵘, 等. 微机械加速度计的研究 [J]. 清华大学学报（自然科学版），1998, 38 (11): 4 - 8.

Design for Electrostatically Levitated Micromechanical Accelerometer

LIU Yun-feng, DING Heng-gao, DONG Jing-xin

(Department of Precision Instruments and Mechanology,
Tsinghua University, Beijing 100084, China)

Abstract A micromechanical accelerometer was designed based on the principle of electrostatic levitation. The device uses a sandwich micro sensor, capacitive position detection, and electrostatic levitation control. The sensor design is described along with the span of the main design parameters, calculations of the electrostatic force, and accuracy limitations. The analysis shows that a levitated accelerometer can be designed with a 5g dynamic range and $10^{-5}g$ accuracy. One ring-shaped sensing unit was designed and fabricated to verify levitation and acceleration detection along one axis. The design scheme will also be used to build a 3-axis levitated accelerometer.

Keywords Navigation equipment; Micromechanical; Electrostatic levitation; Accelerometer; Dynamic range

A Study of Cross – axis Effect for Micromachined Thermal Gas Inertial Sensor

(2007 年)

Abstract　Micromachined thermal gas inertial sensors are novel devices that take advantages of simple configuration, high shock resistance and good reliability in virtue of using gaseous medium instead of mechanical proof mass as key moving and sensing elements. Because of complexity of gas flow movements, the gas inertial sensor generally undergoes a cross – axis problem. In this paper a study of cross – axis sensitivity on the thermal gas inertial sensor based on convection heat transfer is reported. The cross – axis sensitivity of the sensor is induced from the multidimensional coupling movement of the convection flow in the sensor chamber and can be reduced by a tailored structural design. Combining more than two sensors instead of using a particular structure design to build an integrated sensor system incorporating with a fusion methodology is proposed in this paper, which is an effective way to overcome the cross – axis effect and realize decoupling the measurements of the angular – rates around cross – axes.

　　本文发表于 IEEE SENSORS 2007 Conference。合作作者：Rong Zhu, Henggao Ding, Yan Su, Yongjun Yang。

1 Introduction

Thermal gas accelerometer[1] and gyroscope[2] with no proof mass have been demonstrated recently, which take advantages of simple configuration, easy to be fabricated, high shock resistance and good reliability comparing with the conventional inertial sensors. A low-cost, thermo-fluidic micromachined inertial sensor, the configuration of which consists of a small silicon etched cavity, a suspended central heater that heats up and lowers the density of the surrounding gas, and four suspended detecting thermistor wires symmetrically placed on both sides of the heater, all of which are assembled and packaged in a hermetic chamber, was reported by our research group previously[2]. Based on gas thermal convection, this sensor can detect single-axis rotation and dual-axis accelerations[2]. In this paper, we only consider the measurement on the rotation using this thermal inertial sensor.

Like most sensors, such as accelerometers[3] and magnetometers[4], this thermal gas inertial sensor undergoes the problem of the cross-axis effect, which generally arouses serious errors in measurements. In this paper we present an analysis to the cross-axis effect of the sensor by performing a numerical simulation of a three dimensional convection flows in the sensor chamber. Consequently the principle and the affecting factors of the cross-axis sensitivity of the sensor are given. A methodology to overcome the cross-axis effect by integrating more than two sensors to form a compensation system for decoupling cross sensitivity is proposed as well. The methodology can mainly resolve the cross problem and retain the good sensitivities for the thermal gas inertial sensor.

2 Operation and Design

Our homemade prototype of the sensor, the schematic view of the sensor configuration and the signal processing circuit for detecting rotation around Z axis are shown in Fig. 1 (a), (b) and (c). The prototype of the sensor was fabricated using the known fabrication techniques of thermal accelerometers that have been already detailed in [1]. Platinum resistors are used as the heater and thermistors. Air is used as the gaseous medium sealed in the chamber. Define $X \times Y \times Z$ as the coordinate frame of the sensor, where X and Y are the central axes of the cavity surface (define as working plane), and Z is the normal axis pointing from the working plane of the sensor to the outside. The operating principle of the sensor is based on the interaction of inertial and thermal properties in a laminar, internal chamber flow, which has been described detailedly in [2]. The wheatstone bridge circuit shown in Fig. 1 (c) is utilized to acquire the output of the sensor (i. e. rotation around Z), which is proportional to the difference of the resistances among four thermistors and dependent on the opposite symmetric flows in the sensor chamber[2].

The gas flow in the chamber generally behaves as complex multidimensional movements. The multidirectional flows concur thus induces cross-axis effect on the sensor. From experiments, we found that either rotation around Z and X axes of the sensor may induce an

A Study of Cross-axis Effect for Micromachined Thermal Gas Inertial Sensor

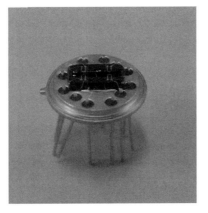

(a) Homemade sensor prototype without package

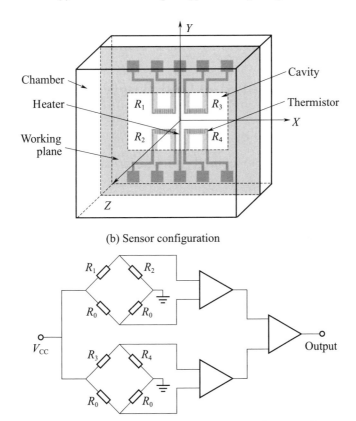

(b) Sensor configuration

(c) Signal processing circuit for detecting rotation around Z

Fig. 1 Homemade prototype of the sensor and schematic view of the sensor configuration and signal processing circuit for detecting angular rate

output response in the electric circuit shown in Fig. 1 (c). As a result the rotation around X induced cross-axis effect on the angular rate sensor whose sensitive axis was supposed to be Z. However the rotation around Y was tested to have little cross effect on the sensor output. The output voltages of the sensor prototype (i.e. the outputs of the electric circuit in Fig. 1 (c)) versus the angular rates applied around X and Z axes respectively while an acceleration of $9.8 m/s^2$ is applied along the corresponding rotation axis are shown in Fig. 2 (a) and (b). It can be seen also that the output of the sensor is almost linear with the applied angular rate, around either of X and Z axes.

3 Numerical Simulation and Analysis

To study the behaviour of the cross-axis sensitivity of the thermal inertial sensor, we performed a numerical simulation to three dimensional convection flows in the sensor chamber by using ANSYS FLOTRAN.

Fig. 3 illustrates the simulated flow vectors in the working plane under an acceleration of $9.8 m/s^2$ applied along X and Z axes respectively. We can see that the orientation of the applied acceleration mainly dominates the gas flow behavior in the chamber. When the acceleration is applied along X, the flow in the working plane is moved along X axis with opposite flow components along Z on two sides of the heater. The Coriolis force derived from a rotation around X will deflect these opposite flows along Z to the opposite directions of Y, which causes opposite cooling effect between two opposite thermistors along Y, and

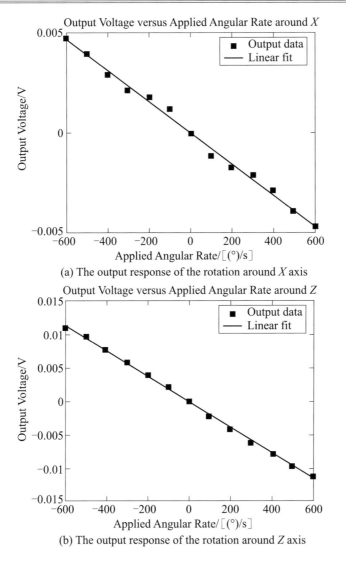

Fig. 2 Output voltages of sensor prototype versus angular rates applied around X and Z axes respectively

as a result, outputs a signal by the Wheatstone bridge circuit in Fig. 1 (c). While the acceleration is applied along Z axis, the flow in the working plane mainly moves along X and be inverse symmetric about Y. The rotation around Z induces opposite Coriolis acceleration along Y

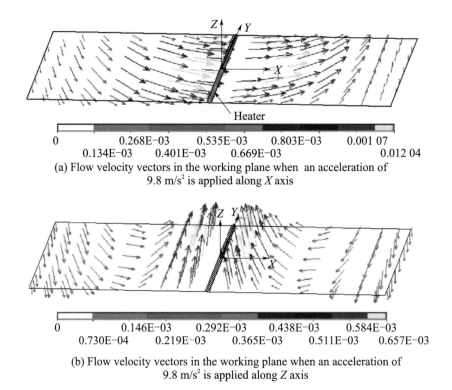

(a) Flow velocity vectors in the working plane when an acceleration of 9.8 m/s² is applied along X axis

(b) Flow velocity vectors in the working plane when an acceleration of 9.8 m/s² is applied along Z axis

Fig. 3　Simulated flow vectors under an acceleration of 9.8 m/s² applied along X and Z axes respectively

between the two sides of the heater so as to move the flows to deflect in opposite directions of Y, which also produces an output signal of the circuit. As a result, either rotation along Z and X axes can induce opposite Coriolos acceleration along Y when the acceleration component along the rotation axis is not null, thus brings an output via the electric circuit. These gyroscope signals at two axes are coupling together thus lead to a cross-axis sensitivity between X and Z axes. The seriousness of the cross-axis sensitivity is dependent on the acceleration magnitude along the other axis other than the detected axis. Since the heater is placed along Y axis and the dimension of the cavity in Y direction has been designed to be much smaller than the dimension in the X direction, the acceleration along Y axis will not cause an effective Coriolos acceleration along Y or X thus bring little cross effect on the sensor, which is in accordance with the measurement result.

4　A Way To Overcome Cross – Axis Effect

The way to diminish the cross – axis effect of the sensor can be carried out by reconstructing the chamber and recomposing the placement of the heater and the thermistors. But in this paper, we present another way to overcome the cross – axis problem by performing a data fusion methodology to an integrated measurement system that is comprised of more than two thermal sensors placed in normal orthogonal directions. One example with two sensors is schematically shown in Fig. 4. Assume that both of the two sensors are the same in structure with the configuration in Fig. 1, but are placed in the orthogonal directions, specifically the Z_1 and X_1 axes of the Sensor 1 are parallelized with the X_2 and Z_2 axes of the Sensor 2, respectively. We also define a referencing orthogonal coordinate system $X - Y - Z$, where X is parallelized with X_1, Y is parallelized with Y_1, and Z is parallelized with Z_1.

Based on the working principle of the sensor, the scale factor of the sensor in detecting a rotation along one certain axis is related with the acceleration component along this axis and the structural orientation. For instance, the scale factor of the Sensor 1 along the Z axis is denoted as

$a_{z1} \cdot k_{z1}$, where a_{z1} refers to the acceleration along the Z_1 axis, k_{z1} refers to the structural factor of the sensor along the Z_1 axis. In the same way, the scale factor of the Sensor 1 along the X axis is denoted as $a_{x1} \cdot k_{x1}$, and the scale factors of the Sensor 2 along Z and X axes are $a_{z2} \cdot k_{z2}$ and $a_{x2} \cdot k_{x2}$ respectively. It must be noted that the Z_1 and X_1 axes of the Sensor 1 are parallelized with the X_2 and Z_2 axes of the Sensor 2, and the two sensors are the same in structure except the placement of the orientation, thus we have

$$\begin{aligned} a_{z1} &= a_{x2} = a_z \\ a_{x1} &= a_{z2} = a_x \\ k_{z1} &= k_{z2} = k_z \\ k_{x1} &= k_{x2} = k_x \end{aligned} \quad (1)$$

Then we obtain the relationship equation as follows for the two sensors

$$\begin{bmatrix} u_1 \\ u_2 \end{bmatrix} = \begin{bmatrix} a_x \cdot k_x & a_z \cdot k_z \\ a_x \cdot k_z & a_z \cdot k_x \end{bmatrix} \begin{bmatrix} \omega_x \\ \omega_z \end{bmatrix} + \begin{bmatrix} e_1 \\ e_2 \end{bmatrix} \quad (2)$$

where u_1 and u_2 refer to the outputs of the Sensor 1 and the Sensor 2, ω_x and ω_z refer to the rotation rates around the X and Z axes, respectively. e_1 and e_2 are the measurement errors of the two sensors.

Equation (2) is derived from the assumption that the output of the sensor is proportional to the acceleration applied along either axis of X and Z. This assumption can be proved by the simulation results of the relationship between the corresponding convection flow velocity and the applied acceleration. Take the acceleration along the Z axis as an example, Fig. 5 gives the result, which shows that the flow velocity along X -

axis is quite linear with the applied acceleration along the Z-axis. Since the corresponding flow velocity is proportional to the Coriolis acceleration, the applied acceleration is proportional to the output of the sensor.

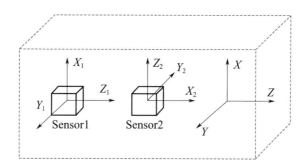

Fig. 4 An integrated system comprised of two thermal sensors for eliminating the cross-axis effect

Equation (2) is a linear algebraic equation, the rotation rates ω_x and ω_z can be conditionally figured out based on the theorem of the least squares estimation[5]. Thus it can be further extracted that the rotations around the two axes X and Z can be decoupled and solved out only if the accelerations along the two axes are not null and $k_x \neq k_z$. The latter condition $k_x \neq k_z$ can be easily fulfilled by the sensor configuration in fact. Consequently, the condition to decouple the two cross sensitivities is that the accelerations along the two detected axes are not null.

Thereby, the estimations of the rotations around X and Z axes are

$$\hat{W} = (K^T \cdot K)^{-1} \cdot K^T \cdot U \qquad (3)$$

Where $\hat{W} = [\hat{\omega}_x \quad \hat{\omega}_z]^T$, $K = \begin{bmatrix} a_x \cdot k_x & a_z \cdot k_z \\ a_x \cdot k_z & a_z \cdot k_x \end{bmatrix}$, $U = [u_1 \quad u_2]^T$.

The error of the estimation is $\Delta W = \hat{W} - W = (K^T \cdot K)^{-1} \cdot K^T \cdot E$,

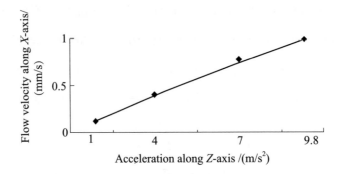

Fig. 5 Flow velocity along X-axis versus the applied acceleration along Z-axis

where $\boldsymbol{E} = [e_1 \quad e_2]^T$.

For the case that the acceleration along one of two cross-axes approaches null, the cross-axis effect of rotations is selfvanished. The output of the sensor yields the angular rate around the single axis, along which the acceleration is not null.

5 Conclusion

The cross-axis sensitivity of the thermal gas inertial sensor is analyzed in the paper. The cross effect is mainly brought from the complex multidimensional motion of the convection flow induced by the external accelerations and temperature differences. The methodology of eliminating the cross-axis effect and decoupling the cross sensitivities among orthogonal axes is developed by combining more than two sensors and employing a fusion data technology. The proposed results are greatly useful for the practical applications of the thermal gas sensors, especially in the multidimensional motion measurements.

References

[1] F Mailly, A Giani, A Martinez, R Bonnot, P Temple - Boyer, A Boyer. Micromachined thermal accelerometer. Sensors and Actuators A, vol. 103, pp. 359 - 363, 2003.

[2] R Zhu, Y Su, H G Ding, Z Y Zhou. Micromachined gas inertial sensor based on convection heat transfer. Sensors and Actuators A, vol. 130 - 131, pp. 68 - 74, 2006.

[3] R Amarasinghe, D V Dao, T Toriyama, S Sugiyama. Design and fabrication of miniaturized six - degree of freedom piezoresistive accelerometer. 18th IEEE International Conference on Micro Electro Mechanical Systems, pp: 351 - 354, 2005.

[4] D Misra, B D Wang. Elimination of cross sensitivity in a three - dimensional magnetic sensor. IEEE Transactions on Electron Devices, vol. 41 no. 4, pp. 622 - 624, 1994.

[5] S Bittanti and G Picci. Identification, Adaptation, Learning: the science of learning models from Data, Berlin ; New York : Springer, 1996, pp. 350 - 351.

Sensor Fusion Methodology to Overcome Cross – Axis Problem for Micromachined Thermal Gas Inertial Sensor

(2009 年)

Abstract Micromachined thermal gas inertial sensors are novel devices that take advantages of simple configuration, large working range, high shock resistance, and good reliability in virtue of using gaseous medium instead of mechanical proof mass as key moving and sensing elements. Basing on multidimentional movements of gas flow in a small chamber, the sensor generally undergoes a cross – axis problem. In this paper, a study on the cross – axis sensitivity of the thermal gas rotation sensor is reported. The cross – axis problem of the sensor is resulted from the multidimensional coupling movement of the convection flow in the sensor chamber and possibly be diminished by a tailored structural design. Unlike using a complex scheme on the mechanical structure, combining more than two sensors to form an integrated compensation system and using a fusion methodology to decouple cross rotations are proposed in this paper. The method helps to

---

本文发表于 IEEE SENSORS JOURNAL，第 9 卷第 6 期。合作作者：Rong Zhu, Henggao Ding, Yongjun Yang, Yan Su。

enhance practical applications for thermal rotation sensors.

Index Terms Cross – axis problem; Data fusion; Micromachined thermal gas inertial sensor

1 Introduction

Most micromachined inertial sensors generally use a mechanical structure including a solid proof mass attached to springs [1]. For example, several micromachined vibrating gyroscopes including vibrating shells[2], tuning – forks[3], and vibrating beams[4] have been utilized. From the end of last century, a kind of micromachined thermal accelerometer without seismic mass has been studied and applied by Zhao and Leung et al. [5-7]. The thermal sensors take advantages of simple configuration, easy to be fabricated, high shock resistance, and good reliability. Through development for one decade, these acceleration sensors have presently been facing rapid commercialisation. Based on the similar thermal principle, a thermal rotation sensor with no proof mass was demonstrated recently[8,9]. The working principle of these thermal inertial sensors is mainly based on the free convection flow of gas in a small chamber. In our previous work[8], we have proposed a low – cost, thermofluidic micromachined inertial sensor, the configuration of which consists of a small silicon etched cavity, a suspended central heater that heats up and lowers the density of the surrounding gas, and four suspended detecting thermistor wires symmetrically placed on both sides of the heater, all of which are assembled and packaged in a hermetic chamber. The proposed sensor can detect single – axis rotation and dual-axis accelerations. The sensor takes merits of simple structure, low cost, wide working range ($>1\ 000\ (°)/s$), and extremely high shock resistance ($20\ 000\ g$)[10].

Like most axis-sensitive sensors, such as accelerometers[11] and magnetometers[12], the thermal gas inertial sensor inevitably undergoes the cross-axis problem[13], which generally causes serious errors in measurements. The common strategy to overcome the cross-axis problem for a micromachined sensor is to adopt a tailored complex structure design to diminish the coupling effect among cross-axes[14,15], and thus brings complex fabrication. In this paper, we will propose a fusion methodology to overcome the cross-axis problem for realizing multiaxial rotation measurements. First, we give an experiment on a thermal inertial sensor prototype to test its cross sensitivity. Consequently, the cross principle and influencing factors of the cross-axis effect are analyzed by using a numerical simulation of three-dimensional convection flows in the sensor chamber. Then, a methodology to overcome the cross-axis problem by combining more than two sensors to form a compensation system and using a data fusion will be proposed. Finally, an example is presented to validate the feasibility of the method. The methodology enable to effectively resolve the cross problem and retain good performances for thermal inertial sensor applications.

2 Operation and Design

Our homemade prototype of the sensor, the schematic view of the sensor configuration, and the signal processing circuit for detecting rotation around Z axis are shown in Fig. 1 (a) – (c). The prototype of the sensor can be fabricated by using similar fabrication techniques that have been used to fabricate thermal accelerometers. First of all, a low stress silicon nitride membrane Si_xN_y is deposited on a silicon substrate by low pressure chemical vapor deposition (LPCVD). Heater and thermistors are made of platinum thin film, which is sputtered and pat-terned on Si_xN_y by liftoff. Then, the Si_xN_y is etched by RIE and the Silicon substrate is etched to obtain the cavity and Pt resistors on Si_xN_y bridges by ethylenediamine pyrochatechol water (EPW) etching. After packaging, air sealed in the chamber is used as the gaseous medium. Define $X-Y-Z$ as the coordinate frame of the sensor, where X and Y are the central axes of the cavity surface ($X-Y$ is defined as working plane), and Z is the normal axis pointing from the working plane of the sensor to the outside. The operating principle of the sensor is based on the interaction of inertial and thermal properties in a laminar, internal chamber flow, which has been described detailedly in[8]. The sensor can detect the rotation around Z and the dual-axis accelerations along X and Y. In this paper, we only consider the rotation measurement. The Wheatstone bridge circuit shown in Fig. 1 (c) is utilized to acquire the output of the sensor (i.e., rotation around Z), which is proportional to the difference of the resistances among four thermistors and dependent on the opposite symmetric flows in the sensor chamber[8]. The gas flow in the chamber generally behaves as complex multidimensional movements. The multidimensional flows concur and thus induce a cross-axis effect on the sensor. For knowing the cross-axis sensitivity, we executed an experimental measurement on the prototype. The output voltages of the sensor prototype [i.e., the outputs of the electric circuit in Fig. 1 (c)] versus the rotation around X and Y axes, respectively, are shown in Fig. 2 (a) and (b), in the case an acceleration of 9.8 m/s² is applied along the corresponding rotation axis. From experiments, we find that either rotation around Z and X axes of the sensor may cause an effective rotation output of the electric circuit shown in Fig. 1 (c). As a result, the rotation around X induces a cross-axis effect on the sensor whose measurement axis is supposed to be Z. The cross degree is dependant on the acceleration along the cross axis. For our prototype, a cross error of about $0.43 \cdot a_x \cdot \omega_x$ from the rotation around X exists in the rotation measurement around Z, which is deduced from experimental data shown in Fig. 2, where a_x and ω_x represent the acceleration and rotation rate around X axis, respectively. Fortunately, it is tested that the rotation around Y has less cross effect on the sensor output. The experimental results also indicate that the output of the sensor is almost linear with the applied angular rate.

3 Numerical Simulation and Analysis

To investigate the cross-axis behavior of the thermal inertial sensor, we perform a primary numerical simulation on three-dimensional convection flows in the sensor chamber by using ANSYS FLOTRAN.

Fig. 3 illustrates the simulated flow vectors in the working plane

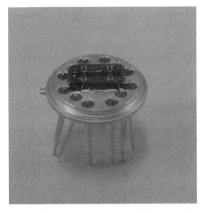

(a) Homemade sensor prototype without package

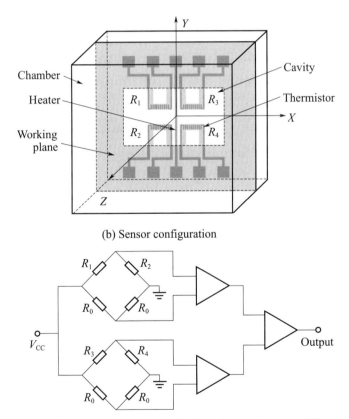

(b) Sensor configuration

(c) Signal processing circuit for detecting rotation around Z

Fig. 1 Homemade prototype of the sensor and schematic view of the sensor configuration and signal processing circuit for detecting angular rate

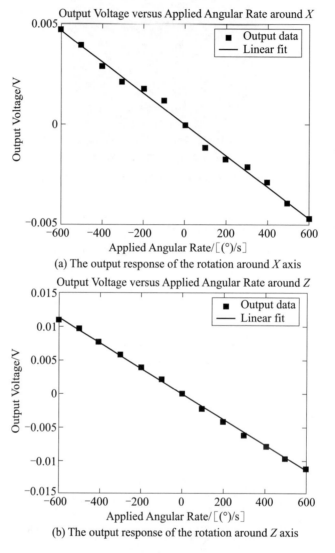

Fig. 2　Output voltages of sensor prototype versus angular rates applied around X and Y axes, respectively

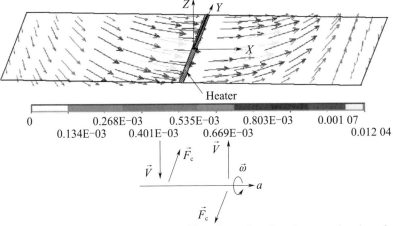

(a) Flow velocity vectors in the working plane when there is an acceleration of 9.8 m/s² along X axis

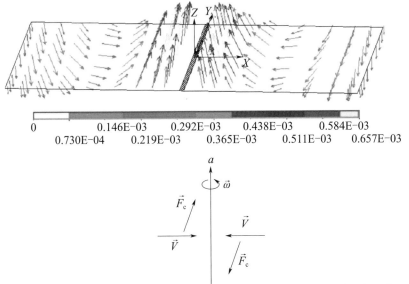

(b) Flow velocity vectors in the working plane when there is an acceleration of 9.8 m/s² along Z axis

Fig. 3 Simulated flow vectors under an acceleration of 9.8 m/s² applied along X and Z axes, respectively

under an acceleration of 9.8 m/s² applied along X and Z axes, respectively. We can see that the orientation of the applied acceleration mainly dominates the gas flow behavior in the chamber. If an acceleration is applied along X as shown in Fig. 3 (a), the flow vector in the working plane donated as $\vec{V}$ moves along X with opposite flow in Z direction on two sides of the heater. The Coriolis force $\vec{F}_c = 2\vec{\omega} \times \vec{V}$ derived from a rotation $\vec{\omega}$ around X will deflect the opposite flows along Z to the opposite directions of Y, which causes opposite cooling effect between two opposite thermistors along Y, and as a result, responses an output signal by using the Wheatstone bridge circuit in Fig. 1 (c). In a word, the rotation around X axis results in an effective output of the sensor as long as the acceleration along X is not null. For the case, there is an acceleration along Z axis, as shown in Fig. 3 (b), the flow in the working plane mainly moves along X and is inverse symmetric about Y. The rotation around Z induces opposite Coriolis forces $\vec{F}_c$ along Y on the two sides of the heater so as to move the flows to deflect in opposite directions of Y, which also produces an output signal of the sensor, as shown in Fig. 3 (b). As a result, either rotation along Z and X axes can induce an opposite symmetric Coriolis acceleration along Y on the two sides of the heater when the acceleration along the rotation axis is not null, thus brings an effective output in the electric circuit. These two rotation signals around Z and X axes are coupling together thus lead to a cross-axis sensitivity. The magnitude of the cross sensitivity between X and Z axes is dependent on the acceleration magnitude along X. Another analysis can be given to the cross effect between Y and

Z. Since the heater is placed along the Y axis, the rotation around the Y axis cannot induce effective Coriolis accelerations along Y (namely, effective Coriolis accelerations along Y should be opposite symmetric on the two sides of the heater), and thus bring no cross effect on the sensor. This analysis is in accordance with the experimental results.

4 A Way to Overcome Cross – Axis Effect

A possible way to diminish or eliminate the cross – axis errors of the sensor can be carried out by well constructing the chamber structure and well composing the placement of the heater and thermistors, which may arouse a complexity in the fabrication. Furthermore, even the mechanical structure has been well designed, but the cross – axis problem cannot be completely eliminated. In this paper, we present another way to overcome the cross – axis problem by performing a data fusion methodology on an integrated compensation system that contains more than two thermal inertial sensors placed in different directions. One example with two sensors is schematically shown in Fig. 4. Assume that the two sensors have the same structure with the same configuration in Fig. 1, but are placed in the orthogonal directions, specifically the Z_1 and X_1 axes of Sensor 1 are aligned with the X_2 and Z_2 axes of Sensor 2, respectively. Y_1 is opposite to Y_2. We also define a reference coordinate system $X - Y - Z$, where X, Y, and Z are aligned with X_1, Y_1, and Z_1, respectively.

Based on the working principle of the sensor, the scale factor of the sensor in detecting a rotation around one specific axis and is

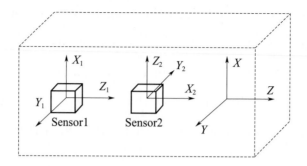

Fig. 4 An integrated compensation system comprised of two thermal inertial sensors for eliminating cross-axis errors

correlated with the acceleration component along this axis and the structural orientation. For instance, the scale factor of Sensor 1 along the Z axis is denoted as $a_{z1} \cdot k_{z1}$, where a_{z1} refers to the acceleration along the Z_1 axis, k_{z1} refers to the structural factor of the sensor along the Z_1 axis. In the same way, the scale factor of Sensor 1 along the X axis is denoted as $a_{x1} \cdot k_{x1}$, and the scale factors of Sensor 2 along Z and X axes are $a_{z2} \cdot k_{z2}$ and $a_{x2} \cdot k_{x2}$, respectively. It must be noted that the Z_1 and X_1 axes of Sensor 1 are aligned with the X_2 and Z_2 axes of Sensor 2, and the two sensors are the same in structure except the orientation placement, thus we have

$$\begin{aligned} a_{z1} &= a_{x2} = a_z \\ a_{x1} &= a_{z2} = a_x \\ k_{z1} &= k_{z2} = k_z \\ k_{x1} &= k_{x2} = k_x \end{aligned} \qquad (1)$$

Then we can establish the relationship equation as follows for the two sensors

$$\begin{bmatrix} u_1 \\ u_2 \end{bmatrix} = \begin{bmatrix} a_x \cdot k_x & a_z \cdot k_z \\ a_x \cdot k_z & a_z \cdot k_x \end{bmatrix} \begin{bmatrix} \omega_x \\ \omega_z \end{bmatrix} + \begin{bmatrix} e_1 \\ e_2 \end{bmatrix} \quad (2)$$

where u_1 and u_2 refer to the outputs of Sensors 1 and 2, ω_x and ω_z refer to the rotation rates around the X and Z axes, respectively. e_1 and e_2 are the measurement errors of the two sensors.

Equation (2) is derived from the assumption that the output of the sensor is proportional to the acceleration applied along either axis of X and Z. This assumption is proved by the simulation on the relationship between the corresponding convection flow velocity and the applied acceleration. Take the acceleration along the Z axis as an example. Fig. 5 give the simulation results of the flow velocity along X at one point in the working plane under the Z axis acceleration of 1, 4, 7, 10 m/s², respectively. The result shows that the flow velocity along the X axis is quite linear with the applied acceleration along the Z axis. Since the corresponding flow velocity is proportional to the Coriolis acceleration that determines the output of the sensor, the applied acceleration is thus proportional to the output of the sensor.

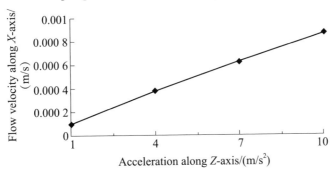

Fig. 5 Flow velocity around X versus the applied acceleration along Z

Equation (2) is a linear algebraic equation, the rotation rates ω_x and ω_z can be conditionally figured out based on the theorem of the Least Squares Estimation[16]. The condition is that only if the accelerations along the two axes are not null and $k_x \neq k_z$, the rotations around the two axes X and Z can be decoupled and estimated from the measurement outputs. The latter condition $k_x \neq k_z$ can be easily fulfilled by the sensor configuration in fact. Consequently, the condition to decouple the two rotations around X and Z is that the accelerations along X and Z are not null.

Thereby, if the condition is satisfied, the estimations of the rotations around X and Z axes can be extracted by

$$\hat{\boldsymbol{W}} = (\boldsymbol{K}^\mathrm{T} \cdot \boldsymbol{K})^{-1} \cdot \boldsymbol{K}^\mathrm{T} \cdot \boldsymbol{U} \tag{3}$$

Where $\hat{\boldsymbol{W}} = [\hat{\omega}_x \quad \hat{\omega}_z]^\mathrm{T}$, $\boldsymbol{K} = \begin{bmatrix} a_x \cdot k_x & a_z \cdot k_z \\ a_x \cdot k_z & a_z \cdot k_x \end{bmatrix}$, $\boldsymbol{U} = [u_1 \quad u_2]^\mathrm{T}$. The error of the estimation is $\Delta \boldsymbol{W} = \hat{\boldsymbol{W}} - \boldsymbol{W} = (\boldsymbol{K}^\mathrm{T} \cdot \boldsymbol{K})^{-1} \cdot \boldsymbol{K}^\mathrm{T} \cdot \boldsymbol{E}$, where $\boldsymbol{E} = [e_1 \quad e_2]^\mathrm{T}$.

For the case that the acceleration along one of two cross-axes approaches null, the cross-axis effect between two rotations around the cross-axes is self-vanished. The output of the sensor only yields the rotation around one axis, along which the acceleration is not null. However, the rotation around the other axis, along which the acceleration is null, cannot be estimated for this case.

For solving the general problem and decoupling the cross rotations for more cases, we consider adding another thermal inertial sensor (defined as Sensor 3) in the compensation system, the X and Z axes of which (define as X_3 and Z_3) bevel with X_1 and Z_1, e. g. , the X_3 meets the X_1 and Z_1 at 45°, as shown in Fig. 6. The placement of the coordinate systems among the $X-Y-Z$, $X_1-Y_1-Z_1$, $X_2-Y_2-Z_2$, and $X_3-Y_3-Z_3$ are illustrated in Fig. 6.

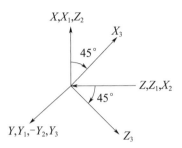

Fig. 6   Placement of coordinate systems of sensors

Considering the geometric transformation of the vectors of the accelerations and rotations among coordinate systems, we have

$$a_{x3} = a_x \cdot \cos 45° + a_z \cdot \sin 45°$$
$$a_{z3} = -a_x \cdot \sin 45° + a_z \cdot \cos 45°$$
$$\omega_{x3} = \omega_x \cdot \cos 45° + \omega_z \cdot \sin 45°$$
$$\omega_{z3} = -\omega_x \cdot \sin 45° + \omega_z \cdot \cos 45°$$
(4)

where a_{x3} and a_{z3} refer to the acceleration along the X_3 and Z_3 axes, respectively. ω_{x3} and ω_{z3} refer to the rotation rates around the X_3 and Z_3 axes, respectively. Then, the output of Sensor 3 responds as

$$u_3 = a_{x3} \cdot k_x \cdot \omega_{x3} + a_{z3} \cdot k_z \cdot \omega_{z3} + e_3$$
$$= 0.5 [k_1 \quad k_2] \begin{bmatrix} \omega_x \\ \omega_z \end{bmatrix} + e_3$$
(5)

where $k_1 = (k_x + k_z) \cdot a_x + (k_x - k_z) \cdot a_z$ and $k_2 = (k_x - k_z) \cdot a_x + (k_x + k_z) \cdot a_z$.

Combine (2) and (5), the relations among the outputs of three sensors are formulated by

$$\begin{bmatrix} u_1 \\ u_2 \\ u_3 \end{bmatrix} = \begin{bmatrix} a_x \cdot k_x & a_z \cdot k_z \\ a_x \cdot k_z & a_z \cdot k_x \\ 0.5 \cdot k_1 & 0.5 \cdot k_2 \end{bmatrix} \begin{bmatrix} \omega_x \\ \omega_z \end{bmatrix} + \begin{bmatrix} e_1 \\ e_2 \\ e_3 \end{bmatrix} \qquad (6)$$

Provided that $k_x \neq k_z$ and at least one acceleration component of a_x and a_z are not null. The rank of the parameter matrix $\begin{bmatrix} a_x \cdot k_x & a_z \cdot k_z \\ a_x \cdot k_z & a_z \cdot k_x \\ 0.5 \cdot k_1 & 0.5 \cdot k_2 \end{bmatrix}$ is always greater than or equal to 2, thus the rotation rate $\hat{W} = [\hat{\omega}_x \quad \hat{\omega}_z]^T$ can be estimated by solving (6).

The above methodology can be further extended to solve the cross-axis problem for decoupling the rotations around three-axis. The following example will demonstrate the solutions.

5 Example

Consider an air vehicle following a trajectory motion from a start point A to a goal point B where the vehicle is rotating around the forward motion axis X_0 at a high-speed ω_1 and the forward direction is rotating at a constant angular rate ω_2 about the horizontal axis $-Y_0$. We define a normal coordinate system $X_D - Y_D - Z_D$ as a local horizontal coordinate system, and definethe system $X_0 - Y_0 - Z_0$ as the initial body system of the vehicle. Consider the actual body system of the vehicle as $X - Y - Z$, where X refers to the forward direction of the vehicle, Y

refers to the right-left direction of the vehicle, and Z incorporating with X and Y forms a Cartesian coordinate. At the beginning, X-Y-Z is aligned with X_0-Y_0-Z_0. The mission scenario is shown in Fig. 7. Assume $\omega_1 = 3\,000(°)/s$ and $\omega_2 = 10(°)/s$.

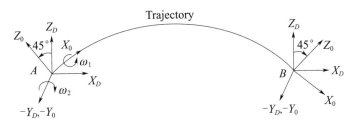

Fig. 7　Mission scenario of vehicle flight

For tracking the flight trajectory of the vehicle and detecting the rotations ω_1 and ω_2 in flight, we combine six thermal inertial sensors and place them in a way shown in Fig. 8, where the X_i-Y_i-Z_i represents the coordinate frame of Sensor i, the X_Z axis is formed by rotating the Z axis to 45° around Y, and the Y_Z axis is formed by rotating the Y axis to 45° around X. Amongst these sensors, Sensors 1, 2, and 6 are combined to figure out the rotation rates around X and Z axes (define as ω_x and ω_z) by solving (6). In a same way, Sensors 3, 4, and 5 are combined to figure out the rotation rates around Y and Z axes (define as ω_y and ω_z). Since the X axis of the vehicle always points to the forward direction X_0, i.e., $\omega_1 = \omega_x$ yields. The rotation ω_2 is derived from the combination of ω_y and ω_z. Under the condition that the measurement errors of the sensors are less than 1 μV, Fig. 9 shows the detected results for the rotations, which indicate that the rotations around the three axes X-Y-Z can been decoupled and

· 351 ·

measured, the rotations ω_1 and ω_2 are detected with the measurement errors of less than 5 (°)/s and 0.5 (°)/s, respectively.

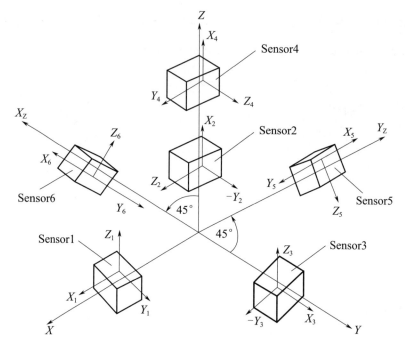

Fig. 8 Placement of thermal gas sensors in an integrated system

6 Conclusion

The cross-axis sensitivity of the thermal gas inertial sensor is studied in this paper. The cross effect is mainly resulted from the complex multidimensional motion of the convection flow induced by the external accelerations and temperature differences. The methodology of eliminating the cross-axis errors and decoupling the cross rotations is developed by integrating multiple thermal sensors and employing a fusion data technology. The proposed method is simple, easy to be realized, and free of structural complexity. The proposed schemes will

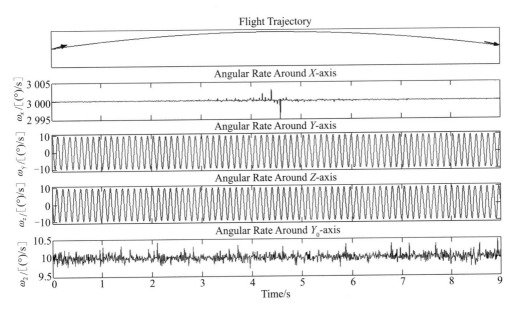

Fig. 9 Results of rotating detection in flight mission, where $\omega_1 = \omega_x$, and $\omega_2 = \sqrt{\omega_y^2 + \omega_z^2}$

helps to enhance the practical applications for thermal gas inertial sensors, especially in the multidimensional rotation measurements.

References

[1] Y Navid, F Ayazi, and K Najafi. Micromachined inertial sensors. Proc. IEEE, vol. 86, pp. 1640 – 1659, 1998.

[2] F Ayazi and K Najafi. A harpss polysilicon vibrating ring gyroscope. J. MEMS, vol. 10, pp. 169 – 179, 2001.

[3] J Bernstein. A micromachined comb – drive tuning fork rate gyroscope. in Proc. IEEE Micro Electro Mech. Syst. (MEMS), Fort Lauderdale, FL, Feb. 7 – 10, 1993, pp. 143 – 148.

[4] K Maenaka, T Shiozawa. A study of silicon angular rate sensors using anisotropic etching technology. Sens. Actuators A, vol. 43, pp. 72 – 77, 1994.

[5] Y Zhao, A P Brokaw, M E Rebeschini, A M Leung, G P Pucci, and A Dribinsky. Thermal convection accelerometer with closed loop heater control. U.S. 6 795 752 B1, Sep. 21, 2004.

[6] A M Leung, J Jones, E Czyzewska, J Chen, and B Woods. Micromachined accelerometer based on convection heat transfer. in Proc. IEEE Micro Electro Mech. Syst. (MEMS), 1998, pp. 627 – 630.

[7] F Mailly, A Giani, A Martinez, R Bonnot, P Temple – Boyer, and A Boyer. Micromachined thermal accelerometer. Sens. Actuators A, vol. 103, pp. 359 – 363, 2003.

[8] R Zhu, Y Su, H G Ding, and Z Y Zhou. Micromachined gas inertial sensor based on convection heat transfer. Sens. Actuators A, vol. 130 – 131, pp. 68 – 74, 2006.

[9] D V Dao, V T Dau, T Shiozawa, and S Sugiyama. Development of a dual – axis convective gyroscope with low thermal – induced stress sensing element. J. Microelectromech. Syst., vol. 16, pp. 950 – 958, 2007.

[10] Y Zhao, R Zhu, X Y Ye, and Y J Yang. Analysis on shock resistance of micromachined angular rate sensor based on convection heat transfer. Chinese J. Sens. Actuators, vol. 21, no. 4, 2008.

[11] R Amarasinghe, D V Dao, T Toriyama, and S Sugiyama. Design and fabrication of miniaturized six – degree of freedom piezoresistive accelerometer. in Proc. 18th IEEE Int. Conf. Micro Electro Mech. Syst., 2005, pp. 351 – 354.

[12] D Misra, B D Wang. Elimination of cross sensitivity in a three – dimensional magnetic sensor. IEEE Trans. Electron Devices, vol. 41, pp. 622 – 624, 1994.

[13] R Zhu, H Ding, Y Su, and Y Yang. A study of cross – axis effect for micromachined thermal gas inertial sensor. in Proc. 6th IEEE Conf. Sensors, 2007, pp. 840 – 843.

[14] K Kwon, S Park. A bulk – micromachined three – axis accelerometer using silicon direct bonding technology and polysilicon layer. Sens. Actuators A, vol. 66, pp. 250 – 255, 1998.

[15] W T Pike, S Kumar. Improved design of micromachined lateral suspensions using intermediate frames. J. Micromech. Microeng., vol. 17, pp. 1680 – 1694, 2007.

[16] S Bittanti, G Picci. Identification, Adaptation, Learning: The Science of Learning Models From Data. New York: Springer, 1996, pp. 350 – 351.

Modeling and Experimental Study on Characterization of Micromachined Thermal Gas Inertial Sensors

(2010年)

Abstract Micromachined thermal gas inertial sensors based on heat convection are novel devices that compared with conventional micromachined inertial sensors offer the advantages of simple structures, easy fabrication, high shock resistance and good reliability by virtue of using a gaseous medium instead of a mechanical proof mass as key moving and sensing elements. This paper presents an analytical modeling for a micromachined thermal gas gyroscope integrated with signal conditioning. A simplified spring – damping model is utilized to characterize the behavior of the sensor. The model relies on the use of the fluid mechanics and heat transfer fundamentals and is validated using experimental data obtained from a test – device and simulation. Furthermore, the nonideal issues of the sensor are addressed from both the theoretical and experimental points of view. The nonlinear behavior demonstrated in experimental measurements is analyzed based on the

---

本文发表于 Sensors 2010 年第 10 期。合作作者：Rong Zhu, Henggao Ding, Yan Su, Yongjun Yang。

model. It is concluded that the sources of nonlinearity are mainly attributable to the variable stiffness of the sensor system and the structural asymmetry due to nonideal fabrication.

Keywords Micromachined thermal inertial sensor; Heat convection; Modeling; Nonlinearity

1 Introduction

The development of micromachined inertial sensors has been widely addressed for many years. Typical inertial sensors are based on the movement of a seismic proof mass caused by an inertial quantity. These sensors utilize different sensing principles: capacitive, piezoresistive and piezoelectric measurements[1-3]. Different from these conventional devices, micromachined thermal gas inertial sensors based on heat convection, such as thermal accelerometers[4] and thermal gas gyroscopes[5], offer the advantages of simple structures, easy fabrication, high shock resistance and good reliability due to their use of a gaseous medium instead of a mechanical proof mass as the key moving and sensing elements. The working principle of these thermal inertial sensors is mainly based on the natural convection of gas in a small sealed chamber. In our previous work[5], we demonstrated a low-cost, thermo-fluidic micromachined inertial sensor, the configuration of which consisted of a small silicon etched cavity, a suspended central heater that heated up and lowered the density of the surrounding gas, and four suspended detectors symmetrically placed on two sides of the heater, all of which were assembled and packaged in a hermetic chamber. The proposed sensor could detect single – axis angular rate and dual – axis accelerations. In this paper, we only consider the angular rate detection using the sensor.

A mechanism analysis along with mathematical modeling is an essential part of the required work in the sensor design and sensor optimization processes, especially for an inertial device. An analytical model often helps to understand the behavior of a device and resolve any concurrent problems. For example, an inertial sensor generally has nonlinear problems that usually lower the sensitivity and narrow the working range of the device. In order to get rid of these problems, many researchers have taken great efforts to investigate the nonlinear mechanisms and identify the nonideal sources by modeling[6]. For a thermal gas inertial sensor, systematic modeling is inevitably important for its design and error analysis[7]. However, the modeling in a fluidic and thermal domain is more complicated than in a seismic – mass – based device due to the complexity of multi – physics coupling among electrical, thermal, fluidic, and mechanical properties. Up to now, the corresponding results of modeling in a system level for thermal gas gyroscopes have been rarely reported.

In this paper, theoretical and experimental studies on characterization of a micromachined thermal gas gyroscope are presented. For the first time, a characterization of the sensor incorporating its signal conditioning using a simplified model of a spring – damping system is proposed and experimental verification is demonstrated. The modeling approach relies on the fundamentals of fluid mechanics and heat transfer, in association with empirical techniques. The proposed compact model is effective to handle the complexity of the device optimization. The experimental data are provided from both of model - based simulations and physical measurements using fabricated prototypes. The nonlinear characteristics of the sensor are analyzed based on the model and the nonideal sources are summarized.

2　Device Operation and Design

A conceptual design of a micromachined gas gyroscope is shown in Fig. 1. Its convection field in region of hermetic chamber is shown in Fig. 2, and the signal transfer and processing strategy are shown in Fig. 3.

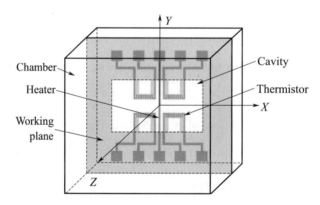

Fig. 1　Conceptual design of a thermal gas gyroscope

The working principle of the device is based on the phenomenon of natural convection. A convectional flow is generated by heating the suspended central heater. For instance, when the central heater heats up and acceleration is applied on the direction of the Z - axis, a gas flow

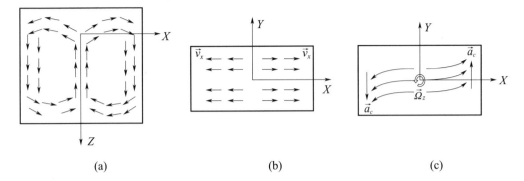

Fig. 2 The convection field in region of hermetic chamber driven by heating the central heater under an acceleration along Z-axis. (a) The convective flow in the plane of $X-Z$; (b) The flow in the working plane of $X-Y$; (c) the flow deflection due to the Coriolis effect

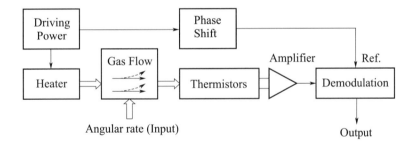

Fig. 3 Block diagram of signal transfers in the thermal gas gyroscope

is generated in the region of the hermetic chamber and depicted in Fig. 2. On the working plane where the detecting thermistors are symmetrically placed, convection flows mainly move along X-axis and are inversely symmetric about the Y-axis. The external inertial rotation $\vec{\Omega}_z$ around the Z-axis will induce a Coriolis acceleration $\vec{a}_c$ and leads the convective flows on the two sides of the heater to deflect in opposite directions of Y, which can be detected by the distributed detectors (thermistors) in a Wheatstone bridgecircuit. Like most vibratory gyroscopes[6], the detection system together with the signal conditioning electronics of the gas gyroscope comprise two orthogonal gaseous oscillators. One of the oscillators, called the primary oscillator or the drive oscillator, is driven by applying an alternating power on the central heater to modulate the convective flow. When the gyroscope rotates about its sensitive axis (i.e., the Z - axis), the Coriolis effect couples the vibration from the primary oscillator to another oscillator in the deflection along the Y - axis, called the secondary oscillator or the sense oscillator. As a result of the Coriolis coupling, the secondary oscillator movement contains the angular rate information, which is the amplitude of the signal modulated around the operating frequency. To obtain the angular rate information, the movement of the secondary oscillator has to be converted into a voltage, and thereafter, be demodulated.

3 Modeling

The entire working process of the sensor consists of multi - physics interactions: electrical - thermal conversion, heat transfer, flow convective movement, and fluid - electrical conversion. A block diagram of the system model, including heating source, gas conduction, gas convection, and sensing, is shown in Fig. 4.

Firstly, we consider the heating source. The electric power supplied to the heating resistor is dissipated by heat transfer toward the ambient fluidic medium and also toward the substrate (heating

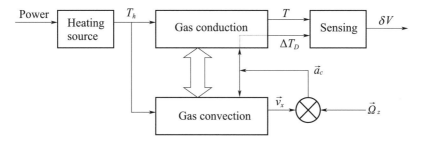

Fig. 4 Block diagram of the sensor model.

resistor), and which leads to a temperature difference between the heater and ambience. According to the Energy Principle[8], the dynamic process of the heating can be modeled by:

$$C \frac{\partial T_h}{\partial t} = P_h - h g_0 \Delta T_h \quad (1)$$

$$\Delta T_h = (T_h - T_a)$$

where T_h and T_a refer to the temperatures of the heater and ambience, C is the thermal capacity of the heater, P_h is the electrical power, h is the heat transfer coefficient, and g_0 is a constant coefficient depending on the geometrical parameters of the heater. According to linear perturbation theory, the heat transfer coefficient h can be considered to be constant. Perform Laplace transform to (1), the transfer function of the heating source can be formulated by a first-order model, where s represents differential operator:

$$G_1(s) = \frac{T_h(s)}{P_h(s)} = \frac{k_1}{1 + \tau_1 s} \quad (2)$$

where $\tau_1 = C/h g_0$, $k_1 = 1/h g_0$.

Then, we analyze the process of gas conduction. The gas

conduction is the heat conduction. The temperature difference between the heater and external ambience leads to a heat transfer in the gaseous medium in the chamber. According to the heat transfer principle[8], the local temperature T at a point in the chamber can be ruled by:

$$\frac{\partial(\rho c T)}{\partial t} = k_2 \cdot \nabla^2(T) \qquad (3)$$

where, ρ, c, and k_2 are the gas density, specific heat, and thermal conductivity, respectively. The vector operator ∇ is defined as $\nabla \equiv i\frac{\partial}{\partial x} + j\frac{\partial}{\partial y} + k\frac{\partial}{\partial z}$. Here we only consider the heat flow within the working plane and define x as characteristic dimension for the device. Therefore equation (3) can be reduced to $\frac{\partial(\rho c T)}{\partial t} = k_2 \cdot \frac{\partial^2 T}{\partial x^2}$. Solving the partial differential equation using a Separation Variable technique[9] together with the boundary conditions T_h at the wall of the heater, we obtain the following first-order transfer relationship:

$$G_2(s) = \frac{T(s)}{T_h(s)} = \frac{1}{1+\tau_2 s} \qquad (4)$$

where $\tau_2 = -\rho c \iint T_0(x) \mathrm{d}x^2 / (k_2 T_0(x))$, and $T_0(x)$ is a normalized shape function of temperature profile.

In the process of gas convection, the gradient pressure is generated by the gradient temperature in terms of the state equation $\rho = \frac{p}{R \cdot T}$, where p and R are the pressure and gas constant, respectively. According to the Navier-Stokes equation[10], the convection flow velocity $\vec{v}$ of the gas in the chamber is ruled by:

$$\frac{\partial(\rho\vec{v})}{\partial t} = -\nabla(p) + \nabla(\mu \cdot \nabla(\vec{v})) \tag{5}$$

where μ is the dynamic viscosity of the gas in the chamber. Solving (5) using the Separation Variable approach and combining the state equation together with the wall condition $\vec{v}_w = 0$, we obtain the transfer function between the temperature T and the flow velocity $\vec{v}$ of the gas given by a first-order expression:

$$G_3(s) = \frac{\vec{v}(s)}{T(s)} = \frac{k_3}{1+\tau_3 s} \tag{6}$$

where $\tau_3 = -\rho v_0(x)/(\mu \nabla^2 v_0(x))$, $k_3 = -Ra_2 \nabla T_0(x)/T_0(x)$, and $v_0(x)$ is a normalized shape function of convection flow.

Following the gas momentum equation and Archimedes's law, an applied acceleration results in a buoyancy force and deforms the temperature profile[8]. When an angular rate $\vec{\Omega}_z$ is applied about the Z-axis, the Coriolis acceleration $\vec{a}_c = 2\vec{\Omega}_z \times \vec{v}$ is generated, which leads to a deformation on the temperature profile that is detected by the thermistors. The temperature deformation has been found to be proportional to the Grashof number G_r determined by a given acceleration[4], which comes a linear relationship between the temperature difference ΔT_D across the thermistor detectors and the given acceleration (here is Coriolis acceleration a_c):

$$\Delta T_D \propto G_r \quad \text{with} \quad G_r = \frac{a_c \rho^2 \eta T_h l^3}{\mu^2} \tag{7}$$

where η is gas coefficient of expansion, l is linear dimension.

Considering the governing transient momentum process[8], the above transformation also corresponds to a first-order response:

$$G_4(s) = \frac{\Delta T_D(s)}{\vec{v}(s)} = \frac{\Omega_z k_4}{1+\tau_4 s} \tag{8}$$

where k_4 and τ_4 are constant coefficients depending on thermal and fluidic properties of gas.

The thermistors convert the thermal signals (local temperatures) into the resistance signals of the resistors. Due to thermal inertia of the thermistors, another first-order transfer function should be considered since thermistors have to be in equilibrium with the local temperature of the gas to convert temperature variations into electrical resistance variations. The first-order transfer function represents the signal transfer from the local temperature difference ΔT_D to the temperature difference on the thermistors ΔT_d:

$$G_5(s) = \frac{\Delta T_d}{\Delta T_D} = \frac{1}{1+\tau_5 s} \tag{9}$$

where τ_5 represents the time constant of the thermal inertia of the thermistors.

Using a Wheatstone bridge circuit, the temperature difference on the detecting thermistors is proportionally converted into a voltage difference δV[5]. This process can be formulated by:

$$\delta V = k' \cdot \Delta T_d \tag{10}$$

where k' is a constant coefficient depending on the parameters of the electronic circuit.

Combining the equations (2), (4), (6), (8), (9), and (10), the entire transfer function from the heating power P_h to the output voltage δV can be given by:

$$H(s) = \frac{Y(s)}{X(s)} = k'G_1(s)G_2(s)G_3(s)G_4(s)G_5(s) \quad (11)$$

where $X(s)$ refers to the Laplace vector of the applied electrical power P_h on the heater, $Y(s)$ refers to the Laplace vector of the output voltage δV. For easing up the analysis for the system and considering the time constant of individual process G is generally small value typically in the order of ms or μs, we ignore the high-order terms in (11) so as to yield a compact simplified spring-damping model formulated by a second-order differential equation:

$$H(s) = \frac{\lambda \Omega_z}{s^2 + cs + k} \quad (12)$$

where $k = 1/\tau''$ and $c = \tau'/\tau''$ denote equivalent stiffness and damping coefficient, $\lambda = k'k_1 k_3 k_4/\tau''$ is a gain representing the sensor sensitivity;

$$\tau' = \sum_{i=1}^{5} \tau_i \text{ and } \tau'' = \tau_1\tau_2 + \tau_1\tau_3 + \tau_1\tau_4 + \tau_1\tau_5 + \tau_2\tau_3 + \tau_2\tau_4 + \tau_2\tau_5 + \tau_3\tau_4 + \tau_3\tau_5 + \tau_4\tau_5.$$

In practice, the coefficients k, c, and λ can be identified through experimental calibration. The response function at a frequency ω is further modeled in the frequency domain:

$$H(j\omega) = \frac{\lambda \Omega_z(j\omega)}{(j\omega)^2 + c(j\omega) + k} \quad (13)$$

Extract the amplitude and phase of the output response as:

$$|H(j\omega)| = \frac{\lambda \Omega_z(\omega)}{\sqrt{(k-\omega^2)^2 + (c\omega)^2}}$$

$$\angle H(j\omega) = -\arctan\frac{c\omega}{k-\omega^2} = \theta \tag{14}$$

As explained in Section 2, the sensor output signal is detected using a synchronous demodulation technique, which can greatly eliminates disturbances and reduces noise level so as to enhance the accuracy and sensitivity of the sensor. The heating power is modulated at the frequency of ω, which leads the corresponding temperature, convection flow, and thermoelectric conversion signals to be the carrier signals at ω. The amplitude $|H|$ of the output voltage signals is extracted using demodulation, i.e., multiplying the detected signal by a local reference oscillator with the same frequence and phase as the carrier of the detected signal to convert the detected signal (incoming signal) into a dc version. After low-pass filtering, the incoming signal consisting of the carrier at ω is retained and others are filtered. For guaranteeing in-phase, the original phase of the local reference oscillator is usually shifted. Define a phase shift $\Delta\theta$, the normalized demodulation output signal is given by:

$$V_{output} = |H|\cos(\theta - \Delta\theta) \tag{15}$$
$$= H_0 \Omega_z \cos(\theta - \Delta\theta)$$

where, $H_0 = \dfrac{\lambda}{\sqrt{(k-\omega^2)^2+(c\omega)^2}}$. Ideally, the phase shift $\Delta\theta$ of the reference oscillator needs to be adjusted to be equal to the phase θ of the incoming signal for guaranteeing synchrony. As a result the normalized demodulation output is $|H|=H_0\Omega_z$. It implifies the ideal output of the sensor is linear with the angular rate Ω_z.

4 Nonideal Factors in Sensors

The preceding analyses are based on the assumption of ideal gas and ideal device - structure. However, the practical conditions are complex and in general not ideal. The considered nonideal aspects affecting the device are mainly as follows: inaccurate phase - shift, asymmetrical structure due to unsatisfied fabrication, nonlinear dependence between temperature differences across detectors and Coriolis acceleration.

The first nonideal factor is an improper phase shift in the local reference oscillator due to improper electronic circuits, which will reduce the scale factor of the sensor (i.e., sensitivity) according to (15). Since the phase θ of the output response is a function of the driving frequency ω according to (14), the compensation - purposed phase shift $\Delta\theta$ of the reference oscillator needs to be carefully adjusted along with the variation of ω.

The second nonideal factor affecting the sensor output is structural asymmetry in the chamber, heater, and detectors (i.e., thermistors). Ideally, the suspending heater beam needs to be located in the centre and the chamber needs to be symmetrical in structure in order to generate symmetrical convection flows; the distributed thermistor wires (four thermistors are used in our device) need to be identical and placed symmetrically on two sides of the heater to detect the deflection of the gas flow[5]. However, these ideal symmetry conditions are difficult to realize in a practical fabrication. These structural asymmetries will induce a parasitical term existing in the output signal, and exhibit as a zero offset voltage depending on the fluidic and thermal inertia of the sensor element. Considering this asymmetrical factor, the model in (15) should be modified as follow, which will be proved in the experiments:

$$V_{\text{output}} = H_0 \cdot [\delta \cdot \cos(\theta' - \Delta\theta) + \Omega_z \cdot \cos(\theta - \Delta\theta)] \quad (16)$$

where δ represents the asymmetrical coefficient, θ' is the phase of the zero-offset output.

The third nonideal source comes from nonlinear dependence of temperature difference across detectors on Coriolis acceleration. A similar nonlinear phenomenon was found in a thermal accelerometer based on heat conduction[11,12], where the sensor output correlation with the temperature is a nonlinear function of the applied acceleration; for a small acceleration there is a linear dependence between temperature and acceleration, where as with increasing acceleration the non-linearity increases. The nonlinear dependence between the acceleration and temperature difference in our devices behaves as a hardening spring, for large impact forces the spring becomes harder than it does for low impact forces. This nonlinearity is attributed to the gas properties with inconstant viscosity, compressibility, slip boundary or even more complicated effects. Especially in a confined space, the thermal and fluidic properties of the gas are variable with inertia[13]. Therefore, the equivalent stiffness k of the system should be a function of the angular rate Ω_z, i.e., $k = k(\Omega_z)$.

5 Experimental Study and Analysis

To validate the effectiveness of the model established above, we conducted experiments using a device prototype shown in Fig. 5, fabricated using micromachining techniques. The detailed fabrication process has been introduced in our previous paper[5]. The sensor was heated by applying an ac power at a given frequency to the heater, and four detectors (i.e., thermistors) in a Wheatstone bridge circuit detected the flow deflection in the chamber that was correlated with the external rotation and exported an output, which was demodulated by a reference signal with the same frequency as the output.

Fig. 5 Fabricated sensor prototype without packaging

The prepared sensor (device A) was mounted on a controlled rotary table. The Z-axis of the sensor was aligned vertically so that the Earth's gravity acceleration was applied on the Z-axis of the sensor. The angular rate ranging from -600 deg/sec up to $+600$ deg/sec was applied around the Z-axis of the sensor. The output voltages of the sensor under a modulation/demodulation frequency of 8 Hz (i.e., the frequency of ac power on the heater) are shown in Fig. 6. A near linear relationship between the output voltage and the angular rate was exhibited. However, it is seen that the linearity for small angular rate is better than that for large angular rate, which is consistent with the theoretical analysis.

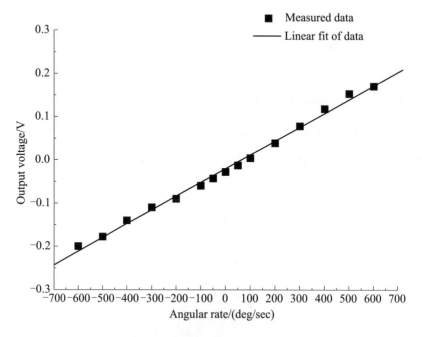

Fig. 6 Output voltage of the sensor versus the angular rate applied around the Z‑axis

To investigate matters of nonlinearity, we further conducted a dual‑phase demodulation measurement on the device and used the established model to simulate the output of the sensors. In the dual‑phase measurement, two orthogonal reference signals with the same frequency and a phase difference of 90° were used to multiply the detected signal to obtain two orthogonal components of the output vector: $V_{\cos}$ and $V_{\sin}$, respectively. According to equation (16), the theoretical formula of $V_{\cos}$ and $V_{\sin}$ are $H_0 \cdot [\delta \cdot \cos(\theta' - \Delta\theta) + \Omega_z \cdot \cos(\theta - \Delta\theta)]$ and $H_0 \cdot [\delta \cdot \sin(\theta' - \Delta\theta) + \Omega_z \cdot \sin(\theta - \Delta\theta)]$. The simulated outputs based on the theoretical model and the real measured data are compared in Fig. 7, which demonstrate a good agreement between the simulated and measured results. The corresponding identification of the model parameters indicated that the equivalent damping coefficient c and the gain λ were about 12 and 0.62, respectively, the equivalent stiffness k increased gradually from 1.72×10^3 to 1.81×10^3 with the increase of the magnitude of angular rate, and an asymmetrical coefficient δ was around 350 for device A.

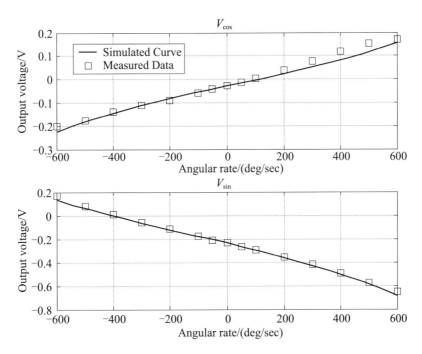

Fig. 7　Tow orthogonal components of the output vector versus the angular rate in a dual-phase measurement for device A

To further test the nonlinearity dependence on the stiffness and structural asymmetry, we used another device (device B) with a serious nonlinear feature to conduct experiments. The experimental setup for the device B was the same as that for the device A. The dual-phase measurements were used once again in this experiment. Fig. 8 demonstrates the measured results and model-based simulation.

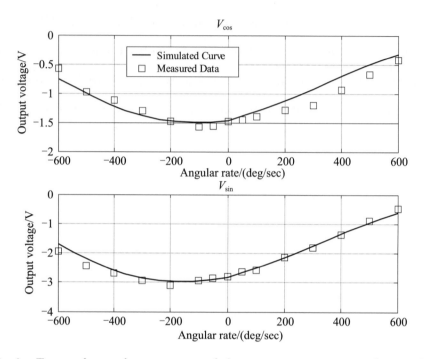

Fig. 8　Tow orthogonal components of the output vector versus the angular rate in a dual-phase measurement for device B

The model parameters were identified by fitting the measurement data. For device B, the equivalent damping coefficient c and the gain λ were 17 and 4, respectively, the equivalent stiffness k varied from 3.95×10^3 up to 6.08×10^3 with the increase of angular rate, and an asymmetrical coefficient δ was as large as 1 300. The larger asymmetry induced a serious nonlinearity, and even produced unilateral warp. For identifying the nonlinear sources, we simulated the sensor output under different conditions: with only variable stiffness or with both of variable stiffness and structural asymmetry. The results are shown in Fig. 9. It was seen that the variable stiffness contributed to the symmetrical nonlinearity shown as dashed lines with x - marks, and the structural asymmetry contributed to the asymmetric warp shown as solid lines with circle - masks.

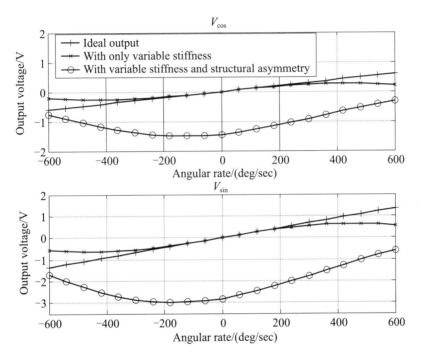

Fig. 9 Simulation results of two orthogonal components of the output vector versus angular rates in three circumstances: ideal state, with variable stiffness, with variable stiffness and structural asymmetry

The structural asymmetry also brings on a zero offset existing in the sensor output as shown in Fig. 7 and Fig. 8. According to the model (16), the zero offset voltage $V_0 = H_0 \delta \cos(\theta' - \Delta\theta)$, which varies with the phase shift $\Delta\theta$ in cosine law. This dependence between the zero offset and the phase shift was confirmed by an experimental measurement on the device B, in which the phase shift $\Delta\theta$ was changed from 0 deg to 360 deg while the device was kept still. The measured results are shown in Fig. 10, where $V_{\cos}$ and $V_{\sin}$ denote two orthogonal components of the zero offset; theoretically they are $H_0\delta\cos(\theta' - \Delta\theta)$ and $H_0\delta\sin(\theta' - \Delta\theta)$, respectively. Fig. 10 indicates that the measured data follow cosine and sine function of $V_{\cos}$ and $V_{\sin}$ very well.

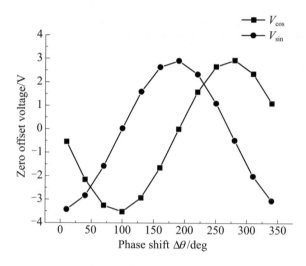

Fig. 10 Two orthogonal components of the zero output versus the phase shift $\Delta\theta$

From the preceding measurements and analyses, it is seen that the model established in the paper can characterize well the performance of the sensor, and is feasible to be used for the optimal design and device improvement. It is also seen that the structural symmetry in the device is crucial for the linearity. The fabrication needs to be improved to amend structural asymmetry for eliminating the nonlinearity of the sensor. Besides the linearity of the sensors, we also tested noise limited resolution of the angular rate for the sensors. We found the noise densities of the sensors were around 1 deg/s/$\sqrt{Hz}$.

6 Conclusions

A mathematical model (simplified as a spring – damping system) is established for a micromachined thermal gas gyroscope based on convection heat transfer to characterize multi – physics interaction processes: electrical – thermal conversion, convection heat transfer, flow convective activity and fluid – electrical conversion. A signal detection process using a modulation/demodulation technique is considered in the modeling, where the heating power is modulated at a given frequency and the angular rate is extracted by demodulation and a low – pass filter. The established model is validated by comparing the simulated results with the real measured data from dual – phase measurements. The theoretical and experimental studies reveal that the nonideal effects in the device are mainly attributable to the structural asymmetry and the variable stiffness of the system; the linearity of the sensor can be improved via amending the structural asymmetry in fabrication.

References

[1] Putty M W, Najafi K. A micromachined vibrating ring gyroscope. In Proceedings of Solid – State Sensor and Actuator Workshop, Hilton Head Island, SC, USA, June 1994; pp. 213 – 220.

[2] Li X X, Chen X M, Song Z H, Dong P T, Wang Y L, Jiao J W, Yang H. A micro

gyroscope with piezoresistance for both high-performance coriolis-effect detection and seesaw-like vibration control. IEEE/ASME J. Microelectromech. Syst. 2006, 15, 1698-1707.

[3] Soderkvist J. Micromachined gyroscopes. Sens. Actuat. A 1994, 43, 65-71.

[4] Mailly F, Martinez A, Giani A, Pascal-Delannoy F, Boyer A. Design of a micromachined thermal accelerometer: Thermal simulation and experimental results. Microelectr. J. 2003, 34, 275-280.

[5] Zhu R, Su Y, Ding H G, Zhou Z Y. Micromachined gas inertial sensor based on convection heat transfer. Sens. Actuat. A 2006, 130-131, 68-74.

[6] Saukoski M, Aaltonen L, Halonen K. Effects of synchronous demodulation in vibratory MEMS gyroscopes: A theoretical study. IEEE Sens. J. 2008, 8, 1722-1733.

[7] Leman O, Chaehoi A, Mailly F, Latorre L, Nouet P. Modeling and system-level simulation of a CMOS convective accelerometer. Solid-State Electr. 2007, 51, 1609-1617.

[8] Lienhard, J. H. A Heat Transfer Textbook, 3rd ed.; Prentice-Hall: Englewood Cliffs, NJ, USA, 1987; pp. 131-250.

[9] Tannehill J C, Anderson D A, Pletcher R H. Computational Fluid Mechanics and Heat Transfer, 2nd ed.; Taylor & Francis Ltd.: London, UK, 1997; pp. 15-22.

[10] Finnemore E J, Franzini J B. Fluid Mechanics with Engineering Applications, 10th ed.; McGraw-Hill Companies Inc.: New York, NY, USA, 2002; pp. 185-189.

[11] Dauderstädt U A, Vries P H S, Hiratsuka R, Korvink J G, Sarro P M, Baltes H, Middelhoek S. Simulation aspects of a thermal accelerometer. Sens. Actuat. A 1996, 55, 3-6.

[12] Dauderstädt U A, French P J, Sarro P M. Temperature dependence and drift of a thermal accelerometer. Sens. Actuat. A 1998, 66, 244-249.

[13] Bairi A, Garcia J M, Laraqi N, Alilat N. Free convection generated in an enclosure by alternate heated bands. Experimental and numerical study adapted to electronics thermal control. Int. J. Heat Fluid Flow 2008, 29, 1337-1346.

A Micromachined Integrated Gyroscope and Accelerometer Based on Gas Thermal Expansion

(2013年)

Abstract This paper reports a novel micromachined integrated gyroscope and accelerometer based on gas thermal expansion. Using gas movement induced by alternating heat expansion in a micro chamber instead of a solid proof mass movement, a new type of micro gyroscope with a wide-range rotation detectability incorporating with 1-axis acceleration sensing is accomplished. Experiments demonstrate a good linearity of the rotation output in response to the angular rate from −3 000 (°)/s to +3 000 (°)/s and the acceleration output in response to the acceleration in the range of ±1 g, which validate the simultaneous measurements of 1-axis angular rate and 1-axis acceleration in one chip. Meanwhile, the coupling effects among 3-axis rotations and accelerations are investigated experimentally to demonstrate low coupling effect via using thermal expansion flow.

Keywords Gyroscope, Accelerometer, Gas expansion, MEMS

本文发表于 IEEE Access。合作作者：Songlin Cai, Rong Zhu, Henggao Ding, YongJun Yang, Yan Su。

1 Introduction

With the rapid market growth of micromachined gyroscopes and accelerometers, a great deal of research attention has been paid to their novel designs for low cost and high performances. Among the wide variety of micromachined gyroscopes and accelerometers, the most common are devices with a solid – proof mass, which are generally driven by electrostatic force, electromagnetic force or piezoelectric force. As for sensing method, piezoresistive, capacitive and piezoelectric methods are widely used. Owing to the existence of the solid proof mass, those gyroscopes and accelerometers cannot afford high – shock vibration and require complicated and costly fabrication process for the proof mass and suspending beams. Another type of accelerometer with no solid proof mass based on gas convective heat transfer has been investigated in the past sixteen years[1-4]. The thermal accelerometer shows great advantages of high – shock resistance and low cost. A gas gyroscope based on convective heat transfer was also reported[5]. Like the gas accelerometer, this gyroscope has simple structure with a heater, four thermistors and a sealed chamber. However, this type of convective gyroscope has great rotation – acceleration coupling effects due to the natural convection flow induced by buoyancy that is proportional to the linear acceleration imposed on the gas[6,7], which limits its applications. In this paper, we present a new micromachined thermal integrated gyroscope and accelerometer based on gas thermal expansion, which demonstrates a wide range of rotation measurement and low coupling effects.

2 Operation principle

The device shown in Fig. 1 consists of three heaters (ht_1, ht_2, ht_3), four thermistors (R_1, R_2, R_3, R_4) and a sealed micro chamber, which is filled with gas medium. Three heaters are suspended over a cavity etched on silicon substrate, while four thermistors are suspended symmetrically between two adjacent heaters. The heaters and thermistors form the working plane.

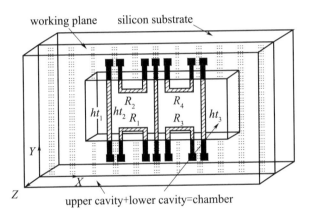

Fig. 1 Schematic diagram of the thermal sensor

Two seismic gas streams based on thermal expansion are generated in the working plane by alternately heating and cooling the heaters. Unlike the natural convection flow induced by buoyancy, the expansion flow is induced by transient temperature change that is independent of accelerations, and thus prevents coupling effect of the acceleration on the rotation measurement. The natural convection in the chamber can be restrained by compressing the chamber height and using an appropriate gas medium. The driving signal is alternating periodic wave. At the first half period, with ht_2 heating up and ht_1, ht_3 cooling down, the gas near ht_2 expands and generates two streams in opposite directions [Fig. 2 (a)]. While at the second half, the two streams reverse [Fig. 2 (b)] with ht_2 cooling down and ht_1, ht_3 heating up. A symmetrical temperature field distribution is generated along the Y - axis [Fig. 4 (a)].

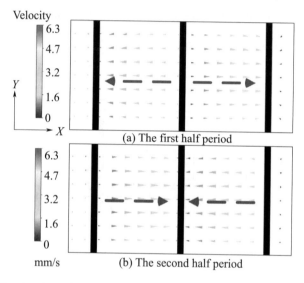

Fig. 2　Simulated periodic gas flows without rotation

When Z - axis rotation occurs, the Coriolis acceleration along the Y - axis ($\vec{a}_c = -2\vec{\omega}_z \times \vec{v}$, where $\vec{\omega}_z$ refers to Z - axis angular rate, $\vec{v}$ refers to gas velocity along the X - axis) is generated and deflects the two streams to the Y - axis direction [Fig. 3 (a), (b)], resulting in the deflection of temperature profile [Fig. 4 (a)] in the Y - axis direction that is detected by the four thermistors symmetrically distributed in the chamber.

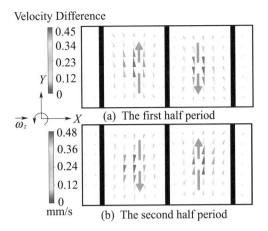

Fig. 3 Simulated relative velocity of periodic gas flows induced by Coriolis acceleration under rotation $\vec{\omega}_z$

The deflected directions of temperature profiles on two sides of the Y – axis are opposite due to the opposite Coriolis accelerations induced by the Z – axis rotation. The angular rate output, Out1, which is proportional to $\vec{\omega}_z$, is formulated as Eq. 1,

$$\text{Out1} = T(R_1) - T(R_2) - T(R_3) + T(R_4) \quad (1)$$

where $T(R)$ is the temperature of the thermistor R.

The temperature profile deflections induced by a linear acceleration on two sides of the Y – axis are in the same direction, Fig. 4 (b). The Y – axis acceleration output can be acquired via formulating the Out2 as Eq. 2.

$$\text{Out2} = T(R_1) - T(R_2) + T(R_3) - T(R_4) \quad (2)$$

Hence, for combining the rotation and acceleration sensing based on thermal principle, we propose an improved conditioning circuit (Fig. 5), with which the Z – axis angular rate and Y – axis acceleration

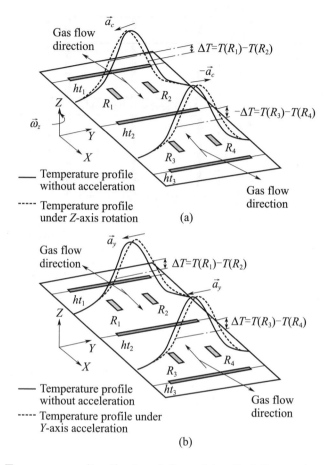

Fig. 4 Temperature distribution deflected by Coriolis acceleration (a) and linear acceleration during the second half period (b)

are detected simultaneously in one chip. Two Wheatstone bridges consist of the four thermistors (R_1, R_2, R_3, R_4) and four reference resistances (R). Since the signal is modulated by the driving signal, the demodulation circuit is utilized to extract the angular rate and acceleration, following the two-level differential amplifiers.

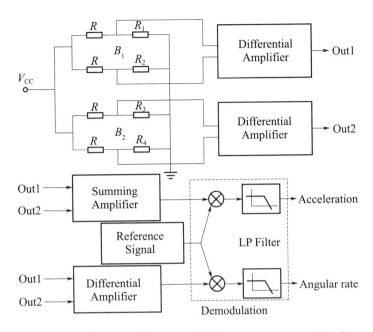

Fig. 5　Schematic diagram of the measurement circuit

3　Numerical simulation

Numerical simulations using software ANSYS/CFX are conducted to validate the device and help to optimize geometric parameters. Nitrogen ideal gas is selected as the gas medium in the chamber and the heaters are driven by square wave power. Simulation results in Fig. 6 verify the seismic gas flow [Fig. 6 (b)], the temperature difference under Z - axis rotation [Fig. 6 (c)] and the temperature difference under Y - axis acceleration [Fig. 6 (d)] by combining the temperatures of the four thermistors, depicted in Eq. (1) - (2). Finally, the demodulation algorithm is used to extract angular rate and acceleration. Fig. 7 and Fig. 8 validate Z - axis angular rate measurement and Y - axis acceleration measurement, respectively.

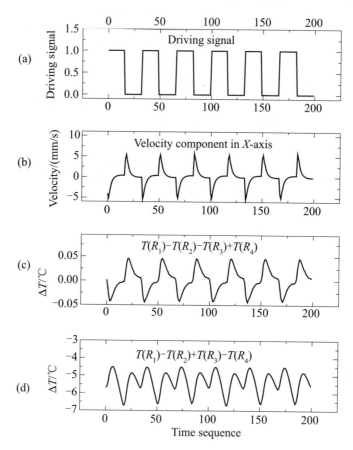

Fig. 6 Timing diagram of (a) driving signal of ht_2, (b) flow velocity at the midpoint between ht_1 and ht_2, (c) angular rate output with positive Z-axis rotation and (d) acceleration output with positive Y-axis acceleration

4 Fabrication process

The fabrication process and prototype are shown in Fig. 9 and Fig. 10. After depositing 1.5 μm thick LPCVD low stress silicon nitride, the metal layer, including 160 nm thick Platinum and Cr-adhesion

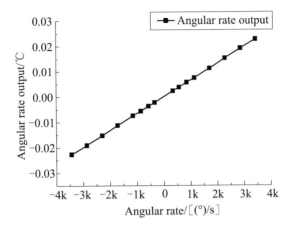

Fig. 7 Simulation result of angular rate output with respect to Z - axis rotation

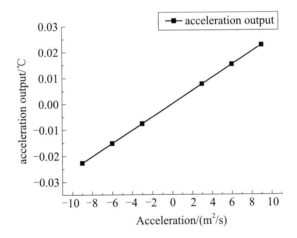

Fig. 8 Simulation result of acceleration output with respect to Y - axis acceleration

layer, is patterned [Fig. 9 (a), (b)]. Then RIE (Reactive Ion Etching) is utilized to shape the etching window of the cavity [Fig. 9 (c)]. Afterwards, as shown in Fig. 9 (d), the cavity is obtained using wet anisotropic etching. Finally, the chip is hermetically packaged in a metal cap with a chamber height of 0.7 mm and filled with nitrogen as gas medium.

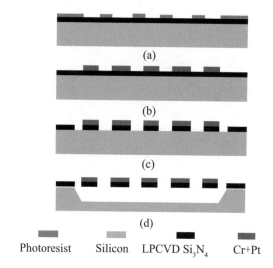

Fig. 9 Outline of the fabrication process

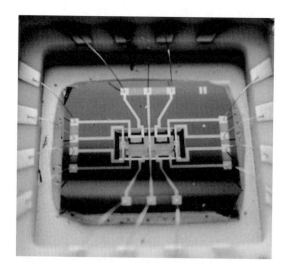

Fig. 10 Prototype of an unpackaged device

5 Experiments

Experiments are conducted to validate the device. Fig. 11 indicates the angular rate output with respect to the Z - axis angular rate ranging from $-3\,000$ (°) /s to $+3\,000$ (°) /s with a nonlinearity of 0.57%.

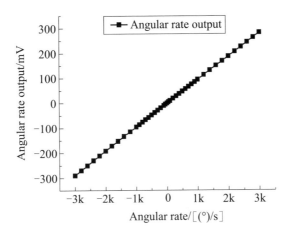

Fig. 11　Experimental result of Z - axis angular rate measurement

The measurements of the cross coupling among the 3 - axis rotations in the range of -300 (°) /s to $+300$ (°) /s are shown in Fig. 12, where the X - axis and Y - axis rotations generate less than 4% cross coupling effect on the output of Z - axis rotation.

The acceleration measurements are conducted using a tilt table. Fig. 13 shows the acceleration outputs of the device when the gravity changes direction in the XY, XZ, and YZ planes respectively, where $Y-XY$ represents the gravity changes direction in the XY plane and the acceleration component along the Y axis is as the abscissa, and so on for the definitions of $X-XZ$ and $Y-YZ$. The measurement results indicate the X - axis and Z - axis accelerations generate less than 6% cross coupling effect on the Y - axis acceleration output.

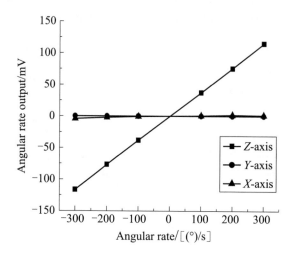

Fig. 12　Experimental results of the angular rate outputs under 3 - axis rotations

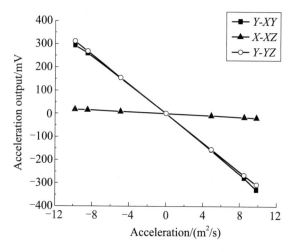

Fig. 13　Experimental results of 3 - axis acceleration measurements

The coupling effects between the gyroscope and accelerometer are also experimentally tested. Fig. 14 presents the coupling effects of 3 -

axis accelerations on the scale factor (sensitivity) of angular rate output. The previous reports[5,6] have shown that the linear acceleration will greatly influence the sensitivity of the convective gyroscope due to the natural convection induced by buoyancy that depends on linear accelerations, while the coupling effects of accelerations on the gyroscope sensitivity are minimized to be about 5% in this new device by using the thermal expansion rather than natural convection. The bias sensitivity couplings between the gyroscope and accelerometer are tested to be less than 10 (°)/s/g and 7×10^{-5} g/(°)/s, which can be further improved by adjusting the conditioning circuit and reducing the structure non-uniformity.

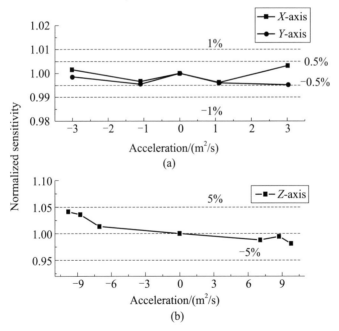

Fig. 14 Coupling effects of the 3-axis accelerations on the sensitivity of the angular rate output

6 Conclusion

An integrated gyroscope and accelerometer based on gas thermal expansion is presented in this paper. Experimental results, which agree with numerical simulations, validate the feasibility of the simultaneous measurements of Y - axis acceleration and Z - axis angular rate in one chip. Besides the advantages of simple structure and high - shock resistibility, the coupling effects of accelerations on the gyroscope sensitivity are inhibited via using gas thermal expansion. The device also exhibits wide - range rotation detectability with good linearity.

7 Acknowledgement

This work is supported by the National High - tech Program '863' of China under the grants 2012AA02A604.

References

[1] A M Leung, J Jones, E Czyzewska, J Chen and B Woods. Micromachined accelerometer based on convection heat transfer. Proceedings of the IEEE Micro Electro Mechanical Systems, Heidelberg, Germany, Jan. 25 - 29, 1998, pp. 627 - 630.

[2] A Garraud, P Combette, J M Gosalbes, B Charlot and A Giani. First high - g measurement by thermal accelerometers. in Digest Tech. Papers Transducers'11 Conference, Beijing, June 5 - 9, 2011, pp. 84 - 87.

[3] X B Luo, Y J Yang, F Zheng, Z X Li, Z Y Guo. An optimized micromachined convective accelerometer with no proof mass. Micromech. Microeng. Journal, vol. 11, pp. 504 - 508, 2001.

[4] F Maily, A Martinez. Design of a micromachined thermal accelerometer: thermal simulation and experimental results. J. Microelectronics, vol. 34 pp. 275 - 280, 2003.

[5] R Zhu, H G Ding, Y Su, Z Y Zhou. Micromachined gas inertial sensor based on convection heat transfer. Sensors &. Actuators: A. Physical, vol. 130 - 131, pp. 68 - 74, 2006.

[6] R Zhu, H G Ding, Y J Yang, Y Su, Sensor Fusion Methodology to Overcome Cross - axis Problem for Micromachined Thermal Gas Inertial Sensor. IEEE Sensors Journal. vol. 9, no. 6, pp. 707 - 712, 2009.

[7] R Zhu, H G Ding, Y Su and Y J Yang. Modeling and Experimental Study on Characterization of Micromachined Thermal Gas Inertial Sensors, Sensors. vol. 10, no. 9, pp. 8304 - 8315, 2010.

A Micromachined Gas Inertial Sensor Based on Thermal Expansion

(2014 年)

Abstract Thermal expansion in volume is a natural phenomenon as a result of change in temperature. However, it is used rarely as an actuating source of movement in macrocosm due to its weak motion. In this paper we propose to utilize a gas thermal expansion in microcosm to generate seismic gaseous mass for sensing inertial quantities including angular rate and acceleration. The expansion – based inertial sensor possesses the advantages of simple structure, low fabrication cost, high shock resistance, large – range rotation gauge, and low coupling between rotation and acceleration. We present the theory principle for applying thermal expansion to inertial sensor, and conduct the validation of simulations and experiments on the new sensor. The results indicate that the sensor with a tailor – made read – out circuit is effective to simultaneously detect both of the Z – axis rotation and the Y – axis acceleration, and exhibits a good linearity of angular rate output with respect to a large – range rotation of $\pm 3\,000$ (°) /s.

本文发表于 Sensors and Actuators A：Physical. 合作作者：Rong Zhu, Songlin Cai, Henggao Ding, Yongjun Yang, Yan Su。

A Micromachined Gas Inertial Sensor Based on Thermal Expansion

Keywords Inertia sensor; Gyroscope; Accelerometer; Thermal expansion; Micro electromechanical system (MEMS)

1 Introduction

Development of micromachined inertial sensors has been widely addressed for many years. Most micromachined inertial sensors generally use a mechanical structure including a solid proof mass suspended on springs[1]. For example, several micromachined vibrating gyroscopes including vibrating shells[2], tuning-forks[3], and vibrating beams[4] have been demonstrated. These inertial sensors, such as gyroscopes are based on the movement of a seismic proof mass caused by an inertial quantity. The mechanical proof mass configuration raises the complexity of structure and fabrication, and especially restricts the high shock resistance of the device. Another concept of inertial sensor using fluid medium instead of mechanical proof mass as key moving and sensing elements has been reported[5-8]. These sensors are called fluid inertial sensors, the sensing principle of which is that the inertial-induced deflection of the gas or liquid flow is detected using the distributed thermistor wires. Based on different driving mode, the sensors can be categorized into the fluid gyroscope using jet flow[5,6], the thermal accelerometer and gyroscope based on heat convection[7-9]. The former sensor uses a jet flow that is driven by piezoelectric diaphragm pump[5] or electroconjugate fluid that needs high DC voltage of several kilovolts applying between the electrodes[6]. For generating an appropriate jet flow, elaborate elements are needed, such as diaphragm, nozzle orifice, or needle electrode, which restricts the miniaturization of the devices. Another inertial sensor based on thermal convection possesses a relative simple structure with one or several heater and thermistor wires, which detect a gaseous flow deflection coming from inertial - induced heat convection surrounding the heater wire[8]. The convective inertial sensors including accelerometer and gyroscope have exhibited great advantages of high - shock resistance and low cost[9]. However, the general convective gyroscope has a coupling effect with external acceleration because the natural convection current originates from buoyancy that is proportional to the external acceleration imposed on the gas [8,9].

In this paper we propose to utilize a different gas flow that is driven from thermal expansion to develop a novel micromachined inertial sensor. Gas expands or contracts in volume as a result of change in temperature. Gas thermal expansion flow exists generally in nature, but it is rarely acted as an actuating source of the movement because it is weak in macro - scale compared with other driving sources, such as forced convection flow generated by external mechanical sources or natural convection flow generated by density differences in the fluid occurring due to temperature gradients. However, if the thermal expansion/contraction is driven in a micro - scale chamber by alternately heating and cooling, the gas flow of which can be greatly enhanced so as to be possible serving as a driving source. Therefore, we propose to use the thermal expansion of gaseous medium in a micro chamber to generate the seismic current along a specific path and detect the deflection of the current induced by the Coriolis acceleration that is proportional to the angular rate. Due to use the same measuring principle with thermal accelerometer to detect the Coriolis acceleration, the sensor can be also acted as an accelerometer. The configuration of the sensor is comprised of three heater wires, distributed four thermistor wires between heaters and a sealed micro chamber, which is filled with gas medium[10]. The expansion flow is independent of external linear acceleration and thus diminishes the coupling effect between the acceleration and the angular rate measurements of the sensor.

2 Theory and operation principle of thermal inertial sensor

Natural convection is a mechanism, or type of heat transport, in which the fluid motion is not generated by any external mechanical source (like a pump, fan, suction device, etc.) but only by density differences in the fluid occurring due to temperature gradients. In natural convection, fluid surrounding a heat source receives heat, becomes less dense and rises. The surrounding, cooler fluid then moves to replace it. This cooler fluid is then heated and the process continues, forming a convective current. The driving force for natural convection is buoyancy, a result of differences in fluid density. Because of this, the presence of acceleration such as arises from gravity, or an equivalent force (arising from linear acceleration, centrifugal force or Coriolis effect), is essential for natural convection. Due to the above reason, the convection flow can be used to detect rotation for an inertial sensor, but the sensor is greatly influenced by the body acceleration occurring due to the gravity, and centrifugal effect. Considering another flow existent in the fluid environment, such as jet flow and thermal expansion flow which is driven and dominated by an external source rather than the body acceleration, the coupling between the flow and acceleration can be diminished. Thermal expansion is the tendency of matter to change in volume in response to a change in temperature[11]. Compared with the jet flow, the thermal expansion flow is easily produced by alternately heating and cooling the two heater sources. As one of the heater heats up, the fluid particles between two heaters move about and flow from the hot heater to the cool heater. As another heater heats up while the previous heating heater cools down, the fluid particles run reversely and generate the back-flow between the two heaters.

For understanding the mechanisms of fluid flows applied in the inertial sensors, a computational fluid dynamics (CFD) model is developed as following to describe the three-dimensional flow (ignoring jet flow) in the sensor chamber, which can be solved to obtain the transient flow velocity, temperature and pressure of gas in the chamber.

$$\frac{\partial \rho}{\partial t} + \nabla \cdot (\rho \vec{U}) = 0 \tag{1}$$

$$\frac{\partial \rho \vec{U}}{\partial t} + \nabla(\rho \vec{U} \otimes \vec{U}) = -\nabla P + \rho \vec{f} + \mu \nabla \cdot \tau \tag{2}$$

$$\frac{\partial T}{\partial t} + \vec{U} \cdot \nabla T = \alpha \nabla^2 T \tag{3}$$

$$P = \rho R_g T \tag{4}$$

where $\tau = \nabla\vec{U} + (\nabla\vec{U})^T - (2/3)\sigma \nabla\cdot\vec{U}$, $\sigma = \begin{pmatrix} 1 & 0 & 0 \\ 0 & 1 & 0 \\ 0 & 0 & 1 \end{pmatrix}$; $\vec{r} = (x, y, z)$ refers to the position vector; $\vec{U} = (U_x, U_y, U_z)^T$ refers to velocity vector; $\vec{f} = (g_x, g_y, g_z)^T$ refers to the acceleration vector; ρ, P, μ, T are gas density, pressure, dynamic viscosity and temperature; $\alpha = \lambda/\rho C_p$, λ is thermal conductivity, C_p is specific heat, R_g is gas constant. The vector operator ∇ is defined as gradient $\nabla \equiv i(\partial/\partial x) + j(\partial/\partial y) + k(\partial/\partial z)$, the operator ∇ is defined as divergence $\nabla \cdot \vec{U} \equiv \partial U_x/\partial x + \partial U_y/\partial y + \partial U_z/\partial x$. $\vec{U} \otimes \vec{U} = \begin{pmatrix} U_x U_x & U_x U_y & U_x U_z \\ U_y U_x & U_y U_y & U_y U_z \\ U_z U_x & U_z U_y & U_z U_z \end{pmatrix}$.

The above momentum Eq. (2) characterizes the kinetic process of the three-dimensional flow. Considering about the rotation $\vec{\Omega} = (\Omega_x, \Omega_y, \Omega_z)$ that refers to angular rate vector, and expanding the two items in the left side of Eq. (2), the momentum equation can be equivalently converted to the following:

$$\rho \frac{\partial \vec{U}}{\partial t} + \rho \vec{U} \cdot \nabla \vec{U} + \vec{U}(\nabla \cdot (\rho \vec{U})) = -\nabla P + \rho \vec{f} - \vec{U}\frac{\partial \rho}{\partial t} - 2\rho \vec{\Omega} \times \vec{U} - \rho \vec{\Omega} \times (\vec{\Omega} \times \vec{r}) + \mu \nabla \cdot \tau$$

(5)

Define the coefficient of thermal expansion as α_v and assume admitting a small variation of density with temperature[12]:

$$\alpha_v = \frac{1}{V}\frac{\partial V}{\partial T}\bigg|_p = -\frac{1}{\rho}\frac{\partial \rho}{\partial T}\bigg|_p \approx -\frac{1}{\rho}\frac{\rho - \rho_0}{T - T_0} \quad (6)$$

$$\nabla P = \rho_0 \vec{f} \quad (7)$$

where T_0 refers to the ambient temperature, ρ_0 refers to the gas density at the ambient temperature. Substitute Eqs. (6) and (7) into the momentum Eq. (5), then there is

$$\frac{\partial \vec{U}}{\partial t} + \vec{U} \cdot \nabla \vec{U} + \vec{U}(\nabla \cdot \vec{U}) = \alpha_v (T - T_0)\vec{f} + \alpha_v \vec{U}\frac{\partial T}{\partial t} - 2\vec{\Omega} \times \vec{U} - \vec{\Omega} \times (\vec{\Omega} \times \vec{r}) + \frac{\mu}{\rho}\nabla \cdot \tau \quad (8)$$

Define $\overline{\vec{U}} = \vec{U}/(v/H) = (U_x/(v/H), U_y/(v/H), U_z/(v/H))^T$, $\overline{\vec{r}} = \vec{r}/H = (x/H, y/H, z/H)^T$, $\Theta = (T - T_0)/(T_h - T_0)$. $\overline{\vec{\Omega}} = \vec{\Omega}/(v/H^2) = (\Omega_x/(v/H^2), \Omega_y/(v/H^2), \Omega_z/(v/H^2))^T$, $\delta = tv/H^2$, where T_h is the heater temperature, $v(v = \mu/\rho)$ is momentum viscosity, H is the feature dimension. Then we put a non-dimensionalization to the momentum Eq. (8) and get a non-dimensional model:

$$\frac{\partial \overline{\vec{U}}}{\partial \delta} + \overline{\vec{U}} \cdot \overline{\nabla}\overline{\vec{U}} + \overline{\vec{U}}(\overline{\nabla} \cdot \overline{\vec{U}}) = \vec{Gr}\Delta\Theta + \alpha_v \frac{\partial T}{\partial \delta}\overline{\vec{U}} - 2\overline{\vec{\Omega}} \times \overline{\vec{U}} - \overline{\vec{\Omega}} \times (\overline{\vec{\Omega}} \times \overline{\vec{r}}) + \overline{\nabla} \cdot \tau \quad (9)$$

where the Grashof number $\vec{Gr} = (Gr_x, Gr_y, Gr_z) = (\alpha_v(T_h - T_0)g_x H^3/v^2, \alpha_v(T_h - T_0)g_y H^3/v^2, \alpha_v(T_h - T_0)g_z H^3/v^2)$, $\overline{\nabla} = (\partial/\partial \overline{x}, \partial/\partial \overline{y}, \partial/\partial \overline{z})^T$, $\overline{\tau} = \overline{\nabla}\overline{\vec{U}} + (\overline{\nabla}\overline{\vec{U}})^T - (2/3)\sigma \overline{\nabla} \cdot \overline{\vec{U}}$. The first item $\vec{Gr}\Delta\Theta$ at the right of Eq. (8) is a driving force to generate the heat convection flow that is correlative with the Grashof number $\vec{Gr}$; the second item $\alpha_v(\partial T/\partial \delta)\vec{U}$ at the right of Eq. (8) is another driving force to generate the expansion flow that is correlative with the transient variation rate of the temperature, which is induced by alternately heating and cooling of the heaters.

From Eq. (9), we can analyze the inertial sensing mechanism for thermal accelerometer and thermal gyroscope. For sensing acceleration, the above equation can be simplified as the following formula by setting the rotation as zero and ignoring the transient variation of the temperature

$$\frac{\partial \vec{U}}{\partial \delta}+\vec{U}\cdot\nabla\vec{U}+\vec{U}(\nabla\cdot\vec{U})=\vec{Gr}\Delta\Theta+\nabla\cdot\bar{\tau} \tag{10}$$

The fluid flow $\vec{U}$ is generated due to the buoyancy, a result of the temperature difference $\Delta T = T_h - T_0$ and the acceleration $\vec{g}=(g_x, g_y, g_z)$. This fluid flow induces the deflection of the temperature profile based on the energy Eq. (3) that is detected by using the distributed thermistors. Tracing to the source, we can see that the output of the thermal accelerometer is dominated by the Grashof number $\vec{Gr}$.

For sensing the angular rate of the rotation, we firstly consider the flow vectors generated by two driving sources: buoyancy and volumetric thermal expansion. Two driving sources result in two flow vectors, defined as $\vec{U}_c$ for the convective flow and $\vec{U}_e$ for the expansion flow. When the rotation $\vec{\Omega}$ occurs, the non-dimensional Coriolis accelerations $-2\vec{\Omega}\times(\vec{U}_c+\vec{U}_e)$ are accordingly engendered that induces the deflection of the temperature field and brings out the output of the gyroscope. Obviously, the flow $\vec{U}_c$ interconnects with the acceleration, and the flow $\vec{U}_e$ is absolutely independent. Accordingly, the thermal convection gyroscope exists a high coupling with the external acceleration, specifically the scale factor (or sensitivity) of the gyroscope is dependent on the acceleration, which has been demonstrated in our previously experiments[8,9]. However, the thermal expansion gyroscope is completely isolated. which is correlated with the expansion number $\alpha_v(\partial T/\partial \delta)$ that is proportional to the transient variation rate of the temperature.

Based on the above analysis, we propose a new type of thermal expansion sensor incorporating gyroscope with accelerometer. For eliminating the influence of external disturbance, we adopt differential structure design by using three heaters to modulate the center heater ht_2 and the two side heaters ht_1 and ht_3 in alternating way so as to generate two flow streams in opposite direction, as shown in Fig. 1. When a rotation around the Z - axis occurs, the flow stream along the X - axis will be deflected to the Y - axis due to the Coriolis acceleration ($\vec{a}_c = -2\vec{\omega}_z \times \vec{v}_x$, where $\vec{\omega}_z$ refers to Z - axis angular rate, $\vec{v}_x$ refers to gas velocity along the X - axis), resulting in the deflection of temperature profile in the Y - axis direction that is detected by the four thermistors symmetrically distributed in the chamber.

The temperature profiles in the Y - axis direction are shown in Fig. 2, where Fig. 2 (a) represents the deflection of the temperature

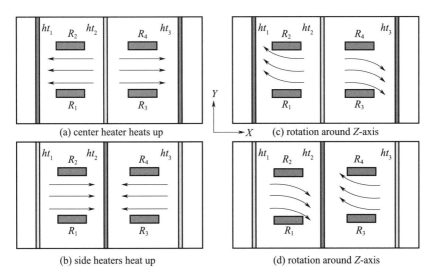

Fig. 1 Configuration of the sensor and flow stream diagram. (a) Center heater heats up, (b) rotation around Z-axis, (c) side heaters heat up and (d) rotation around Z-axis

profile induced by the Coriolis acceleration due to the rotation $\overline{\omega}_z$ occurs, Fig. 2 (b) represents the deflection of the temperature profile induced by an external linear acceleration. The Coriolis accelerations on two sides of the Y-axis are opposite and thus deflect the temperature profiles to the opposite directions along the Y-axis. Therefore the Z-axis rotation can be read out by the following formula assuming the gas velocity along the X-axis is fixed

$$\text{Output}_{GyroZ} \propto [T(R_1) - T(R_2)] - [T(R_3) - T(R_4)] \propto \omega_z \tag{11}$$

where $T(R_1)$, $T(R_2)$, $T(R_3)$, $T(R_4)$ are the temperatures of the four thermistors. However, the external linear acceleration deflects the temperature profiles on the two sides of the Y-axis in the same direction, and thus the Y - axis acceleration can be read out by the following formula assuming the gas velocity along the X - axis is fixed.

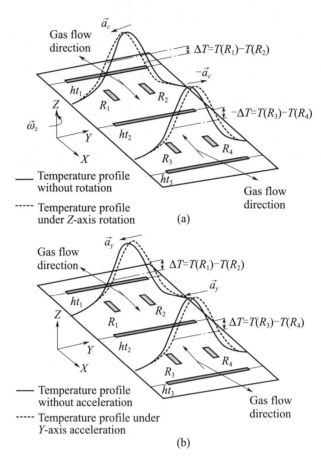

Fig. 2 Temperature distribution deflected by Coriolis acceleration (a) and a linear acceleration (b)

$$\text{Output}_{\text{AccY}} \propto [T(R_1) - T(R_2)] + [T(R_3) - T(R_4)] \propto a_y \quad (12)$$

Thermistors can be operated either in constant voltage (CV) and constant current (CC) modes as shown in Fig. 3, where R is the reference resistances for balancing the Wheatstone bridge. Assuming the gas velocity along the X - axis is fixed, the gyroscope outputs under CV and CC respectively can be deduced by

$$\text{Output}_{\text{GyroZ}}^{\text{CV}} \propto \frac{U_C R(R_1 - R_2)}{(R+R_1)(R+R_2)} - \frac{U_C R(R_3 - R_4)}{(R+R_3)(R+R_4)} \approx \frac{\alpha R_0 U_C \Delta T_G}{4R} \tag{13}$$

$$\text{Output}_{\text{GyroZ}}^{\text{CC}} \propto \frac{I_C R(R_1 - R_2)}{2R + R_1 + R_2} - \frac{I_C R(R_3 - R_4)}{2R + R_3 + R_4} \approx \frac{I_C \alpha R_0 \Delta T_G}{4} \tag{14}$$

where U_C and I_C are the constant voltage and constant current applied into the bridge in the CV and CC modes. R_0 is the resistance of the thermistors at 0 ℃ and α is the temperature coefficient of resistance (TCR) of the thermistors. $\Delta T_G = [T(R_1) - T(R_2)] - [T(R_3) - T(R_4)]$. Assuming the gas velocity along the X - axis is fixed, the accelerometer outputs under CV and CC respectively can be deduced by

$$\text{Output}_{\text{AccY}}^{\text{CV}} \propto \frac{U_C R(R_1 - R_2)}{(R+R_1)(R+R_2)} + \frac{U_C R(R_3 - R_4)}{(R+R_3)(R+R_4)} \approx \frac{\alpha R_0 U_C \Delta T_A}{4R} \tag{15}$$

$$\text{Output}_{\text{AccY}}^{\text{CC}} \propto \frac{I_C R(R_1 - R_2)}{2R + R_1 + R_2} + \frac{I_C R(R_3 - R_4)}{2R + R_3 + R_4} \approx \frac{I_C \alpha R_0 \Delta T_A}{4} \tag{16}$$

where $\Delta T_A = [T(R_1) - T(R_2)] + [T(R_3) - T(R_4)]$.

From Eqs. (13) - (16), it can be seen that the sensitivity of the gyroscope and accelerometer depends on the applied voltage or current, and also depends on the TCR of the thermistors.

3 Numerical simulation and analysis

The behaviors of the expansion flow in a micro chamber with

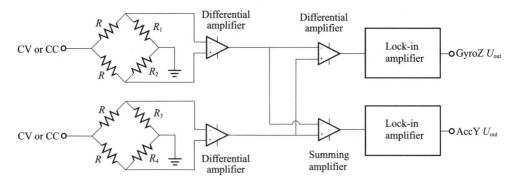

Fig. 3　Schematic diagram of the read-out circuit of the four thermistors

dimension of 2 600 μm×1 200 μm×1 000 μm ($X \times Y \times Z$) filled with nitrogen are simulated by using ANSYS/CFX software. The origin of the coordinate system is located at the lower left corner of the chamber. Considering the gap distance of 300 μm from the side heater to the sidewall, the flow and temperature simulation results are only shown between two side heaters, i. e. from 300 μm to 2 300 μm in the X-axis. The simulated flows actuated in the first and second half periods are shown in Fig. 4. Fig. 4 (a) and (b) depicts the X-axis velocity of the flow at the first and second half period, Fig. 4 (c) and (d) depicts the Y-axis velocity, Fig. 4 (e) depicts the temperature profile across the thermistors R_1 and R_2 (or R_3 and R_4). When a rotation occurs, the flow and temperature profiles deflect along the Y-axis as shown in Fig. 5. Fig. 6 further demonstrates the time sequences of the seismic gas flow [Fig. 6 (b)] driven by the square wave power of the heaters [Fig. 6 (a)], the temperature difference under Z-axis rotation [Fig. 6 (c)] and the temperature difference under Y-axis acceleration [Fig. 6 (d)] by combining the temperatures of the four thermistors, depicted in Eqs. (11) and (12).

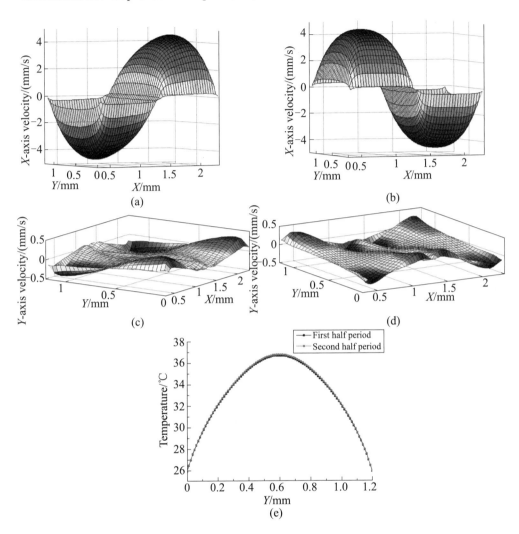

Fig. 4 Simulations of the expansion flow in a micro chamber. X-axis velocity, (a) at the first half period when the center heater heats up and (b) at the second half period. Y-axis velocity, (c) at the first half period and (d) at the second half period. (e) Temperature profiles across the thermistors R_1 and R_2 at the first and second half period

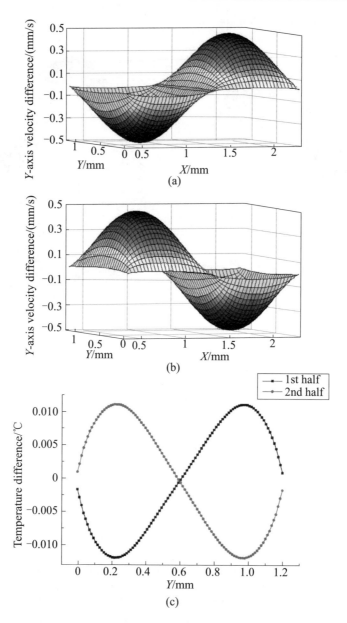

Fig. 5 Simulation of the difference of Y-axis velocities and temperature profile between rotation case and non-rotation case. (a) Difference of Y-axis velocities between rotation case and non-rotation case at the first half period, (b) difference of Y-axis velocities between rotation case and non-rotation case at the second half period and (c) temperature differences between rotation case and non-rotation case at the first and second half period

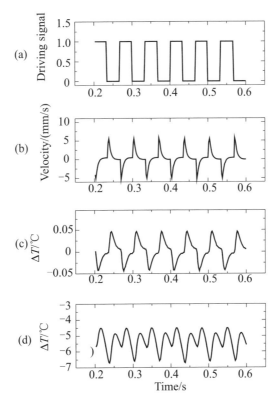

Fig. 6 Timing diagram of (a) driving signal of the center heater, (b) X - axis flow velocity at the midpoint between ht_1 and ht_2, (c) the temperature difference under Z - axis rotation $[T(R_1)-T(R_2)]-[T(R_3)-T(R_4)]$ and (d) the temperature difference under Y - axis acceleration $[T(R_1)-T(R_2)]+[T(R_3)-T(R_4)]$

As seen in Fig. 6, the temperature differences under rotation/acceleration are periodic signals due to the alternating gas velocity in the chamber modulated by the square wave power of the heaters. The periodic output signals need to be demodulated to deduce the rotation and acceleration outputs. The demodulation algorithm of extracting the amplitude of the modulated periodic signals is used to extract the angular rate and the acceleration from the signals of Fig. 6 (c) and (d).

Figs. 7 and 8 demonstrate the demodulated output proportional to angular rate and the demodulated output proportional to acceleration respectively. The actual demodulator adopted in the device is realized by using lock-in amplifier shown in Fig. 3.

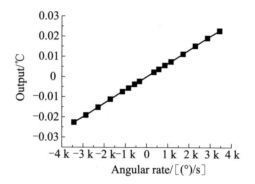

Fig. 7　Simulation result of demodulated output proportional to angular-rate with respect to Z-axis rotation.

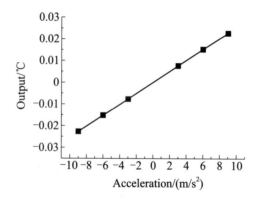

Fig. 8　Simulation result of demodulated output proportional to acceleration with respect to Y-axis acceleration

4　Fabrication and primary experiments

The sensor prototypes are fabricated using the same fabrication techniques of thermal gyroscope and accelerometers that have been already detailed in[8-10]. The process starts with depositing 1.5 μm thick LPCVD low stress silicon nitride and 160 nm thick platinum and Cr – adhesion metal layer on a 400 μm thick silicon wafer. The Cr/Pt film is etched by using photolithography to form the heaters and thermistors. Then RIE (reactive ion etching) is utilized to shape the etching window of the cavity. Afterwards, the silicon is etched using wet anisotropic etching to form a cavity with 2 600 μm length × 1 200 width × 200 μm depth and suspend the heaters and thermistors. The chip is hermetically packaged in a metal cap with a chamber height of 0.7 mm and filled with nitrogen as gas medium. Platinum resistors are used as the heater and thermistors. The resistances of the heaters are around 600 Ω and the resistances of the thermistors are around 900 Ω. The operating heater power is about 10 mW with the voltage of 4 V_{pp} at the frequency of 15 Hz. The four thermistors are operated in CC modes. The supplied current to each thermistor in the Wheatstone bridge is 1.5 mA. The fabricated prototypes of unpackaged and packaged devices are shown in Fig. 9.

Experiments are conducted to validate the device at room temperature around 20 ℃. The rotation test is performed using a turntable that has an angular rate range of ±3 000 (°)/s with error less than 0.01 (°)/s. The device is mounted at the center of the hori‑zontal plane of the turntable. The turntable is rotated around the vertical axis from −3 000 (°)/s to +3 000 (°)/s at 100 (°)/s per step (in the range of −100 (°)/s to +100 (°)/s at 50 (°)/s per

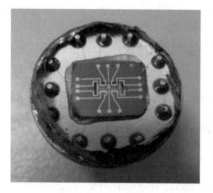

Fig. 9 Prototypes of unpackaged and packaged devices

step), while the readouts of the sensor with an amplification of about 36 000 are transferred to a PC for data collection at each step. Fig. 10 demonstrates the angular rate output over the Z - axis angular rate ranging from $-3\ 000\ (°)/s$ to $+3\ 000\ (°)/s$. The nonlinearity in full - scale range is less than 0.57%.

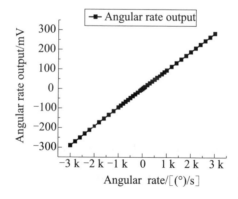

Fig. 10 Experimental result of Z - axis angular rate measurement

The cross coupling test is conducted by adjusting the Z, Y and X axes of the device along the vertical axis (i.e. rotating axis) of the turntable respectively and exporting the readouts of the sensor with an amplification of about 144 000. The measurements of the cross coupling among the 3-axis rotations in the range of −300 (°)/s to +300 (°)/s are shown in Fig. 11. The results indicate the X-axis and Y-axis rotations generate less than 4% cross coupling errors on the output of Z-axis rotation. The cross coupling effects among the rotation axes may be aroused by the misalignment of the sensor in the test.

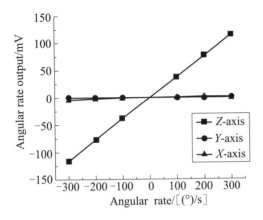

Fig. 11　Experimental results of the angular rate outputs under 3-axis rotations

The acceleration measurements are conducted using a rotary dividing table with dividing precision better than 0.02°. The sensor is mounted on the vertical plane of the table, in which the direction of the gravity turns around. Fig. 12 shows the acceleration outputs of the device under an amplification of about 10 000 when the gravity changes direction in the XY, XZ, and YZ planes respectively, where Y-XY represents the gravity changing direction in the XY plane and the acceleration component along the Y axis is as the abscissa, and the same definitions for X-XZ and Y-YZ. The measurement results indicate the X-axis and Z-axis accelerations generate less than 6% cross coupling errors on the Y-axis acceleration output. Similarly, the cross coupling effects among the acceleration axes may be aroused by the misalignment of the sensor in the test.

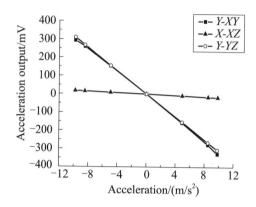

Fig. 12 Experimental results of 3-axis acceleration measurements

The coupling between the rotation output and acceleration is further tested using the centrifugal method and tilting method. Firstly the sensor is mounted eccentrically on the turntable with the Z-axis of the sensor along the rotating direction that is also the direction of gravity. When the turntable is rotating, a centrifugal acceleration along the radial direction is imposed onto the sensor. According to the sensor axis along the radial direction, the centrifugal acceleration can be modulated along the X-axis or the Y-axis of the sensor respectively. Fig. 13 (a) and (b) demonstrates the results and shows the coupling effects of the X-axis and Y-axis accelerations to the scale factor (sensitivity) of the Z-axis rotation output are less than 0.5%. The coupling effect between the Z-axis rotation output and the Z-axis acceleration is tested using the tilting method where the rotating axis is tilted (i. e. the gravity component along the Z – axis varies with tilt). Fig. 13 (c) demonstrates the testing results, where the coupling effect of the Z – axis acceleration to the scale factor of the angular rate sensing is less than 5%. The low coupling between the rotation sensing output and the external acceleration validates the fact that the thermal expansion rather than the heat convection dominates the gas flow in the sensor chamber.

The experiments also indicate that the sensitivity of the angular rate sensor is about 95 μV/ (°) /s under the amplification of 36 000 and the equivalent angular rate noise is around 1 (°) /s RMS, and the sensitivity of the acceleration sensor is about 300 mV/g under the amplification of 10 000 and the equivalent acceleration noise is around 1 mg RMS. In addition, the sensor sensitivity variation depending on the ambient temperature is tested to be less than 0.025/℃ from −40 ℃ to 60℃ and the temperature drift is less than 1.3 (°) /s/℃, which can be compensated by using temperature calibration method. More study works need to be conducted in the future for further improving the performances of the device.

5 Conclusions

The gas thermal expansion is a fluid flow dominated by temperature variation. Compared with the heat convection, the thermal expansion is non – originated from buoyancy and thus isolated from body acceleration. We demonstrate the inertial sensing mechanism using the convection and expansion flow. Based on the fact that the thermal expansion flow

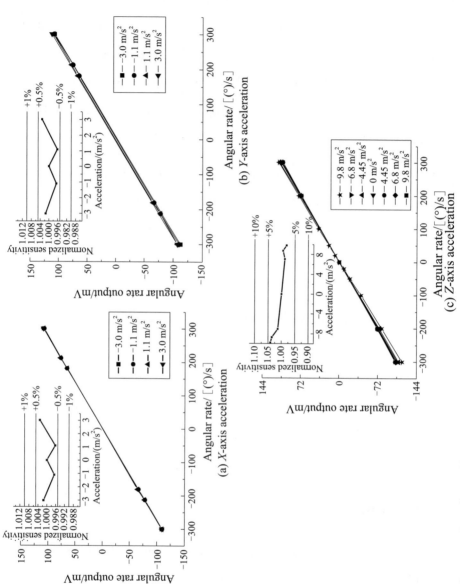

Fig. 13 Coupling effects of the 3-axis accelerations on the scale factor of the angular rate output. (a) X-axis acceleration. (b) Y-axis acceleration and (c) Z-axis acceleration

is independent of the buoyancy, we propose a new type of inertial sensor based on thermal expansion to simultaneously detect one-axis angular rate ($\pm 3\,000$ (°)/s) and one-axis acceleration ($\pm 1\,g$) and demonstrates less coupling effect between the rotation output and external acceleration. The sensor design, fabrication, and performance testing are presented, which verify the effectiveness of the sensor.

Acknowledgment

This work is supported by the National High-tech Program '863' of China under the grants 2012AA02A604 and National Natural Science Foundation of China Project (91123017).

References

[1] Y Navid, F Ayazi, K Najafi. Micromachined inertial sensors. Proc. IEEE 86 (1998) 1640-1659.

[2] F Ayazi, K Najafi, A HARPSS polysilicon vibrating ring gyroscope. J. MEMS 10 (2001) 169-179.

[3] J Bernstein. A micromachined comb-drive tuning fork rate gyroscope. in: Proceedings of the IEEE Micro Electro Mechanical Systems (MEMS), 1993, pp. 143-148.

[4] K Maenaka, T Shiozawa. A study of silicon angular rate sensors using anisotropic etching technology. Sens. Actuators A 43 (1994) 72-77.

[5] T Shiozawa, V T Dau, D V Dao. A dual axis thermal convective silicon gyroscope. in: Proceedings of the 2004 International Symposium on Micro-Nano Mechatronics and Human Science, 2004, pp. 277-282.

[6] K Takemura, S Yokota, M Suzuki, K Edamura, H Kumagai, T Imamura. A liquid rate

gyroscope using electro-conjugate fluid, Sens. Actuators A 149 (2009) 173-179.

[7] A M Leung, J Jones, E Czyzewska, J Chen, B Woods. Micromachined accelerometer based on convection heat transfer. in: Proceedings of the IEEE Micro Electro Mechanical Systems (MEMS), 1998, pp. 627-630.

[8] R Zhu, H G Ding. Y Su, Z Y Zhou. Micromachined gas inertial sensor based on convection heat transfer, Sens. Actuators A 130-131 (2006) 68-74.

[9] R Zhu, H G Ding, Y J Yang, Y Su. Sensor fusion methodology to overcome cross-axis problem for micromachined thermal gas inertial sensor, IEEE Sens. J. 9 (2009) 707-712.

[10] S L Cai, R Zhu, H G Ding, Y J Yang, Y Su. A micromachined integrated gyroscope and accelerometer based on gas thermal expansion. in: The 17th International Conference on Solid-State Sensors, Actuators and Microsystems, 2013. pp. 50-53.

[11] P A Tipler, G P Mosca. Physics for Scientists and Engineers, vol. 1, 6th ed., New York, NY, Worth Publishers, 2008, pp. 666-670.

[12] J H Lienhard IV, J H Lienhard V. A Heat Transfer Textbook, 4th ed., Phlogiston Press, Cambridge, MA, 2012, pp. 402-405.

A Temperature Compensation Method for Micromachined Thermal Gas Gyroscope

(2015 年)

Abstract This paper presents a novel temperature compensation technology to overcome the problem of temperature drift for micromachined thermal gas gyroscopes. The compensation methodology utilizes an alternating constant – temperature – difference (CTD) operation circuit to thermally drive gas motion for stabilizing the sensor scale – factor and reducing the bias temperature sensitivity of the gyroscope. Compared with other compensation methods for gyroscopes, which generally rely on temperature calibration and eliminate the temperature drift through algorithm – based compensation, the proposed method can realize self – sustained temperature compensation through an analog circuitry, which is easily operated and with good stability. Experimental results validated that the temperature dependences of the scale – factor and bias of the thermal gas gyroscope were effectively reduced.

Keywords Temperature compensation; Self – sustained; Thermal gas gyroscope; Micromachined inertial sensor

本文发表于 IEEE。合作作者：Shiqiang Liu, Rong Zhu, Henggao Ding。

1 Introduction

Thermal errors, often behaving as temperature drifts, generally exist in all micromachined inertial sensors, such as gyroscope and accelerometer. For micromachined gyroscopes, the sources of thermal errors commonly include electrical - thermal error resulted from random motion of electrons in a conductor and mechanical - thermal error resulted from molecular agitation and mechanical - thermal distortion[1-4]. In our previous work, we have proposed a micromachined thermal gas gyroscope with simple structure, high shock resistance, and large - range rotation gauge. The gyroscope used the thermal expansion of gaseous medium driven by alternately heating and cooling the heaters in a micro chamber to generate the seismic gas streams along a specific path and detected the deflection of the streams induced by the Coriolis acceleration that is proportional to the angular rate[5]. However, we found the thermal gas gyroscope exhibited a large temperature drift, i. e. the variance of environment temperature had a great influence on the output of the gyroscope. The temperature drift largely damages the performance of the gyroscope, therefore, it is very important to investigate an effective temperature compensation method to ensure sensor accuracy. Conventional temperature compensations are usually based on a temperature calibration to establish a temperature model and eliminate temperature drift via a compensation algorithm, as in [6 - 9], which are time - consuming, costly, and sometimes invalid. In this paper, we propose an advanced thermo - driving operation technology by utilizing an alternating constant-temperature-difference (CTD) circuit instead of previous constant-voltage (CV) circuit to maintain a constant temperature difference between the working temperature of the heated heater and the ambient temperature. It is experimentally validated that the scale-factor temperature sensitivity was reduced from 4 903 ppm/℃ to 960 ppm/℃ and the bias temperature sensitivity was reduced from 5.4 (°)/s/℃ to 1.4 (°)/s/℃ in virtue of the proposed temperature compensation.

2 Mechanism of temperature dependance

2.1 Structure of Thermal Gas Gyroscope

As illustrated in Fig. 1, the configuration of the proposed thermal gas gyroscope comprises three heater wires (ht_1, ht_2, ht_3) and four thermistor wires (R_1, R_2, R_3, R_4) which are distributed in the working plane and suspended over a sealed micro chamber filled with gas medium. In the sensor's body coordinate frame $x-y-z$ (shown in Fig. 1), the $x-y$ plane is the working plane of the sensor and the z-axis is perpendicular to the working plane.

2.2 Working Principle

The thermal gas gyroscope was operated previously by using a CV circuit illustrated in Fig. 2[5]. By alternately heating and cooling the central heater (ht_2) and the two side heaters (ht_1 and ht_3), two seismic gas streams in opposite directions along the x-axis are generated in the working plane and the temperature profiles (red solid

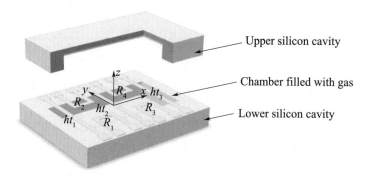

Fig. 1   Schematic diagram of thermal gas gyroscope

lines in Fig. 3) are formed in the chamber. The z - axis rotation deflects the temperature profiles (blue dotted lines in Fig. 3) on the two sides of the y - axis in the opposite direction along the y - axis due to the opposite Coriolis accelerations on the two sides. These deflections of the temperature profiles are detected by using the distributed four thermistors, outputs of which are used to deduce the z - axis angular rate[5].

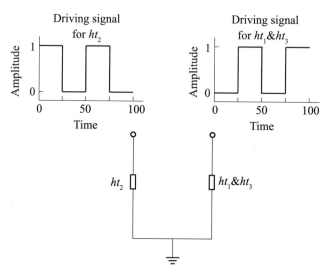

Fig. 2   Constant - voltage (CV) drive circuit

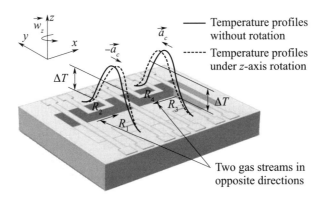

Fig. 3 Working principle of thermal gas gyroscope

Theoretically, the deflections of the temperature profiles detected by the thermistors (shown in Fig. 3) can be expressed as equation (1), assuming that the structure and distribution of the heaters and thermistors are ideally symmetrical.

$$\Delta T = T(R_1) - T(R_2) = -[T(R_3) - T(R_4)] \tag{1}$$

where $T(R_1)$, $T(R_2)$, $T(R_3)$, $T(R_4)$ are the temperatures of the four thermistors. Thus, the z-axis rotation can be deduced by

$$\text{Gyro} \propto [T(R_1) - T(R_2)] - [T(R_3) - T(R_4)] \propto \omega_z \tag{2}$$

Furthermore, the gyroscope output demodulated from a read-out circuit shown in Fig. 4 can be formulated as[5]

$$\text{Gyro} \propto \frac{\alpha R_0 U_C \Delta T_G}{4R} \tag{3}$$

where R is the reference resistances for balancing the Wheatstone bridge, U_C is the constant voltage applied into the bridge, $\Delta T_G = [T(R_1) - T(R_2)] - [T(R_3) - T(R_4)] = 2\Delta T$, assuming that the thermistors have the same R_0 (the resistance at 0 ℃) and the same α (the temperature coefficient of resistance, TCR).

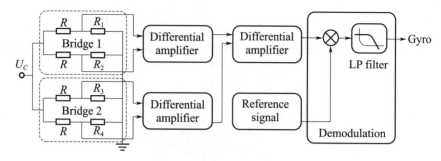

Fig. 4 Schematic diagram of conditioning circuit of the gyroscope

2.3 Influence of Ambient Temperature

Inevitably, the output of the gyroscope is influenced by the temperature. However, the thermal gas gyroscope has exhibited a larger sensitivity dependence on the ambient temperature than conventional mechanical gyroscopes. Therefore, it is vital to investigate mechanism of temperature influence and propose an effective temperature compensation scheme to improve the measurement accuracy of the thermal gas gyroscope.

In the thermal gas gyroscope, joule heat is generated by heating the heaters with a drive circuit and dissipated into the environment through the gas medium. Therefore, the temperature profile in the chamber depends on the Joule power of the heaters, the ambient temperature and the thermal properties of the gas medium. Previously, a CV mode drive circuit was used to drive the heaters as shown in Fig. 2. According to Newton's law of cooling, the temperature balance can be formulated as equation (4) without considering of heat radiation and heat conduction.

$$Q = \frac{U^2}{R_{h0}(1+\alpha T_h)} = hA(T_h - T_a) \qquad (4)$$

where Q is the Joule power of the heater, U is the magnitude of the driving voltage, R_{h0} is the resistance of the heater at 0 ℃, α refers to the TCR of the heater, T_h is the temperature of the heater, T_a is the ambient temperature, h refers to the convection heat transfer coefficient, A is the surface area of the heater.

According to equation (4), variation of the ambient temperature will result in a change of the heater temperature under a CV mode. And therefore, the temperature profiles in the chamber will be changed correspondingly. Assume an ambient temperature T_1 is lower than an ambient temperature T_2. As illustrated in Fig. 5 (a), without rotation, the temperature profiles at the lower ambient temperature T_1 (the red solid lines) is different from that at the higher temperature T_2 (the blue dotted lines). When a rotation around the z-axis is applied to the sensor, a deflection of the temperature profile induces a temperature difference ΔT_1 between R_1 and R_2 (or between R_3 and R_4) under the ambient temperature T_1, while ΔT_2 between R_1 and R_2 (or between R_3 and R_4) under the ambient temperature T_2. It is obvious that ΔT_1 is larger than ΔT_2. The change of the temperature difference between the thermistors corresponding to the variation of the ambient temperature generates the temperature drift in the gyroscope output.

Furthermore, equation (3) also indicates that the gyroscope output is proportional to the temperature difference of the thermistors

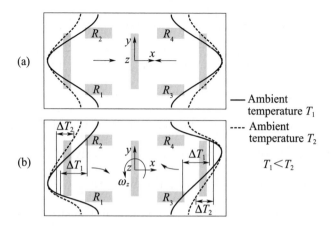

Fig. 5 Ambient temperature influences the temperature profile under CV driving mode, (a) shows the temperature profiles under different ambient temperature without rotation, (b) shows the temperature profiles under different ambient temperature with z - axis rotation

ΔT_G which is dependent on the temperature difference between the heater and the environment. The temperature difference between the heater and the environment plays a crucial role in the outputs of the thermal gas gyroscope.

3 Temperature compensation

3.1 Thermo - driving Operation Circuit for Temperature Compensation

An advanced thermo - driving operation circuit shown in Fig. 6 is proposed to compensate the temperature drift. The circuit utilize an alternating CTD circuit (illustrated in Fig. 6) instead of the previous CV mode (illustrated in Fig. 2) to maintain a constant temperature difference between the working temperature of the heated heater and the ambient temperature. A temperature resistor R_t (PT1000 platinum resistor) is employed to measure the ambient temperature. The center heater ht_2 and two side heaters ht_1 and ht_3 are alternately heated and cooled by alternately plugging the center heater and two side heaters into the Wheatstone bridge of the CTD circuit through an analog switch. The overheat ratios of the heaters (the center heater and the side heaters) are adjusted by regulating the resistors of Rtb2 and Rtb1&3 respectively (Rtb2 for the center heater and Rtb1&3 for the side heaters). R_a and R_b are two constant resistors used to balance the bridge.

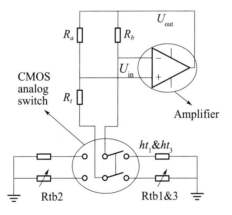

Fig. 6　The constant-temperature-difference (CTD) drive circuit

The CTD operation mode is achieved by incorporating a feedback differential amplifier into the driving circuit to obtain a rapid variation in the heating power to compensate for the instantaneous changes of the ambient temperature. Assuming that ht_2 and Rtb2 (or ht_1&ht_3 and Rtb1&3) are plugged into the bridge of the CTD circuit, as the ambient temperature varies, an error voltage U_{in} proportional to the corresponding change of R_t forms the input of the operational amplifier and generates the output voltage U_{out} of the amplifier. Feeding U_{out} back to the top of the bridge will restore the heater resistance ht_2 (or $ht_1 \& ht_3$) to the value for maintaining a balance between R_t and the heater resistance via adjusting the Joule heating. Thus the constant temperature difference ΔT_C between ht_2 (or $ht_1 \& ht_3$) and the ambient temperature can be maintained, which is determined by the criterion[10] formulated in equation (5).

$$\Delta T_C = \frac{1}{\alpha_t}\left(\frac{\alpha_h - \alpha_t}{\alpha_h} + \frac{R_{tb}}{R_{t0}}\right) \quad (5)$$

where R_{tb} is the resistance of Rtb2 or Rtb1&3, R_{t0} is the resistance of R_t at 0 ℃, α_t and α_h refer to the TCRs of R_t and the heater (ht_2 or $ht_1 \& ht_3$), respectively.

Under the CTD operation mode, the temperature profiles in the chamber for two ambient temperature T_1 and T_2 are illustrated in Fig. 7. The two temperature profiles for T_1 and T_2 have similar shape. And thus the rotation – induced temperature difference ΔT_1 between R_1 and R_2 (or R_3 and R_4) under the ambient temperature T_1 is equal to the temperature difference ΔT_2 between the thermistors under T_2. As a result, the temperature dependence of the gyroscope output is eliminated.

3.2 Configuration of The CTD Driving Circuit

The parameters of the resistors in the CTD circuit can be determined by the criterions[10] formulated in equations (6) and (7).

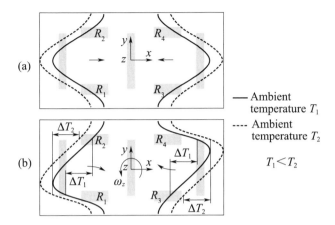

Fig. 7 Ambient temperature influences the temperature profile under CTD driving mode, (a) shows the temperature profiles under different ambient temperature without rotation, (b) shows the temperature profiles under different ambient temperature with z - axis rotation

$$\frac{R_a}{R_b} = \frac{R_{t0} \cdot \alpha_t}{R_{h0} \cdot \alpha_h} \tag{6}$$

$$\frac{R_a}{R_b} = \frac{R_t + R_{tb}}{R_h} \tag{7}$$

where R_h refers to the resistance of the heater (ht_2 or $ht_1 \& ht_3$), R_{h0} refers to the resistance of the heater (ht_2 or $ht_1 \& ht_3$) at 0 ℃.

4 Experimental results

Experiments were conducted to demonstrate the effectiveness of temperature compensation by using the proposed thermo - driving operation method. Fig. 8 shows the tested scale - factor of the thermal gas gyroscope under the ambient temperature varying from 15 ℃ to 55 ℃, where the blue line indicates the results without compensation (operated in CV mode) while the red line indicates the results with the compensation (operated in the proposed alternating CTD mode). The scale-factor temperature sensitivity was reduced from 4 903 ppm/℃ to 960 ppm/℃ by using the proposed temperature compensation.

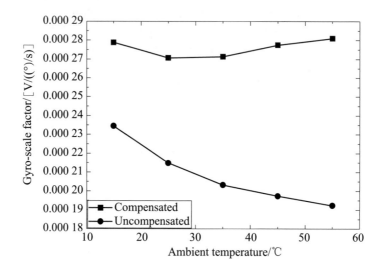

Fig. 8 Gyro-scale factor at different ambient temperature, the red line is with compensation, and the blue line is without compensation

Fig. 9 shows the zero bias of the gyroscope under the ambient temperature varying from 15 ℃ to 55℃, where the blue line indicates the results without compensation (operated in CV mode) while the red line indicates the results with compensation (operated in the proposed alternating CTD mode). The zero bias temperature sensitivity was reduced from 5.4 (°)/s/℃ to 1.4 (°)/s/℃ in virtue of the temperature compensation.

The experimental results validate that the temperature drifts on scale-factor and zero bias of the thermal gas gyroscope are greatly reduced by using the temperature compensation based on an alternating CTD operation circuit. However, due to the actual structural asymmetry and unideal heating of the operation, the temperature difference between the thermistors was not ideally constant during each operation period of the alternating CTD circuit, which restricted the performance of the temperature compensation.

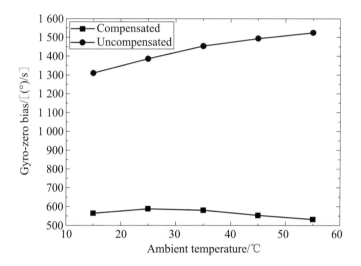

Fig. 9　Gyro－zero bias at different ambient temperature, the red line is with compensation, and the blue line is without compensation

5　Conclusion

A novel temperature compensation technology based on an alternating CTD operation circuit is proposed in this paper to overcome the temperature drift for micromachined thermal gas gyroscopes. The proposed compensation method uses an analog circuitry to realize self－sustained temperature compensation which is easy operation, fast and good stability. It is experimentally validated that the scale – factor temperature sensitivity of the thermal gas gyroscope was reduced from 4 903 ppm/℃ to 960 ppm/℃ and the bias temperature sensitivity was reduced from 5.4 (°)/s/℃ to 1.4 (°)/s/℃ in virtue of temperature compensation.

References

[1] F Jiancheng and L Jianli. Integrated model and compensation of thermal errors of silicon microelectromechanical gyroscope. IEEE Trans. Instrum. Meas., vol. 58, no, 9, pp. 2923 – 2930, 2009.

[2] T B Gabrielson. Mechanical – thermal noise in micromachined acoustic and vibration sensors. IEEE Trans. Electron Devices, vol. 40, no. 5, pp. 903 – 908, 1993.

[3] Z Djuric. Noise sources in microelectromechanical systems. 2000 22nd Int. Conf. Microelectron. Proc. (Cat. No. 00TH8400), vol. 1, 2000.

[4] R P Leland. Mechanical – thermal noise in MEMS gyroscopes. IEEE Sens. J., vol. 5, no. 3, pp. 493 – 500, 2005.

[5] R Zhu, S Cai, H Ding, Y J Yang, and Y. Su. A micromachined gas inertial sensor based on thermal expansion. Sensors Actuators A Phys., vol. 212, pp. 173 – 180, 2014.

[6] K Shcheglov, C Evans, R Gutierrez, and T K Tang. Temperature dependent characteristics of the JPL silicon MEMS gyroscope. 2000 IEEE Aerosp. Conf. Proc. (Cat. No. 00TH8484), vol. 1, 2000.

[7] I P Prikhodko, A A Trusov, and A M Shkel. Compensation of drifts in high – Q MEMS gyroscopes using temperature self – sensing. Sensors Actuators, A Phys., vol. 201, pp. 517 – 524, 2013.

[8] J K Bekkeng. Calibration of a novel MEMS inertial reference unit. in IEEE Transactions on Instrumentation and Measurement, 2009, vol. 58, no. 6, pp. 1967 – 1974.

[9] Q Zhang, Z Tan, and L Guo. Compensation of temperature drift of MEMS gyroscope using BP neural network. in Proceedings - 2009 International Conference on Information Engineering and Computer Science, ICIECS 2009.

[10] R Y Que, R Zhu, Q Z Wei, and Z Cao. Temperature compensation for thermal anemometers using temperature sensors independent of flow sensors. Measurement Science and Technology, 22.8: 085404, 2011.

Thermal Characteristics of Stabilization Effects Induced by Nanostructures in Plasma Heat Source Interacting with Ice Blocks

(2022 年)

Abstract Thermal stability is a crucial issue for the dielectric barrier discharge (DBD) plasma deicing. However, the ice blocks tend to invoke strong nonlinear heating processes, finalized with the devastating instability events, e. g. , the arcing and sparking. In this paper, it is proposed that the periodic field enhancing microscale structures with nanowires could be used to reduce the thermal damage risk. The printing circuit board (PCB) based low profile DBD samples of sawtooth configuration with ZnO nanowires (SN) are prepared, and the controlled groups are the parallel – electrode configurations (P), the parallel – electrode configurations with nanowires (PN), and the sawtooth configurations without nanowires (S) . It is found that: 1) The thermal stability is strongly correlated with the locally enhanced streamers, namely, the LE – streamers, due to the ice blocks. 2) The

---

本文发表于 International Journal of Heat and Mass Transfer。合作作者：Xiaoxu Deng, Henggao Ding, Zhongyu Hou。

ZnO nanowires are shown to play dominant but opposite roles in the LE-streamer dynamics for the PN and SN electrode configurations, i. e. , the enhancement of the streamer channel in the former, but the attenuation in the latter. 3) Uniform pattern of co-existence of the multi-LE-streamers only exists in the SN configurations. 4) The heating behavior of such a pattern is shown to be self-modulated. It heats the fastest at lower voltages but demonstrates the best thermal stability at the higher voltages when the LE-streamers in all the other three groups are ready to evolve into devastating arcs or sparks. To interpret the underlying physical mechanism, it is argued that the nonlinear effects during deicing could be effectively compressed by the decentralization of the plasma heating energy into more structure-based field enhancement sites constructed by the nano-micro electrodes. Such an illustration has been supported by further analyzing the spatial-temporal evolution of the temperature field during deicing, where linearity and uniformity are correlated to play the dominant roles in thermal stabilization. Thus, a statistical factor, ST, is suggested as an index to evaluate and predict the thermal stability of a deicing DBD plasma heat source, which is 23.9% for the SN configurations, the least compared to all the controlled groups.

1 Introduction

Protecting the aircraft from the adverse impacts of ice could date back to the early stage of flights[1,2]. In 2012, anti-icing and de-icing technology by the dielectric barrier discharge (DBD) actuator was proposed for the aircraft by the plasma flow control team of Northwestern Polytechnical University[3-5]. DBD had an additional effect compared with conventional electrical heating methods for its capability to induce airflow[6-9]. We found that the premise that DBD could improve the anti-icing and de-icing efficiency is that the ice block covers the exposed electrode. As the location of ice formation on the DBD surface is random, the case when the ice block is near the exposed electrode rather than covering it is also worth research. When the ice block is located close to the exposed electrode, the locally enhanced streamers and even spark are highly possible to generate between the electrode and the melting ice[10]. Even though the discharge intensity of locally enhanced streamers (LE-streamers) is stronger than the DBD plasma, which is conducive to ice melting, once the spark is formed, accompanied by thousands of degrees[11,12], the surface of the DBD actuator will be devastated. Research on the suppression of these locally enhanced streamers and even streamer-to-spark transition could be necessary to improve the stability of the DBD surface, but it still lacks research currently.

Much of the previous work has been done to verify the feasibility of this anti-icing and de-icing method and improve its efficiency. For example, the DBD actuator installed on a cylinder could anti-ice and de-ice successfully in an icing-wind tunnel[13]. Van den Broecke performed a de-icing device by a nanosecond-pulsed-DBD (ns-DBD) plasma actuator[14]. Chen et al.[15] proved that the DBD actuator driven by higher frequency and lower voltage achieved a better performance of anti-icing at the same applied power. The increasing frequency of the duty-cycled plasma actuator could further enhance anti-/de-icing performance. This performance also strongly relied on the surrounding parameters, such as airflow velocity, gas temperature, angle of attack of the airfoil model, and aerodynamic[16]. Other methods such as the conventional electrical heating method performed no better than the DBD method to anti-ice[17]. The DBD actuator could combine with a super-hydrophobic surface to reduce the surface adhesion and assist anti-ice and de-ice[18]. Zhu et al.[19] simulated the ns-DBD by a numerical model. The simulation result presented that the fraction of the discharge power transferred to fast gas heating was about 6%. Simultaneously, the presence of water or ice on the electrode/dielectric surfaces had a great impact on the electrical characteristics and equivalent capacitance of the DBD actuator[20]. Wei et al.[21,22] successfully de-iced by the NS-DBD actuator and proposed a "stream-wise plasma heat knife" configuration for getting better anti-icing performance. The latest review could be referred to the reference[23] where the working principles and technological developments of the anti/de-icing application of the SDBDs have been well documented in detail.

Many external factors could impact the DBD plasma and the surface temperature such as the thickness of the dielectric, the dielectric permittivity of the material, the frequency, the waveform of the applied signal, the boundary condition, and the geometric parameter of the exposed electrode. PLA and Kapton dielectric of actuators assisted in a higher thermal generation[24]; When the electrical power increased its frequency or the voltage amplitude, the surface temperature increased linearly; A turbulent boundary condition presented better heat dissipation compared with a laminar one[25]; Thinner dielectric had a higher saturation temperature; The increase of the surface temperature was higher When the applied waveform is square or positive ramp[26]. Plate-to-plate DBD and a thin wire-to-plate DBD were two common structures to research the discharge characteristics. Tirumala et al.[27] measured the temperature of the dielectric surface and observed the streamer regime was suppressed in the wire, but the two configurations exhibited a similar range of temperature rise. The glow regime was considered the primary source of thermal energy. However, as a strong nonlinear heat source, DBD plasma is hard to apply to aerodynamic vehicles, if the thermal stability issue is not properly resolved. Although many works have been dedicated to the relationship between the structural features and the actuator performance[28,29] with the concern about the streamer-to-spark transition[30,31], it is still an open question for the improvement of the thermal stability of the DBD deicing devices based on the electrode-structure-induced electric field effects.

In this paper, a novel DBD electrode system is designed with sawtooth structures incorporating ZnO nanowires on the electrode surface. It is well known that the introduction of rough structures on the surface of low surface energy materials can further improve the hydrophobicity of the surface, and even achieve superhydrophobicity[32,33] with excellent anti-icing and anti-fogging properties. An important application of superhydrophobic surfaces (SHS) with nanostructures is the anti-icing of the leading edge of aircraft wings and wind turbine blades[34]. Kreder et al.[33] summarized the anti-icing properties of nanostructured superhydrophobic surfaces as the superior pressure stability and the improved humidity tolerance with jumping condensates[35], but the poor durability. The improved pressure stability and the improved humidity tolerance of the SHS with nanostructures are for its lower ice nucleation probability to resist micron-scale droplet frosting when the spacing between nanowires is at least below 100 nm[36]. Hao et al.[37] found the SHS with nanostructures with jumping condensates could delay condensation icing at a relative humidity (RH) of 60%, the icing is mainly caused by the edge effect or tiny dusts. Hence the collision droplets can bounce off cold surfaces before freezing. However, frost formation could also be an important inflight ice accretion factor on aircraft surfaces, so the durability is an important issue to be addressed as the robustness of icephobic surfaces can be significantly compromised due to frost formation[38]. Many works dedicated to delaying the freeze and improving robustness, e.g., micro/nanostructured surfaces had a more robust anti-icing property which could last for a long delay time of ~7 000 s for ice formation by deep freezing of -25 ℃ in high-humidity condition of 60%[39]; the nanowires with smaller diameters and larger spacing are found to have a better anti-frost performance[40,41]; liquid-infused nanostructured[42] could also highly reduce sliding droplet sizes resulting from the extremely low contact angle hysteresis with low ice adhesion for a promising candidate for developing robust anti-icing materials. Even if the nanostructured SHS surface loses the anti-icing property, some proactive methods of de-icing are required, such as electrical and DBD heating methods. Strobl et al.[34] combined the nanostructured hydrophobic surface and electro-thermal and mechanical method to be able to reduce power consumption by up to 95% compared to conventional ice protection system (IPS).

In addition, the nanostructures could be used to modify the field distributions and theoretically create sub-micrometer scale periodic distribution of the field enhancement sites[43], and the effect could be controlled by the structural features such as spacing and length[44]. Furthermore, a periodic sawtooth structure is introduced to further separate the intensive field regions created by nanostructures within the characteristic length of 10^2 μm, so that the streamer originated from the nano-electrodes could be controlled in some microscale sites, rather than randomly develop across the discharging region. Based on the chemical synthesis method, ZnO nanowires have been deposited on a printing circuit board (PCB) based low profile (0.55 mm in thickness) DBD system (LP-DBD), powered by an alternative current (AC) high voltage source. The experiments show that the methodology is effective to improve thermal stability at higher voltages and operational efficiency at lower voltages. The self-modulation behavior induced by the nanoelectrode-induced field enhancement effect and the microscale sawtooth electrode-induced field isolation effect could be served as a novel mechanism for the understanding and the design of DBD deicing heat source with high stability.

2 Experimental setup

2.1 Configuration preparation and characterization

2.1.1 Nanostructure depositions

There were various methodologies to grow nanowires, such as evaporation and condensation processes, MOCVD schemes, hydro-thermal process, etc[45]. In the paper ZnO nanowires are grown by a simple two-step process: the electroplating method to form Zn coating on copper and the hydro-thermal process to grow ZnO nanowires. The solution used for electroplating is 1.4 g/L $C_{16}H_{33}(CH_3)_3NBr$ which has been prepared in advance and can be reused.

The electroplating method can be completed at room temperature. A 10 cm $\times$ 10 cm $\times$ 5 cm Zn plate is connected to the anode, and the copper surface of the LP-DBD actuator is connected to the cathode and waits to be galvanized. The Zn plate and the LP-DBD actuator are placed in the conductive solution with a 9 cm interval. The constant current source provides a stable 0.03 A current. If small bubbles appear on the surface of the copper with black Zn attaching when we power on the current source, it indicates that the electroplating is in progress. After about 30 min of electroplating, the copper surface is completely covered by black. The Zn plate and the LP-DBD actuator should be rinsed with de-ionized water after the electroplating process.

As the growth conditions of the nanowires are harsh, unnecessary impurities should be minimized. Therefore, put the LP-DBD actuator vertically in the container and rinse it with deionized water and ethanol. Then, add 49 mL deionized water and 1 mL ammonia (the concentration of ammonia is larger than 28%) to the container, and quickly close the lid to prevent the ammonia from volatilizing. Place the container in a 90 ℃ incubator for 12 h to grow nanowires. After the growth, take out the reactor, rinse off the surface ammonia with deionized water, and wait to dry naturally. The morphology of the ZnO nanowires is observed and measured by field emission scanning electric microscopy (SEM, Carl Zeiss Ultra 55, Germany).

2.1.2 Low profile DBD samples

In the experiments, two different designs of the LP-DBD actuators based on the printed circuit board (PCB) technology have been prepared: the parallel structure and the sawtooth structure. ZnO nanowires have been grown on the exposed electrodes of these two groups of the LP-DBD samples. The electrodes are in black after the growth and the results are shown in Fig. 1. The LP-DBD actuators of parallel electrodes, sawtooth electrodes, parallel electrodes with nanowires, and sawtooth electrodes with nanowires are labeled as P, S, PN, and SN, respectively. In Fig. 1 (c), the PCB employs FR4 as the dielectric material with ~4.5 in relative dielectric constant (ϵ_r) and 0.55 mm in gross thickness. The upper exposed electrodes on P and PN configurations are 30 mm in length, 2 mm in width, and 0.05 mm in thickness. The sawtooth is added to one side of each parallel

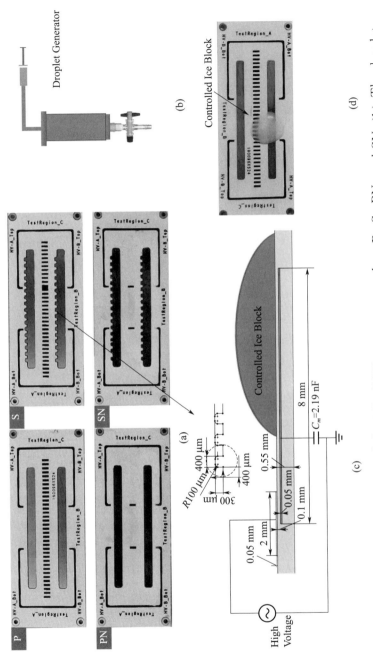

Fig. 1 (a) The four structures of the LP−DBD actuator samples: P, S, PN, and SN. (b) The droplet generator. (c) The structural parameters of the LP−DBD actuators with an ice block under test, and electrode system has been frozen in a 3 m×3 m×2.5 m cold storage with around −20 ℃ kept constant. (d) Ice block on P. The samples should be placed vertically during discharge as the image shows

electrode. The width, height, spacing, and fillet radius of the sawtooth are 0.4 mm, 0.3 mm, 0.4 mm, and 0.1 mm. The lower electrodes, encapsulated by insulating layers, are 35 mm in length, 8 mm in width, and 0.05 mm in thickness, respectively. The deicing experiments have been mainly performed in the vicinity of one of the exposed electrodes, which is connected to the high voltage AC source, the buried electrode below is connected to the ground through a capacitor for the electric measurements. The other pair of the identical backup electrode system of the same structure is suspended, when one row is under test.

2.2 Thermal characterization methods during discharges

2.2.1 Preparation of ice block

To generate the ice block samples in a quantitative and repro-ducible manner at the specified positions on the LP-DBD actuators, the alignment and the generation of the water droplets have been controlled and measured as accurately as possible. As shown in Fig. 1 (b), a droplet generator as is specially designed for this experiment. The droplet generator is made up of two injectors: the larger injector is used for storing water to keep the volume and gravity force approximate constant; sliding the smaller injector at the end of the pipe can slowly change the air pressure inside the pipe, thereby accelerating or slowing down the falling speed of the droplets with better precision. The position of the droplet generator and the number of droplets are fixed for each experiment to obtain the ice blocks with the specified position and quality. At normal temperature above 0 ℃, ten droplets of tap water are dropped onto the LP – DBD actuator by the droplet generator. The conductivity of the tap water is about 280 μS/cm. The total weights of the dropped droplets are about 0.04 g. At last, we place the LP – DBD actuators with the droplets into the cold storage where the temperature is kept around −20 ℃ inside which freezes the droplets to form ice blocks of 10 mm in diameter. The separations between the exposed electrode and the ice block have been controlled within 1.2 mm at 4 kV and 1.5 mm at 5 kV based on the calculated result of Eq. (8). Under the conditions of the DBD deicing experiments, to study the complete LE – streamer development processes, the LP – DBD actuators are placed upright and plasma is above the ice block to avoid the melted water choking the plasma [Fig. 1 (d)].

The general process of deicing is as follows: When the LE – streamers are generated, the ice block will gradually melt and vaporize. The distance between the electrode and the ice block may decrease initially, and the increase in volume induced by melting ice is responsible for the decrease in distance. As melting progresses, melted ice will vaporize and ultimately flow downward when its weight exceeds the carrying capacity of the bottom ice. Further experiments will consume more melted water and the gap began to lengthen. When the expanding gap between the electrode and the ice block exceeds the critical distance that the LE – streamers can reach, the influence of the changes will gradually disappear if the surface does not burn.

2.2.2 Thermal characterization system setup

The discharge between the electrode and the ice block also needs to be carried out in the cold storage. A high voltage capacitor ($C_m = 2.19$ nF) is connected between the ground terminal of the LP − DBD actuator and the ground. Channel 1 of the oscilloscope (Keysight InfiniiVision 1 000 X series) gathers the voltage of C_m by a high voltage probe of 1 000 ∶ 1 attenuation ratio. Channel 2 of the oscilloscope is connected to the voltage output terminal of the voltage power source. The characteristics of the Voltage − Voltage Lissajous graph changing with time are monitored and saved. When we measure the current, the capacitor is substituted with a 61.8 kΩ resistance. After the voltage across the resistance is obtained, dividing by the resistance can get the current. A high voltage power source (TREK, 615 − 10) supplies the sinusoidal wave with voltage amplitudes of 4 kV and 5 kV and a frequency of 10 kHz. An IR thermal camera (FLIR, A6700SC), controlled by a computer, maps the temperature distribution on the surface throughout the discharge process. White protective paint film is used on PCB. Paint is an important coating to protect substrates, and their surface properties can be modified, for example by controlling emissivity or magnetic properties. White paint can be formulated to reflect up to 90% of solar radiation whilst maintaining an emissivity of at least 0.9[46]. Therefore, the emissivity of the dielectric surface is approximated to 0.9. A high speed camera (i-speed 716) records the discharge process with a sampling frequency of 60 fps. In this paper, three replicate de − icing experiments were performed for every LP − DBD sample to recognize the characteristics of the physical processes.

3 Results and discussion

3.1 Structure characterization and field enhancement effect

3.1.1 Nanostructures

The SEM images of ZnO nanowires are shown in Fig. 2. Their lengths are about 500 nm~1 μm and their diameters are about 30~100 nm. In Fig. 2 (a), the distribution of nanowires is relatively dense and sufficient growth length is conducive to further discharge research.

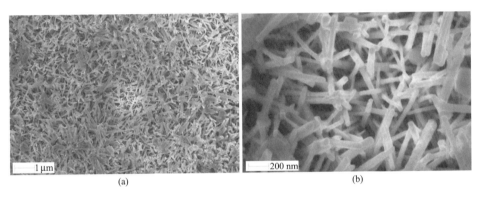

(a)　　　　　　　　　(b)

Fig. 2　SEM images of ZnO nanowires about 30~100 nm diameters and 500 nm~1 μm lengths at the scales of (a) 1 μm and (b) 200 nm

3.1.2 Electric field non-uniformity and ionization localization during deicing process

To characterize the influence of the architecture of different scale-size in the designed DBD electrode system on the discharge and deicing processes, it is important to note that the field enhancement effect exists in all the configurations due to the following structures: 1) sawtooth corner of about 100 μm in the radius of curvature,

2) sharp-pointed protrusion defects of several micrometers in the radius of curvature and 20~50 μm spacing distances due to the chemical etching process of PCB, and 3) nanowires of 30~100 nanometers in diameter with ~50 nm in average. The field enhancement effect induced by the protrusion and the nanowires on the parallel configurations and the sawtooth configurations could be partially measured by the electric field enhancement factor β, $\beta = E_{max}/E_0$, where E_{max} is typically defined as the maximum electric field intensity of the sharp-tip structure and E_0 is defined as background field intensity, i.e., the electric field intensity in the case of parallel structure without protrusions. The field distributions have been numerically calculated using finite-element-method (FEM), and the results are shown as the relative value (E/E_0) distribution to capture the impact of different functional enhancement structures on the field distribution in Fig. 3. According to the case studies about the structural parameter sensitivity using FEM simulations, β is shown to be more sensitive to the protrusion base angles and weakly related to the ratio of the protrusion height for the chemical etching induced sharp-tips[47]. The vertex angle of the protrusion is likely to be obtuse angles, which are larger than the vertex angle in the model, and thus actual β may be slightly smaller than the result shown in Fig. 3 (b). Whereas for nanowires, $\beta \approx L/R$ is often used as a rough approximation for the characterization of the post-like structure when $L \gg R$, where L is the length of the nanowire and R is the base radius[48]. If multiple nanowires overlap each other or are distributed vertically, β will

decrease. It may be a normalized distribution function suggested by curve fitting with a form of the second or third order of exponential decay[43]. The size of the grid has a strong influence on the calculation of β. The models have two scopes of micrometer and nanometer at the same time. Therefore, to compare β more clearly, the same and appropriate grid size for each structure is designed to compare the combined effect of multiple structures on β. Both the sawtooth and nanowires can increase β. Consequently, ZnO nanowires have the strongest effect on β among all the three kinds of field enhancement functional structures, as shown in Fig. 3 (e) and (f).

3.2 Fundamental characteristic and electric characteristics of the plasma heat source interacting with the ice blocks

3.2.1 Fundamental characteristic of the streamer enhancement effect induced by ice blocks and its qualification

To study the influence of the sawtooth and the nanowires on the ice-free streamers (IF-streamer) and the locally enhanced streamers (LE-streamers), we capture the discharge images at the moment when the surface temperatures reach the maximum during the plasma heating process when $V_a = 5$ kV for comparison, and particularly process their LE-streamers and quantify their brightness. The captured images are 8-bit gray-scale images. The processed images of LE-streamers I_s are the results of subtracting the average maximum brightness of IF-streamers $\overline{I}_p$ from the original discharge images I in

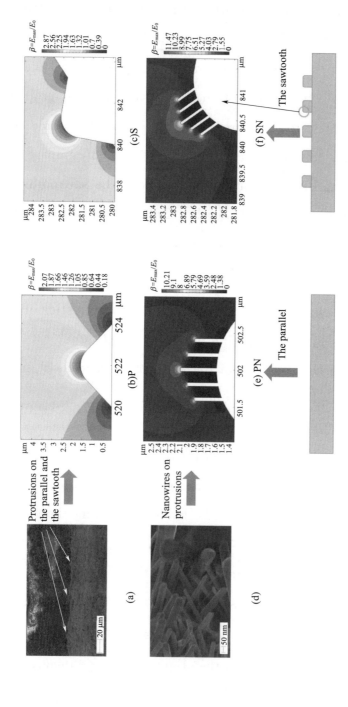

Fig. 3 (a) SEM images of protrusions at the edge of the electrode. The electric field enhancement factor β contours of the portrusions as observed based on SEM images on (b) the parallel electrode, and (c) the sawtooth electrode. (d) SEM images of nanowires at the edge of the electrode. β contours of the nanowires on the protrusions of (e) the parallel electrode, (f) the sawtooth electrode

the image region P

$$I_s = \sum_{i,j=1}^{P}(I(i,j) - \bar{I}_p) \quad (1)$$

Except for the LE-streamers between the electrode and the ice block, some IF-streamers present brighter filaments at some special discharge positions and rounded corners, but they are not involved in the calculations. The background has been normalized across different images and quantified as the brightness at the surface before discharge I_{0d}. The quantified brightness of streamers is calculated by

$$I_b = \sum_{i,j=1}^{P}(I(i,j) - I_0(i,j) + I_{0d}) \quad (2)$$

where I_b is the quantified brightness images of IF-streamers and LE-streamers, I is the image of discharging, and I_0 is the image at the moment before discharge. $I(i,j) - I_0(i,j)$ will cause the brightness of the LE-streamers to decrease, so I_{0d} needs to be added to restore it to the original value.

As shown in Fig. 4, the brightness of the streamers increased in the vicinity of the ice blocks during the deicing process, significantly. In the processed images of 2D and 3D plots, the locally enhanced streamers, namely, the LE-streamers, could be identified with much better morphological details. The images shown in Fig. 4 are selected to characterize the features of the LE-streamers in different electrode configurations.

1) The brightness of LE-streamers is always superior to the average brightness of the IF-streamers during the temporal development of the heating process. There are also some LE-streamers

appearing as much thinner discharge filaments, and their brightness is similar to that of the IF - streamers, also they hardly cause a higher measured temperature [Fig. 10 (h)].

The secondary emission coefficient γ_i in the criteria for Townsend discharge $\gamma_i(e^{\int_0^D \alpha ds} - 1) = 1$ [49] may be crucial for LE - streamer formation between the electrode and the ice block as electron loss could be supplemented by the secondary emission electrons, where, D is the discharge separation, and integration of semi - empirical first ionization factor α is along the line of the electric field from the electrode edge to the ice block surface. Also, surface charge density ρ_{sc} was proposed to be critical for the formation of LE - streamers or even the streamer - to - spark transition. The accumulation of the electric charge on the ice block could increase the voltage drop and the enhancement of the LE - streamers[10,50].

2) The ZnO nanowires and the sawtooth microscale structure could lead to the streamer enhancement effect and the spatial isolation effect, respectively. First, the sawtooth structure is effective to focus the streamer energy onto the corner regions as shown in Fig. 5 (b). Second, the nanostructures on the surface of the PN configuration are shown to be capable of fusing the streamers into one single LE - streamer channel as shown in Fig. 5 (c), which is within expectation because the most intensive field enhancement effect is converged within the submicron scale and about to distribute uniformly across the electrode surface. Besides, the streamers in the ice - free regions outside the ice block, referenced as IF - streamers for short, are more

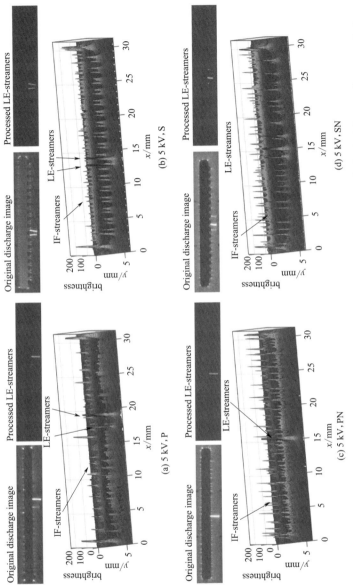

Fig. 4 The original discharge images I, processed LE-streamers images I_s, and the brightness images of the discharge around the electrode I_b. The sampling times are the same as the times in Fig. 10 when the surface temperatures reach the maximum during the plasma heating process on (a) P (b) S (c) PN (d) SN. $V_a = 5$ kV, $f = 10$ kHz, where V_a is the applied voltage amplitude, f is the applied frequency, the locally enhanced streamers is abbreviated to LE-streamers, and ice-free streamers is abbreviated to IF-streamers

intensive than those on the P configurations. Third, the morphology of the LE and LF streamers is somewhat an over-lapped pattern of those shown in Fig. 5 (b) and (c).

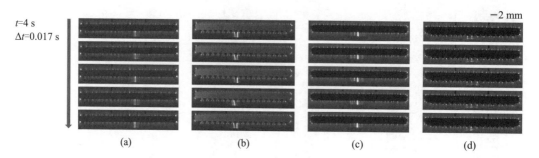

Fig. 5 Repetitive testing results on how the LE-streamers development is related with the ice block in five times of controlled experiments with 0.017 s inter-framing time from $t = 4$ s when $V_a = 5$ kV on (a) P (b) S (c) PN (d) SN, where t is the discharge time

3.2.2 Electric characteristics of the discharges with and without nanostructures

As V_a increases, the stepwise growth of the voltage channels is associated with short current impulses in the external circuit[51]. The sampling frequency of the oscilloscope is 2×10^7 Hz and the sampling interval is 50 ns, much less than the measured intervals of impulses (micro-discharge interval $\tau_{md} = 0.5 \sim 2$ μs in Fig. 6 (a) when $V_a = 5$ kV on P). Therefore, the oscilloscope is accessible to collect every micro-discharge at protrusions and nanowires, same as the principle of successive photographs in Soloviev[52], which could capture these impulses with dozens of ns or several μs. Formula given in Soloviev[52] estimates interval time τ_{md} between each micro-discharge

$$\tau_{md} = \delta V / 4 V_a f \tag{3}$$

$$\delta V \approx 0.3\Delta V_c \quad (4)$$

where δV is voltage increase, and normal fall down of cathode potential ΔV_c is about 600 V for atmospheric air. δV do not depend on the current voltage and τ_{md} is fixed when V_a and f is constant[53]. The number of micro-discharges at each protrusion or nanowire N_0 remained the same if V_a is constant since τ_{md} is inversely proportional to f. $\tau_{md} \approx 0.9$ μm when $V_a = 5$ kV, approximate to the measured result.

In the paper, τ_{md} for the four LP-DBD samples are also different. Micro-discharges mainly concentrate in the positive voltage rising phase ($N_0 \approx 17$ on P). The current impulse amplitude is much smaller during the negative voltage rising phase. In Fig. 6 (c) and (d), the sawtooth and the nanowires cause N_0 to increase, but their current impulse amplitudes are moderate. While on SN in Fig. 6 (e), the nanowires have little effect on the micro-discharges when few new current impulses are generated compared with S.

The active power P_a calculation principle of the LP-DBD actuator based on the Voltage-Voltage Lissajous graph is introduced as follows. The relationship among current i_c, charge q_c, and voltage v_c at two ends of C_m are given by

$$i_c = dq_c/dt = C_m dv_c/dt \quad (5)$$

The equivalent capacitance of the LP-DBD actuator in the magnitude of several pFs is 1/1 000 of the external capacitance. Therefore, the equivalent resistance of the LP-DBD actuator is about 1 000 times that of C_m, as the capacitor's capacitance is inversely proportional

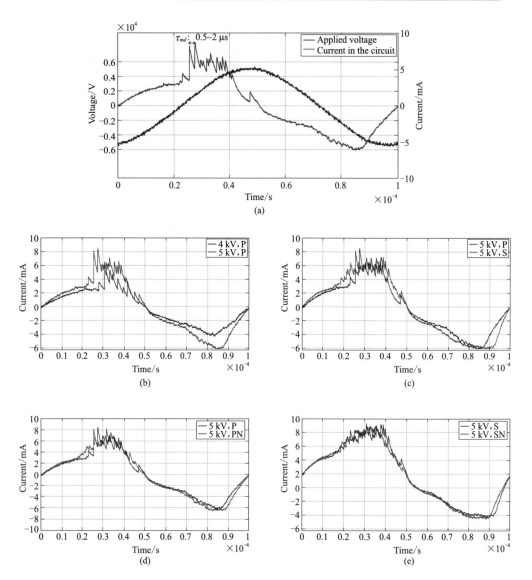

Fig. 6 (a) Plots of the applied voltage v_0 and the current i_c versus time on P when $V_a = 5$ kV. Micro-discharge interval τ_{md} is about the 0.5~2 μs on the current curve. Contrasts of the currents: (b) $V_a = 4$ kV and 5 kV on P; (c) on P and S when $V_a = 5$ kV; (d) on P and PN when $V_a = 5$ kV; (e) on S and SN when $V_a = 5$ kV

to its equivalent resistance

$$P_a = 1/T_s \int_0^{T_s} v_d i_c \, dt \approx 1/T_s \int_0^{T_s} v_0 C_m (dv_c/dt) \, dt \tag{6}$$

$$\approx f C_m \int_0^{T_s} v_0 \, dv_c$$

The formula for calculating P_a of the LP–DBD actuator is expressed in Eq. (6), where v_d is the voltage of the LP–DBD actuator, v_0 is the applied voltage from TREK, and T_s is the time of one cycle. P_a can be obtained by multiplying f, C_m, and the area of the enclosed Lissajous graph.

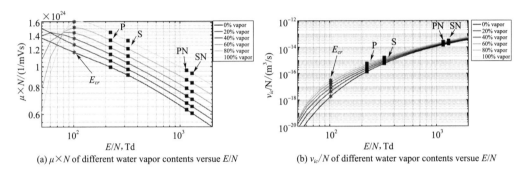

(a) $\mu \times N$ of different water vapor contents versue E/N

(b) v_{ic}/N of different water vapor contents versue E/N

Fig. 7 (a) Electron mobility $\mu \times N$. (b) Ionization frequency v_{ic}/N, where N is the gas particle number density. The E/N positions corresponding to typical initiation field intensity E_{cr} are marked by red hexagons of different water vapor contents. The relative increment ratio for every specific case over that of E_{cr} on the ideal parallel structure are also marked in the figure based on the electric field calculation results as shown in Fig. 3. (For interpretation of the references to colour in this figure legend, the reader is referred to the web version of this article.)

As the differences in active powers of different electrode structures are not large, it is necessary to take the average value of Voltage–

Voltage Lissajous results of multiple cycles for calculation to reduce random errors. In this paper, we perform dozens of active power calculations on 10 cycles of Lissajous data and obtain the average active power $\overline{P_a}$ by the Voltage – Voltage Lissajous method. The results are given in Table 1. It seems that the sawtooth and the nanowires have a gradually increasing influence on $\overline{P_a}$ when V_a rises. Even though the reduction of the discharge points on the sawtooth cause $\overline{P_a}$ of S lower than P when $V_a = 4$ kV, they are almost equal when $V_a = 5$ kV. Similarly, PN consumes the most $\overline{P_a}$ when $V_a = 5$ kV. It is worth noting that even when $V_a = 5$ kV, SN still consumes the least $\overline{P_a}$ which may be caused by its least discharge points.

Table 1　The averaged active power, $\overline{P_a}$, W

Voltage/kV	Structure			
	P	S	PN	SN
4	3.27	2.95	3.24	2.67
5	6.02	6.05	6.67	5.75

3.3　Stabilization effect induced by nanostructures during deicing

3.3.1　Spatial morphology and temporal evolution of the enhanced streamers

The initiation of the LE – steamers is strongly related to the field distribution and thus related to the separation between the electrode and the ice block, which should not exceed the distance of the streamer length (I_{3d}) that energetic free electrons can reach. Otherwise, the ice block melts slowly by the convective heating from the streamers.

The resulted LE – streamer channel length at any moment during deicing is the result of the superposition of multiple successive micro-discharges. A semi-2D discharge is excited by a nano-seconds over-voltage pulse, displaying as a form of flat plasma. When the applied voltage is a sine wave, a 3D discharge is formed and propagates zigzag when near-threshold breakdown discharges (shown in Appendix A) initiates at the slowly rising voltage[53]. The length of a semi-2D discharge l_{2d} is

$$l_{2d} \approx \frac{2h_d v_{ic} V_a}{\mu E_c^2} \tag{7}$$

where characteristic electric field E_c is 277 kV/cm proportional to gas density at atmospheric pressure, streamer layer thickness h_d is about 50 μm determined by experimental or simulated results, the recommended value for ionization frequency v_{ic} is 3.5×10^{11} Hz[53], and for electron mobility μ is 0.06 m²/(Vs) on parallel electrode[54]. In Fig. 7, the mobility μ and ionization frequency v_{ic} have been calculated based on the two-term approximation of the Boltzmann equation using BOLSIG+[55], and the results for the different electrode configurations of the four groups of the DBD samples are compared in the dimension of the reduced electric field (E/N), ranging above the typical initiation field intensity, say, E_{cr} for short, for a self-sustainable discharge in the air. For an electrode system in the atmospheric air of the standard condition, E_{cr} is about 3 kV/mm[56] if the surface condition is carefully treated. and it is given by 3.15 kV/mm as the reference case, marked in red hexagons in Fig. 7, The field enhancement effect for the four groups of the samples under scrutiny can lead to a stronger field under the same applied voltage, and the relative increment ratio for every specific case over that of the field intensity of the P samples could be characterized based on the electric field calculation results as shown in Fig. 3. Such a relationship in both of the dimensions of E/N and water concentrations have been marked as blue for a specific moment when E is assumed to be 3 kV/mm on the ideal parallel electrode (no protrusion), where the reduced field is remarkably more intensive than E_{cr} and self-sustainable discharge may be expected for all the samples. As shown in Fig. 7, firstly, it is shown that the increase of the input voltage, μ will decrease, and v_{ic} will increase. Secondly, the structural features for all four samples tend to increase the field intensity at the same applied voltages, and the largest increment level is expected to exist in the SN samples. Third, the impact of the increment of the water concentration from 0% to 100% tends to induce an increase in electron mobility and ionization frequency. According to the theoretical relationship shown in Eq. (7), it is concluded that the field enhancement effect induced by the structural configurations all lead to an increase in the length of the discharge, and the introduction of the sawtooth structure and the nanowires may induce the largest increment in l_{2d}.

l_{3d} is proportional to l_{2d}, relevant to dielectric thickness d and h_d

$$l_{3d} \approx \frac{l_{2d}(5\epsilon_r + 4)\ln(1 + 2d/h_d)}{6(1+\epsilon_r)(2+\epsilon_r)} \tag{8}$$

The calculated lengths of the streamer for parallel electrodes according to Eqs. (7) and (8) are 1.21 mm and 1.52 mm when $V_a =$

4 kV and 5 kV, respectively. Consequently, there are two important deductions based on the analysis: First, the incorporation of the nanostructures and the micro - scale structures for the stronger local field enhancement tends to initiate streamers reaching the ice blocks with a larger separation between the ice block and the exposed electrode. This may be considered as an enhancement mechanism for a faster deicing process in the applications, due to the formation of the LE - streamers. Second, it is important to note that the difference in the l_{3d} for the examined sample types leads to the necessity of choosing a proper range of the separation between the ice block and the exposed electrode, so that the LE - streamers could be initiated for all the tested cases, which is determined to be within 1.2 mm at 4 kV and 1.5 mm at 5 kV.

The LE - streamer morphology and the rising temperature T at the sampling lines versus Δt are plotted in Fig. 8. The ice melting states are different when $V_a = 4$ kV and 5 kV. When $V_a = 4$ kV, the LE - streamers hardly let the melted ice fall off automatically within minutes of discharge. External factors have greater influences on the melting time, e.g., the width and the quality of the ice block, the separation between the electrode and the ice block, and the unstable peripheral temperature which can only be controlled within a certain range ($-20 \sim -15$°C), etc. It is difficult to compare the ice melting rate of different samples from the melting time. Therefore, we only draw the temperature evolutions in 60 s (Fig. 8). However, it is found that the surface temperature on SN rises the fastest. Hence it is supposed that the heat source of LE‑streamers on SN has the fastest heating rate. While at $V_a = 5$ kV, the ice melting rates are much faster. The maximum temperatures on P, PN, and S are approximate. The ice melting times are almost within 1 min, but on SN, its maximum temperature is much lower with the longest de‑icing time of more than 1 min [Fig. 8 (j)].

Comparing the streamer morphology of the same structure when V_a increases from 4 kV to 5 kV, we find that the discharge states have transformed from diffuse discharge to concentrated discharge. Simultaneously, the temperature evolution can effectively reflect the duration and intensity of these stronger LE‑streamers. When $V_a = 4$ kV, the morphology of LE‑streamers at the ice blocks is uniform and present as diffuse discharge state as Fig. 8 (c) – (f) shows. The maximum temperatures T_{max} of LE‑streamers are approximate. The temperature distribution on P is flat, similar to their uniform LE‑streamer morphology, whereas it is much more volatile at the nanowires on PN. On S, the LE‑streamers at the rounded corner of the sawtooth have steeper temperature distributions, and the weaker LE‑streamers near the two stronger ones slow down the temperature drop in Fig. 8 (d). On SN, no weaker LE‑streamers near the two stronger ones at the nanowires cause the temperature away from them to drop faster.

When $V_a = 5$ kV, there are also multi‑LE‑streamers generated, but the spatial morphology of these LE‑streamers is quite different on

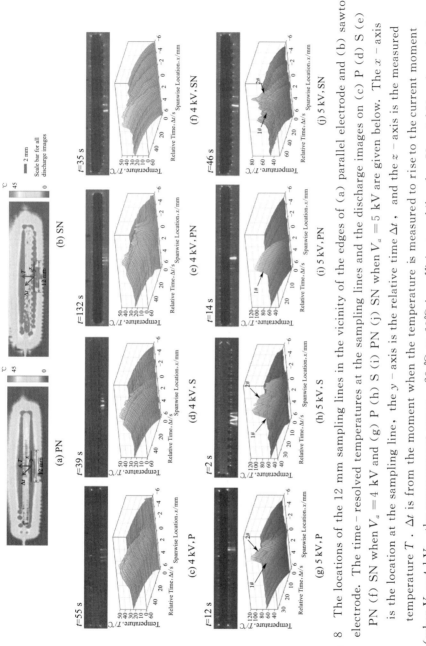

Fig. 8　The locations of the 12 mm sampling lines in the vicinity of the edges of (a) parallel electrode and (b) sawtooth electrode. The time-resolved temperatures at the sampling lines and the discharge images on (c) P (d) S (e) PN (f) SN when $V_a = 4$ kV and (g) P (h) S (i) PN (j) SN when $V_a = 5$ kV are given below. The x-axis is the location at the sampling line, the y-axis is the relative time Δt, and the z-axis is the measured temperature T. Δt is from the moment when the temperature is measured to rise to the current moment (when $V_0 = 4$ kV, the measurement range $-20\ ^\circ\text{C}$ to $60\ ^\circ\text{C}$ is sufficient, while when $V_0 = 5$ kV, the real-time temperature exceeds $100\ ^\circ\text{C}$, so the measurement range should be selected from $30\ ^\circ\text{C}$ to $150\ ^\circ\text{C}$, where the lower temperature boundary is higher than ambient temperature, causing the actual discharge inception time cannot be obtained directly). 1 # and 2 # refer to areas above the temperatures of the IF-streamers

the same structure and presents as a concentrated state on the LP-DBD samples except for SN: a single brighter and longer-lived LE-streamer may obtain most discharge power, producing stronger brightness and longer duration, accompanied by a much higher temperature, while the left thinner or short-lived LE-streamers hardly raise the surface temperature. Ultimately, only a limited number of regions have significant temperature changes. The introduction of the sawtooth or the nanowires can also change the number and positions of these regions. For example, on P, multi-LE-streamers transform into a state of power concentration of a LE-streamer, but this concentrated LE-streamer can be replaced by other LE-streamers at any time as Fig. 8 (g) shows. However, SN could maintain the two uniform LE-streamers at the same time as the ones when $V_a = 4$ kV. The surface temperature rises induced by LE-streamers are much lower than those of the other parts of the sample structures.

3.3.2 The "thermal complementary" behavior and ST factor

The brightness of the LE-streamers increase with the increase in the applied voltages. When $V_a = 5$ kV, the surface temperature of the LE-streamer heating region is significantly higher than other regions, as shown in Fig. 8. The temperature evolution of different portions of the LE-streamers is shown in Fig. 9, where strong evidence of "thermal complementary" behavior could be observed once there are LE-streamer channels more than one. This complementary behavior is not a result of the external power-limiting circuitry, because the AC high voltage source has been operated much less than its compliance condition so that the spark or arc transitions could be tolerated by the same output setup. It is argued that the observed phenomenon results from the competition between the temporal nonlinear heating process and the spatial spreading cooling process. For the first process, the gas in the LE – streamer channels is increasingly heated by the plasma with a higher ionization degree resulting from the nonlinear enhancement effect induced by the increase of the surface charging of the ice block [10] and the introduction of the water molecules (Fig. 3). For the second process, the LE – streamers have originated and are constrained within the as – designed sawtooth structure (S), or the chemical etched protrusions (P). It is clearly shown in Fig. 9 that the nanostructures are crucial to the modification of the general behavior in both processes. In the PN configurations, nanostructures tend to enhance the first process without limitation, so that it is the case most close to the spark transitions; while in the SN configuration, the nanostructures effect is constrained within sawtooth microstructures, so that the LE – streamer channels spreads onto more areas of the electrode surface, resulting in the most uniform temperature distribution.

As shown in Fig. 9, the temperature distributions of the LE – streamers have been further analyzed based on the statistical features considering the relative variations with and without the influence of the ice blocks. The time – resolved temperature difference ($\Delta T = T_{max} - T_{avgIF}$) between the maximum surface temperature T_{max} at each LE – streamer covering region and the average surface temperature T_{avgIF} of the region not affected by the ice block have been calculated for every time step of the temperature records. The resulted curves have been referenced as 1 # and 2 # for the cases of P, S, and SN, and 1# for PN as only one LE - streamer channel is formed during all the test runs (marked in Fig. 8). Furthermore, we plot the $\Delta T - \Delta t$ numerical sum and the absolute value of the difference between the highest two LE - streamer channel covering regions, referenced as 1 # + 2 #, and |1# − 2#| for short, alongside with their original curves. It is obvious that the more effective the self - modulation behavior impact the de - icing process, the less deviation of the 1 # and 2 # curves, i. e., the lower level of |1# − 2#| curve, should be expected. To fur - ther quantify the differences across the tested configurations, an - other statistical index has been introduced as follows: For PN of a single LE - streamer channel, $A_{s1} = A_{s1\#}/A_{s1\#}$. For P, S, and SN of several LE - streamer channels, $A_{s2} = A_{s2\#}/(A_{s1\#} + A_{s2\#})$, and $A_{s1-2} = |A_{s1\#} - A_{s2\#}|/(A_{s1\#} + A_{s2\#})$, where A_s is the numerical integration of the $\Delta T - \Delta t$ curves in the respective cases. It is shown that the A_{s1-2} factor of the SN configurations is shown to be the smallest, indicating the best temperature distribution uniformity of the LE - streamer regions, resulted from the most effective self - modulation. On the contrary, the A_{s1-2} factor is the largest for the PN configurations, indicating the strongest non - uniformity and the tendency of fusing into a strong heating channel until arcing initiates, which is in accordance with the observations.

Consequently, A_{s1-2} could be considered as a temporal statistical

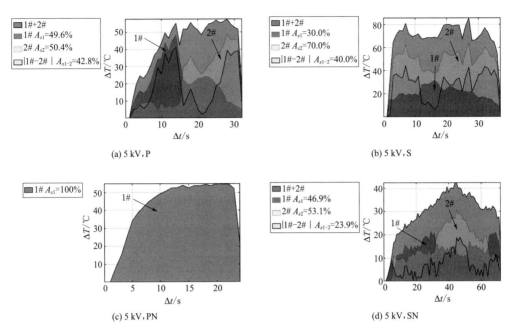

Fig. 9 The sum, A_{s1}, A_{s2} and A_{s1-2} of $\Delta T = T_{max} - T_{avgIF}$ for each LE - streamer channel region's temporal evolution curves in the cases of (a) P (b) S (c) PN (d) SN when $V_a = 5$ kV. 1# and 2# can be referred to the "mountain" areas in Fig. 8. $A_{s1} = A_{s1\#} / A_{s1\#}$, $A_{s2} = A_{s2\#} / (A_{s1\#} + A_{s2\#})$, and $A_{s1-2} = |A_{s1\#} - A_{s2\#}| / (A_{s1\#} + A_{s2\#})$, where A_s is the numerical integration of the $\Delta T - \Delta t$ curves in the respective cases

index to measure the self-modulation behavior of the LE-streamers, which is a significant index for the evaluation of the tendency of the thermal instability of the DBD deicing plasma heat source, renamed as ST, standing for 'Stability of Thermal condition', whose definition is detailed as follows

$$ST = \frac{1}{N_{max}} \sum_{i=1}^{N_{max}} \frac{|\Delta T_{i\_\#1} - \Delta T_{i\_\#2}|}{\Delta T_{i\_\#1} + \Delta T_{i\_\#2}} \quad (9)$$

where N_{max} is the number of sampling points on the temperature axis,

$\Delta T_{i\_\#1}$ and $\Delta T_{i\_\#2}$ correspond to ΔT_i of sampling points at positions of 1# and 2#, respectively. $|\Delta T_{i\_\#1} - \Delta T_{i\_\#2}|$ can reflect the difference between two LE-streamers, then normalized by dividing by their sum, making it easier to compare the uniformity among different samples. It is shown that the calculated result of the SN configuration is the least 23.9% in Table 2, indicating that the self-modulation effect is most effective in the SN configuration.

Table 2 Comparison of the calculated parameters of the four LP-DBD samples when $V_a = 5$ kV

Parameter	P		S		PN	SN	
	1#	2#	1#	2#	1#	1#	2#
T_{max}/℃	100.4	110.7	77.7	112.5	110.7	69.5	77.9
A_s	49.6%	50.4%	30.0%	70.0%	100%	53.1%	46.9%
τ_{cr}/s	31		36		23	71	
σ_T	15.6		12.8		13.9	9.0	
ST	42.8%		40.0%		100%	23.9%	

3.4 Discussion on the physical mechanisms

3.4.1 Different modes of plasma-ice interactions induced by nanostructures

It is a crucial point to realize that the DBD plasma heat source is a distributive heat source with high density heating channels, randomly distributed in space, but tends to centralize non-linearly within transient time whenever a field distortion is under development. Consequently, the fundamental aspect to increase the thermal stability for a DBD deicing system is to break down such a temporal evolution tendency everywhere possible that every 'com - mon' streamer with a small scale can grow into some violent heating electron channel, ready to evolve into arcs. Our approach is to 'saw' the electrode into smaller periodic regions with high and low field intermittent segments on a sub-millimeter scale, significantly smaller than the ice block size, aiming to invoke the self - limiting behavior of the streamers when facing the distortion fields resulting from the melting and charging of the ice blocks, so that the solid-structure-induced field enhancement of any converging features could be limited in the spatial scale as small as possible. However, the unique field enhancement effect of the ZnO nanowires film is proved to be vital to increasing or decreasing the spatial density of the streamers that non - linearly interact with the distortion field of the melting ice block. In what follows, the inter - acting mechanism of the nanostructures that influence the modes of the heat inflow and outflow of the DBD streamers have been illustrated.

Let us begin with the plasma thermal instability mode with strong thermal (translational) - vibrational interaction[57]. TVI for short, which is most likely to be initiated when the strong interacting additives, such as H_2O molecules, have been added to the discharging gas. The mechanism of the plasma thermal instability of the TVI process is a positive feedback loop where the electron density fluctuation, δn_e, once excited in any control volume, will increase the collision frequency and thus the relaxation rate via the vibrational degree within the molecules. Further, the translational degree of freedom of the system acquires more energy so that the gas temperature, T_{gcon}, is increased, the pressure of the control volume, p_{con}, and the gas particle number density of the control volume, n_{gcon}, decreases, so that the reduced electric field strength, E_{con}/n_{gcon}, is intensified in-situ and leads to a positive updated electron density over original fluctuation, say, $k_{pos}\delta n_e$ with k_{pos} significantly larger than unity. The minimum total temporal scale could be as small as $10^{-9} \sim 10^{-8}$ s[58]. It should be emphasized that the plasma is over-heated and compressed into the streamer with a narrower channel due to the increment of the ionization degree, and therefore could increase the deicing rate temporarily before arcing transition. According to the theoretical treatments with quantum mechanics, it is given that the kinetic rate of the vibrational relaxation is related to the translational[58]

$$\tau_{V con} = \frac{1}{p_{con}} [A \exp(T_{gcon}^{-\frac{1}{3}} - 0.015 \mu^{\frac{1}{4}}) - B] \tag{10}$$

where p_{con} is in atm herewith, $\tau_{V con}$ is the vibrational relaxation time in the control volume, μ is the reduced mass of the molecules in atomic units, and A and B are constants related to the molecular species[58]. It is well known that the relaxation rate coefficient of the H_2O molecules is several orders of magnitude larger than that of the N_2 or O_2[58,59]. As a result, according to Eq. (10), the evaporated H_2O gas molecules within the vicinity of the melting ice block are a strong origin to initiate the thermal instability of the deicing plasma streamer channels. The direct consequence should be the contraction of the interacting streamer bridging the exposed electrode and the ice block surface if the deicing DBD has been operated in a power-limiting mode. However, the operational power is far lower than the external power limit and the thermal instability will end up only when the sparking criterion is satisfied, or the positive feeding inflow is balanced by other power-consuming outflow processes.

First, we shall show that the negative feedback loop is hard to be initiated in a quasi-uniform field DBD deicing system, and the local field enhancement only intensifies thermal instability conditionally. For an atmospheric air dielectric barrier discharge in a perfectly uniform electric field, the discharge pattern is multiple streamers with uniform power distribution within those ionizing channels. However, theoretical perfection is not a typical scenario for the low-profile electrode systems, where the lithography-based technologies, such as PCB fabrication processes, tend to generate micro sharp-tips as field enhancement sites, instead of the sub-nanometer uniform surface morphology. As a result, the number of the discharge channels that could consume the external electric power is limited to some selective regions of the stronger field, because the ionization frequency is much higher and the streamers in the weak field regions can even harder to be initiated due to the field screening effect imposed by the strong field region streamers with high electric conductivity. It is shown that the as-deposited ZnO nanowires on the electrode surface tend to enhance the electric field based on the original surface morphology, i.e., the original strong field region tends to be stronger due to the adding of nanowires. Thus, the basic interaction mode between the heat inflow and outflow processes in the DBD plasma of both the P and PN configurations is governed by a collective behavior of every enhanced streamer channel with equal probability for nonlinear evolution if the identical external distortion is presented. The analysis above-mentioned is vital for the prediction of the reaction of an electrode system with nanostructures: Considering the fluctuation posed by the additives of evaporated H_2O has caused an increase of the electron density, δn_{econ}, the perturbation tends to propagate along the longitudinal direction of the streamer due to the field enhancement effect of the nanostructures and multiplied by a factor of, say, k_{local}, when it arrives at the local area affected by the nanostructures. The peaking electric field fluctuation of the ionizing wave reads[60-62]

$$\frac{\delta T_e}{T_e} \approx \frac{1}{(k_s \lambda / \sqrt{\delta \epsilon_k})^2} \frac{\delta E}{E} \tag{11}$$

where T_e is the electron temperature, $\delta \epsilon_k$ is the kinetic energy transferred during one collision, k_s is the wave number of the ionizing wave, and λ is the mean free path. Consequently, given that the factor could be treated approximately as a constant[63], and that the field enhancement resulting from the nanostructures is spatially limited as shown in Fig. 3, the field distortion δE originated from the TVI mode tends to be intensified by the increased electron temperature and the increased field. Supposed that the elastic collisions between electrons and molecules dominate the electron energy losses, the mean free path could be considered constant, and the electron energy distribution follows the well-known Margenau distribution[64], an expanded Druyvesteyn distribution[65] in the frequency domain. The averaging of electron drift velocity over the whole energy spectrum could yield a relation between T_e and the reduced electric field strength, E/n_{gcon}, as follows[65]

$$T_e = \frac{E}{n_{gcon}} \frac{e}{\langle \sigma_{en} \rangle} \sqrt{\frac{\pi}{12\delta\epsilon_k}} \quad (12)$$

where e is the elementary charge, σ_{en} is the cross section of the electron–molecular collision. Taking Eq. (12) into Eq. (11), the local electron temperature increment invoked by the instability of TVI mode reads

$$\delta T_e \approx \frac{e}{k_s^2\lambda} \sqrt{\frac{\pi\delta\epsilon_k}{12}} \delta E \quad (13)$$

Eq. (13) shows that the impact of the thermal–vibrational instability due to the introduction of the evaporated H_2O molecules into the ice–contacting streamers is not directly related to the field enhancement effect induced by the micro–nano structures, and they are related in another way. Let us examine further the instability increment factor, Ω, which characterizes the TVI vibrational mode and is strongly related to the local field strength near the melting ice block[66]. At the high rate of vibrational excitation with $v_{Tp}\tau_{Vcon} \gg 1$, where v_{Tp} is the energy transfer frequency from the electric current to the molecules' translational degree of freedom, which is strongly related with the field strength and the electric conductivity (σ_{cd})[67-70], and Ω could be formulated with local field strength, E, explicitly as follows

$$\Omega \approx \frac{D(QE + 2V_{ip})}{2(FE^2 - \gamma p_{con})} \quad (14)$$

where $Q = e\lambda\sqrt{\pi/12\delta\epsilon_k}$ for the case of Druyvesteyn distribution with λ

as the mean free path, V_{ip} is the effective ionization potential of the ionizing gas, γ is the gaseous adiabatic index, and the other symbols will be discussed after further evaluations. Based on the expression of Eq. (14), the electric field enhancement factor, β, that can intensify the instability increment could be formulated as

$$\frac{\Omega_\beta}{\Omega} \approx \frac{\beta(Q\beta E + 2V_{ip})}{\beta FE^2 - \gamma p_{con}} \cdot \frac{FE^2 - \gamma p_{con}}{QE + 2V_{ip}} > 1 \qquad (15)$$

It is easy to see that Eq. (15) could be expressed as a second-order hyperbolic equation with β as the argument, and inequality could be deduced based on one of the analytical solutions which is larger than zero

$$\beta > \beta_{cr} \equiv \frac{-b + \sqrt{b^2 + ac}}{a} \qquad (16)$$

where β_{cr} is the critical factor, $a = (FE^2 - \gamma p_{con})QE$, $b = -2V_{ip}\gamma p_{con} - FQE^3$, and $c = \gamma p_{con}(QE + 2V_{ip})$, where $F = (\hat{k}_{VT} - 2)(\gamma - 1)\sigma_{cd}\tau_{Tp}$, with τ_{Tp} as the electric current heating relaxation time and $\hat{k}_{VT} = \partial \ln k_{VT}/\partial \ln T_{gcon}$ as the logarithmic sensitivity of the vibrational relaxation rate coefficient to the translational gas temperature. Besides, it is obvious that there are two other conditions to make β a positive real number, so that two inequations about the background field strength must be satisfied

$$\begin{cases} -FQE^3 - 2\gamma p_{con} QE < 6V_{ip}\gamma p_{con} \\ \dfrac{-\gamma p_{con} QE(FE^2 - \gamma p_{con})}{(2V_{ip}\gamma p_{con} + FQE^3)^2} < \dfrac{1}{(QE + 2V_{ip})} \end{cases} \qquad (17)$$

The solutions of the algebraic equations of Eqs. (16) and (17) are mathematically trivial using some routine numerical technique, but quantitative physical appreciations could be given based on the arguments of the parameter range feasible for the atmospheric discharges. It is shown that β_{cr} is about $2 \sim 10$, when E ranges in $5 \times 10^5 \sim 5 \times 10^7$ V/m, that is to say, both the nanostructures and the PCB processing generated protrusions could intensify the thermal instability in the vicinity of the exposed electrode. However, such intensification is a secondary process of the TVI mode originating in the charged melting ice block, whose ionizing wave front arrives and the β effect of nanostructures imposed on the field will further deteriorate the stability and fasten the arcing transition. Consequently, although we cannot predict the exact behavior of the nanostructure's role in the TVI mode, quantitatively, it is clearly shown in Fig. 10 that the heating of the PN streamers is very similar to that of the P group at the lower field cases, and the self-compression features of the thermal instability are only observable in the cases of higher voltages. The results also suggest that the incorporation of the nanostructures should be considered as a method to increase the heating capability without intensifying the vibrational relaxation-induced instability at lower voltages, although the efficiency may not be acceptable.

Second, we shall show that the negative feedback mode is possible with further control of the electric field distribution proportional to the direction of the streamer propagation, namely, the proportional direction. Before we dive into the subject of how the self-modulation

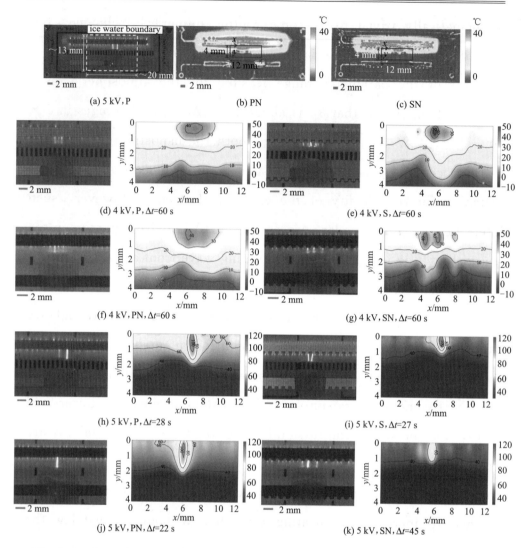

Fig. 10 (a) The locations of the about 13 mm ×20 mm sampling regions of the discharge image. The ice water boundary is marked in this image. Other images all have ice-water boundaries. The locations of the 4 mm×12 mm sampling regions of the surface temperature in the vicinities of the edges of (b) parallel electrode and (c) sawtooth electrode. Melting states, the LE-streamer morphology and temperature distributions at the sampling regions on (d) P (e) S (f) PN (g) SN at $\Delta t=60$ s when discharge states are stable at $V_a = 4$ kV. When $V_a = 5$ kV, the sampling moments are at the maximum temperatures on (h) P (i) S (j) PN (k) SN

mechanism works to function in the negative feedback loops, it is necessary to emphasize that neither the free diffusion nor the attachment of electrons in plasma usually cannot stabilize an intensive positive feedback loop when the thermal instability modes initiate[71]. The attachment is always accompanied by detachment and the high electron temperature resulting from the overheating tends to compensate for the attachment loss of free electrons, and the diffusion of electrons in an intensive field is always not free but coupled with drifting kinetics. It is argued that the time lag of the avalanche – to – streamer development between the adjacent LE – streamers could form an interacting couple[49], and result in the remarkable potential gradient, significant electric field component proportional to the streamer channel, and lateral electron conduction outflow from both of the high electron density channels. By an appropriate design of the inter – distance between the corners, both inside and outside of the sawtooth region, the kinetic energy of electrons mainly deposit their energy into the weak discharge region in between due to the rather long collisional pathway and large volume to dilute the resulted increment of ionization and heating frequency. It is well known that the streamer could develop without the emission processes on the electrode surface, and the electron and photon inflow from the LE – streamers could induce more intensive electron avalanches and increase the spatial density of the streamer in between. The averaged electron density and the averaged electron temperature including the two interacting LE – streamers respectively read $\langle n_e \rangle$ and $\langle T_e \rangle$, the electron current density could be formulated by

$$J_\perp = \sqrt{\frac{3\delta\epsilon_k \langle T_e \rangle e^2}{2m_e}} \langle n_e \rangle \quad (18)$$

where m_e is electron mass. Because both the increase in $\langle T_e \rangle$ and $\langle n_e \rangle$ could cause the increase in the electron loss current density from the major LE – streamer channel, the following equation of the negative feedback loop should be appreciated

$$\begin{cases} \delta n_e^{LE} \Uparrow \to \langle n_e \rangle \Uparrow \to J_\perp \Uparrow \to \delta n_e^{LE} \Downarrow \\ \delta T_e^{LE} \Uparrow \to \langle T_e \rangle \Uparrow \to J_\perp \Uparrow \to \delta n_e^{LE} \Downarrow \end{cases} \quad (19)$$

Consequently, as shown in Fig. 3 where the corner of the saw – tooth micro – nano structures in the SN configuration could enhance the local field intensity, and expand the volume affected by both the structure – induced effect and the ice – surface charging effect[10]. Beyond, the nanostructures deposited on the electrode surface in between could help to increase $\langle n_e \rangle$ to elevate the net loss current density, so that the instability of the LE – streamers could be effectively stabilized. It is also easy to deduce another consequence that the temperature gradient across the discharging and deicing region should tend to be uniform due to the interchanges between the discharging channels, even for the higher electric field cases. The discharge patterns shown in Fig. 10 are in accordance with those theoretical characteristics, where the temperature is the most uniform for the SN configuration at both the lower and the higher operational voltages, although the LE – streamers still possess the dominant positions.

3.4.2 Significance of statistical index of uniformity ST in the characterization of plasma heat source stability

According to the modeling of the underlying mechanisms, it is straightforward to realize that the temperature distribution that characterizes the heat flux amplitude and direction should be important. Therefore, the statistical characteristics may be useful to quantify our understanding and predictions, in some linear fashion. In Table 2, some important parameters related to the thermal stability of the LP-DBD actuators under the condition of $V_a = 5$ kV are summarized. The average duration time, τ_{cr}, is defined as the average duration time for the repeated processes of complete deicing, and V_a is selected for 5 kV because this is close to the arcing event of the PN configurations. Neither the exact critical temporal duration nor the critical voltage of a sample's arcing cannot be defined in theories due to the thermal instability being strongly dependent on the stochastic plasma heating processes[72], so the voltage of 5 kV is selected based on the experiments. Although it is empirical, the comparison of τ_{cr} among different tested samples is shown to be effective to predict the difference in the thermal instability tendency of the discharging plasma streamers, the shorter τ_{cr} is, the more likely the LE-streamers evolve into arcs at a higher applied voltage or a shorter inter-distance between the ice block and the exposed electrode. As shown in Table 2, τ_{cr} of the PN configuration is the shortest, even deteriorates in comparison with the P configuration without nanostructures, while τ_{cr} of S configuration is shown to be longer than the P configuration as the sawtooth structure improves the stability although in a very limited scale, and τ_{cr} of the SN configuration is the longest, indicating the best thermal stability. Empirically speaking, τ_{cr} is in best accordance with the thermal instability arcing experimental observations across all the tested groups and could be used as a ground – truth indicator for all the other parameters. However, another directly measurable parameter, the maximum temperature of the LE – streamers, is shown to be not a linear mapping of the τ_{cr}, actually, T_{max} of the S configuration is even higher than that of the P, and T_{max} of the P and PN are accidentally equal. This implies that the maximum temperature of the LE – streamers of a particular transient time is a quantity sampling the effective volume to be limited to deduce useful information about the thermal stability condition of the whole system.

In Table 2, there are another three statistical indices being introduced to characterize the relative instability tendency. First, A_{s1} or A_{s2} is not a significant index to compare the thermal instability tendency, even within the same micro – scale configurations, unless an extreme situation happens, such as the case of PN configuration, where only a single LE – streamer channel initiates so that A_s indicator of the PN group is significantly larger than that of the P group. However, the A_s index is in a random place to be compared for the other three groups where the index of the LE – streamers more than one have to be compared. Second, the classical statistical parameter of the standard deviation σ_T has been defined as follows

$$\sigma_T = \sqrt{\frac{1}{N_{str}} \frac{1}{N_{spn}} \sum_{i=1}^{N_{spn}} \sum_{j=1}^{N_{str}} (T(i,j) - T_{avg})^2} \qquad (20)$$

where, $T(i,j)$ is the temperatures of sampling points, N_{str} is the number of the sampling point at the streamwise direction in the 2 mm sampling region (enable to cover the LE - streamers), N_{spn} is the number of the sampling point at the spanwise direction, and T_{avg} is the average temperature on the spanwise line. The calculated results are given in Table 2. The meaning of the σ_T factor should be an index of uniformity of temperature distribution during deicing. However, it is shown that σ_T of all of the three controlled groups are very similar, and their relationships are controversial with the observations, for example, σ_T of the PN configuration is smaller than that of the P configuration, indicating a better uniformity and weaker instability tendency, which is contrary to the observations. The reason why σ_T is not a significant indicator lies in the fact that the discharge region influenced by the ice block is only a minor portion, and the statistics about the deviation of the whole sample from the average temperature are not explicitly relevant to the heat transfer features of the LE - streamer region. For the ST index discussed above, it is shown in Eq. (9) that the minimum temperature has been used as the standard index so that the physical meaning of the ST index is the approximation to the temperature gradient from the LE - streamers to the regions that are ice - free or not fully developed to involve in the interactions. Consequently, the ST factor could be served as the statistical index for the evaluation of the thermal instability tendency of a DBD system during the deicing process, with explicit physical significance.

4 Conclusion

To reduce the thermal damage risk from the streamer – to – spark transition between the electrode and the ice block of the DBD plasma de –icing heat source. A novel design of the electrode system of the microscale sawtooth structure incorporated with ZnO nanowires (SN) is scrutinized. The discharge visible and infrared morphology, electric characteristics, and the temperature field evolution patterns are studied and compared with the other three controlled groups: P, PN, and S. The main conclusions are given below:

1) Creation and isolation of multiple channels of the LE – streamer filaments are found to be the valid routes to decentralize the discharge power induced by the nonlinear effects during de – icing so that thermal stability could be significantly improved, and the micro – nano – electrode structures could be used to realize such conditions through the control of electric field distribution. Besides, a non – intuitive result lies in that the direct introduction of similar nano – structured electrodes deposited through the very same protocols into a parallel electrode system may deteriorate the thermal stability, dramatically. These are supported by the following observations. First, the thermal stability is strongly correlated with the localization morphology of the discharge during de – icing, which is characterized by the DBD streamer filaments locally enhanced by the ice block, referenced as the LE – streamers. Second, the ZnO nanowires are shown to play opposite roles

in different electrode configurations, i. e., the enhancement of the brightness and discharge power in the PN configurations, but the attenuation in the SN configurations. Third, the discharge morpho- logical pattern in the SN configurations is unique for its uniform co- existence of the multi-LE-streamers, compared with the other three controlled samples, where heat is much more strongly concentrated. Fourth, the heating behavior of such a discharge morphology in the SN configurations is shown to be self-modulated: It heats the fastest at lower voltages but demonstrates the best thermal stability at the higher voltages when the LE-streamers in all the other three configurations are ready to evolve into the devastating arcs or sparks. This could be considered as the exercise of both thermal stability and efficiency in safety deicing operation, given that the SN configuration possesses the lowest power consumption.

2) The self-modulation effect of the micro-nano electrode struc- tures is suggested to model the underlying mechanism for the thermal stabilization behavior in the proposed SN-DBD system. Firstly, there are three kinds of electric field enhancement features, shown to be vital to realize such behavior, which are induced by three different characteristic structures: i) the as-designed sawtooth corner of $\sim 10^2$ micrometers level, ii) the protrusion defects of $\sim 10^0$ micrometers level due to the chemical etching of the PCB, and iii) the ZnO nanowires of $\sim 10^1$ nanometers level. Secondly, according to the numerical calcu- lations of different structural models based on the SEM observations, the electric field enhancement factor (β) is about $10 \sim 14$ for the as-

deposited nanowires, about 2~4 for the defects, and less than 1.5 for the sawtooth corners. This has led to β in the SN configuration being generally higher than that of the PN configuration counterparts by 10%~20%, both considering the process defects. Thirdly, as a result, the onset voltage of the DBD system is lowest for the SN configuration[73], while the ionization frequency is the highest[43] so that the probability distribution of streamer initiation across the edge of the exposed electrode is the most uniform. Alternatively speaking, the nanostructures result in that the DBD micro streamer channels in any given region ease to fuse into stronger ionization channels so that the LE-streamer in the PN configurations once triggered by the nonlinear processes ease to initiate sparks. This is consistent with the phenomena shown in the PN system, and also in accordance with the phenomena in S and P configurations without nanowires, where sparking or acing is always developed through the nonlinear evolution of one single LE-streamer, although there may coexist two LE-streamer channels prior to the breakdown. Fourthly, the isolation effect resulting from the sawtooth structures can restrict the fusion processes within the region of every single period due to the periodic field screening. This is consistent with the multiple channels of the LE-streamers in the SN cases that exhibit mutually constrained tendencies with the lowest temperature difference, referenced as the self-modulation behavior herewith, under the condition of strong nonlinear effects of phase changing and surface charging.

3) Finally, the following instructions may be useful for the appli-

cation of the proposed methodology: First, for the design of the micro-scale isolation structures, the characteristic length (L_c) between every two adjacent corners is less than the diameter of the ice block (D_c), and the radius of curvature (R_c) is less than that of the ice block so that the separation of the fusion channels could be effective and the ionization localization on the electrodes may survive the competition of the surface charging on the ice blocks. For the field-enhancing nanostructures, the larger the β factor and the better uniformity of distribution, the more fused channel sites may be expected. Second, for the evaluation of certain designs or the prediction of the thermal stability of a DBD deicing device's practical operation, the ST factor extracted from the statistical properties of the temperature field during deicing processes may serve as a useful index.

Declaration of competing interest

The authors declare that they have no known competing financial interests or personal relationships that could have appeared to influence the work reported in this paper.

Credit authorship contribution statement

Xiaoxu Deng: Data curation, Investigation, Methodology, Visualization, Validation, Writing - original draft. Henggao Ding: Conceptualization, Funding acquisition, Supervision, Writing - review & editing. Zhongyu Hou: Conceptualization, Funding acquisition, Supervision, Writing - review & editing.

Data availability

No data was used for the research described in the article.

Acknowledgments

This work was financially supported by the National Natural Science Foundation of China (60906053, 61204069, 61274118, 61306144, 61504079, and 11605112), Scientific and Innovative Action Plan of Shanghai (15DZ1160800 and 17XD1702400)

Appendix A. The characteristic thermal instability phenomenon and the arcing LP – DBD sample during deicing

The LP – DBD sample of P was burnt by devastating arcs or sparks when $V_a = 7$ kV, $f = 10$ kHz. The captured images of LE – streamers and the burnt LP – DBD sample are given in Fig. 11.

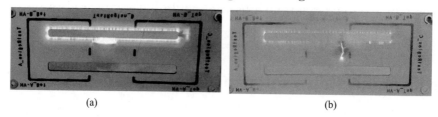

(a) (b)

Fig. 11 (a) LE – streamers on P. (b) Surface burnt by devastating arcs or sparks. High – temperature arcs or sparks cause ice or water to sublime or evaporate rapidly. When the ambient temperature is lower than 0 ℃, the water vapor condenses and a large amount of water mist is produced surrounding the LP – DBD sample

References

[1] G L Dillingham. Aviation safety: preliminary information on aircraft icing and winter operations, GAO Rep. (2010) 1-24. Report/Paper Numbers: GAO-10-441T.

[2] R W Gent, N P Dart, J T Cansdale. Aircraft icing, Philos. Trans, R. Soc. A 358 (1776) (2000), doi: 10.1098/r rsta.2000.0689.

[3] X Meng. AC- and NS-DBD plasma flow control research (2014).

[4] J Cai, Y Tian, X Meng. Q Zhai. Device and method for anti-icing by dielectric barrier discharge plasma, https://patents.google.com/patent/CN104890881A/zh.

[5] J Cai, S Xu, Y Tian, X Meng. Device and method for de-icing by dielectric barrier discharge plasma, https://wenku.baidu.com/view/ b1c656dab5360b4c2e315727a5e9856a561226a8?fr=xueshu.

[6] D F Opaits. Dielectric Barrier Discharge Plasma Actuator for Flow Control, Princeton University, 2010Ph. D, thesis.

[7] E Moreau. Airflow control by non-thermal plasma actuators, J. Phys. D 40 (3) (2007) 605-636, doi: 10.1088/0022-3727/40/3/s01.

[8] N Benard, E Moreau. Electrical and mechanical characteristics of surface ac dielectric barrier discharge plasma actuators applied to airflow control, Exp. Fluids (2014), doi: 10.1007/s00348-014-1846-X.

[9] B Dong, J M Bauchire, J M Pouvesle, P Magnier, D Hong. Experimental study of a DBD surface discharge for the active control of subsonic airflow, J. Phys. D 41 (15) (2008) 155201, doi: 10.1088/0022-3727/41/15/155201.

[10] X Deng, Z Hou. Thermal characteristic and spatial morphology between electrode and phase changing ice during de-icing process of dielectric barrier discharge and critical behavior of the surface charge density. Int. J. Heat Mass Transf. 190 (2022) 122556, doi: 10.1016/j.jiheatmasstransfer.2022.122556.

[11] A Nekahi, M Farzaneh, C Volat, W Chisholm. Electron density and temperature

measurement of arc on an ice surface, 2008, pp. 658 – 661. doi: 10.1109/CEIDP. 2008.4772815.

[12] S Ganesh, A Rajabooshanam, S K Dhali. Numerical studies of streamer to arc transition, J. Appl. Phys. 72 (9) (1992) 3957 – 3965, doi: 10, 1063/1.352248.

[13] J Cai, Y Tian, X Meng, X Han, D Zhang, H Hu. An experimental study of icing control using DBD plasma actuator, Exp. Fluids 58 (8) (2017) 102, doi: 10.1007/s00348 – 017 – 2378 – y.

[14] J Van den Broecke. Efficiency and de – icing capability of nanosecond pulsed dielectric barrier discharge plasma actuators (2016).

[15] J Chen, H Liang, Y Wu, B Wei, G Zhao, M Tian, L Xie. Experimental study on anti – icing performance of NS – DBD plasma actuator, Appl. Sci. 8 (10) (2018), doi: 10.3390/app8101889.

[16] Y Liu, C Kolbakir, H Hu, A Starikovskiy, R B Miles. An experimental study on the thermal characteristics of NS – DBD plasma actuation and application for aircraft icing mitigation, Plasma Sources Sci. Technol. (2018), doi: 10.1088/1361 – 6595/aaedf8.

[17] Y Liu, C Kolbakir, H Hu, H Hu. A comparison study on the thermal effects in DBD plasma actuation and electrical heating for aircraft icing mitigation, Int. J. Heat Mass Transf. 124 (SEP) (2018) 319 – 330, doi: 10.1016/j.ijheatmasstransfer.2018.03.076.

[18] C Kolbakir, H Hu, Y Liu, H Hu. A hybrid anti -/de – icing strategy by combining NS – DBD plasma actuator and superhydrophobic coating for aircraft icing mitigation, AIAA Scitech 2019 Forum, 2019.

[19] Y Zhu, Y Wu, B Wei, H Xu, H Liang, M Jia, H Song, Y Li. Nanosecond – pulsed dielectric barrier discharge – based plasma – assisted anti – icing: modeling and mechanism analysis, J. Phys. D 53 (14) (2020) 145205, doi: 10.1088/1361 – 6463/ab6517. (18pp)

[20] M Abdollahzadeh, F Rodrigues, J Pascoa. Simultaneous ice detection and removal based on dielectric barrier discharge actuators, Sens. Actuators, A 315 (2020) 112361, doi: 10.1016/j.sna.2020.112361.

[21] B Wei, Y Wu, H Liang, J Chen, G Zhao, M Tian, H Xu. Performance and mechanism

analysis of nanosecond pulsed surface dielectric barrier discharge based plasma deicer, Phys. Fluids 31 (9) (2019) 091701, doi: 10.1063/1.5115272.

[22] B Wei, Y Wu, H Liang, Y Zhu, J Chen, G Zhao, H Song, M Jia, H Xu, et al., SDBD based plasma anti-icing: a stream-wise plasma heat knife configuration and criteria energy analysis, Int. J. Heat Mass Transf. 138 (2019) 163-172, doi: 10.1016/j.ijheatmasstransfer.2019.04.051.

[23] A Y Starikovsky, N L Alexandrov. Gasdynamic flow control by superfast local heating in a strongly nonequilibrium pulse plasma (2020).

[24] F Rodrigues, J Páscoa, M Trancossi. Heat generation mechanisms of DBD plasma actuators, Exp. Therm. Fluid Sci. 90 (2018) 55-65, doi: 10.1016/j.expthermflusci.2017.09.005.

[25] R Joussot, V Boucinha, R Weber-Rozenbaum, H Rabat, D Hong. Thermal characterization of a DBD plasma actuator: dielectric temperature measurements using infrared thermography, in: Fluid Dynamics Conference and Exhibit, 2010, doi: 10.1016/j.expthermflusci.2017.09.005.

[26] R Tirumala, N Benard, E Moreau, M Fenot, G Lalizel, E Dorignac. Temperature characterization of dielectric barrier discharge actuators: influence of electrical and geometric parameters, J. Phys. D 47 (25) (2014) 255203, doi: 10.1088/0022-3727/47/25/255203.

[27] R Tirumala, N Benard, E Moreau, M Fenot, G Lalizel, E Dorignac. Temperature characterization of dielectric barrier discharge actuators: influence of electrical and geometric parameters, J. Phys. D 47 (25) (2014) 255203, doi: 10.1088/0022-3727/47/25/255203.

[28] C L Enloe, T E McLaughlin, R D VanDyken, K D Kachner, E J Jumper, T C Corke, M Post, O Haddad. Mechanisms and responses of a dielectric barrier plasma actuator: geometric effects, AIAA J. 42 (3) (2004) 595-604, doi: 10.2514/1.3884.

[29] A R Hoskinson, N Hershkowitz. Differences between dielectric barrier discharge plasma actuators with cylindrical and rectangular exposed electrodes, J. Phys. D 43 (6) (2010) 065205, doi: 10.1088/0022-3727/43/6/065205.

[30] S Achat, Y Teisseyre, E Marode. The scaling of the streamer–to–arc transition in a positive point–to–plane gap with pressure, J. Phys. D 25 (4) (1992) 661, doi: 10.1088/0022-3727/25/4/012.

[31] M Janda, Z Machala, A Niklová, V Martišovitš. The streamer–to–spark transition in a transient spark: a dc–driven nanosecond–pulsed discharge in atmospheric air, Plasma Sources Sci. Technol. 21 (4) (2012) 045006, doi: 10.1088/0963-0252/21/4/045006.

[32] M Liu, S Wang, L Jiang. Nature–inspired superwettability systems, Nat. Rev. Mater. (2017), doi: 10.1038/natrevmats.2017.36.

[33] M J Kreder, J Alvarenga, P Kim, J Aizenberg. Design of anti–icing surfaces: smooth, textured or slippery? Nat. Rev. Mater. (2016), doi: 10.1038/natrevmats.2015.3.

[34] T Strobl, S Storm, D Thompson, M Hornung, F Thielecke. Feasibility study of a hybrid ice protection system, J. Aircr. 52 (6) (2014) 2064–2076.

[35] L Mishchenko, B Hatton, V Bahadur, J A Taylor, T Krupenkin, J Aizenberg. Design of ice–free nanostructured surfaces based on repulsion of impacting water droplets, ACS Nano 4 (12) (2010) 7699–7707.

[36] K Lum, D Chandler, J D Weeks. Hydrophobicity at small and large length scales, J. Phys. Chem. B 103 (22) (1999) 4570–4577.

[37] Q Hao, Y Pang, Y Zhao, J Zhang, J Feng, S Yao. Mechanism of delayed frost growth on superhydrophobic surfaces with jumping condensates: more than interdrop freezing, Langmuir 30 (51) (2014) 15416–15422.

[38] K K Varanasi, T Deng, J D Smith, H Ming, N Bhate. Frost formation and ice adhesion on superhydrophobic surfaces, Appl. Phys. Lett. 97 (23) (2010) 268.

[39] Icephobic/anti–icing properties of micro/nanostructured surfaces, Adv. Mater. 24 (19) (2012) 2642–2648.

[40] L Cao, A K Jones, V K Sikka, J Wu, D Gao. Anti–icing superhydrophobic coatings, Langmuir 25 (21) (2009) 12444–12448, doi: 10.1021/la902882b.

[41] H Min, J Wang, H Li, Y Song. Super–hydrophobic surfaces to condensed micro–droplets at temperatures below the freezing point retard ice/frost formation, Soft Matter 7 (8) (2011) 3993–4000.

[42] P Kim, T S. Wong, J Alvarenga, M J Kreder, W E Adorno-Martinez, J Aizenberg. Liquid-infused nanostructured surfaces with extreme anti-ice and anti-frost performance, ACS Nano 6 (8) (2012) 6569-6577.

[43] Z Hou, B Cai, H Liu. Mechanism of gas breakdown near Paschen's minimum in electrodes with one-dimensional nanostructures, Appl. Phys. Lett. 94 (16) (2009) 163506, doi: 10.1063/1.3123170.

[44] S H Jo, Y Tu, Z P Huang, D L Carnahan, D Z Wang, Z F Ren. Effect of length and spacing of vertically aligned carbon nanotubes on field emission properties, Appl. Phys. Lett. 82 (20) (2003) 3520-3522, doi: 10.1063/1.1576310.

[45] L E Greene, M Law, J Goldberger, F Kim, J C Johnson, Y Zhang, R J Saykally, P Yang. Low-temperature wafer-scale production of ZnO nanowire arrays, Angew. Chem. Int. Ed. 42 (26) (2003) 3031-3034, doi: 10.1002/anie.200351461.

[46] T R Bullett, J L Prosser. Paint: a surface modifier, Phys. Technol. 14 (3) (1983) 119-125.

[47] A Venkattraman. Electric field enhancement due to a saw-tooth asperity in a channel and implications on microscale gas breakdown, J. Phys. D 47 (2014) 425205, doi: 10.1088/0022-3727/47/42/425205.

[48] R G Forbes, C Edgcombe, U Valdrè. Some comments on models for field enhancement, Ultramicroscopy 95 (2003) 57-65, doi: 10.1016/s0304-3991 (02) 00297-8.

[49] Y P Raizer. Gas Discharge Physics, Gas Discharge Physics, 1991.

[50] M Farzaneh, C Volat, A Gakwaya. Electric field modelling around an ice-covered insulator using boundary element method (2002).

[51] V I Gibalov, G J Pietsch. The development of dielectric barrier discharges in gas gaps and on surfaces, J. Phys. D 33 (20) (2000) 2618-2636, doi: 10.1088/0022-3727/33/20/315.

[52] V R Soloviev. Analytical estimation of the thrust generated by a surface dielectric barrier discharge, J. Phys. D (2011), doi: 10.1088/0022-3727/45/2/025205.

[53] V R Soloviev. Analytical model of a surface barrier discharge development, Plasma Phys. Rep. 45 (3) (2019) 264-276, doi: 10.1134/S1063780X19020119.

[54] Y Zhu, Y Wu. The secondary ionization wave and characteristic map of surface discharge

plasma in a wide time scale, New J. Phys. 22 (10) (2020) 103060, doi: 10.1088/1367-2630/abc2e7. (8pp)

[55] PHELPS database, www.lxcat.net, retrieved on June 7, 2021.

[56] J M Meek. The electric spark in air, J. Inst. Electr. Eng. - Part I 89 (1942) 335-356.

[57] O V Skrebkov. Vibrational nonequilibrium in the hydrogen-oxygen reaction at different temperatures, J. Mod. Phys. 5 (16) (2014) 1806, doi: 10.4236/jmp.2014.516178.

[58] R C Millikan, D R White. Systematics of vibrational relaxation, J. Chem. Phys. 39 (12) (1963) 3209-3213, doi: 10.1063/1.1734182.

[59] A B Mikhailovskii. Theory of Plasma Instabilities, Theory of Plasma Instabilities, 1974.

[60] A V Nedospasov, Y B Ponomarenko. Ionization striations in the discharge positive column, in: P. Hubert, E. Crémieu-Alcan (Eds.), Phenomena in Ionized Gases, VI International Conference, Volume II, 1963, p. 223.

[61] J R Roth. New mechanism for low-frequency oscillations in partially ionized gases, Phys. Fluids 10 (12) (1967) 2712.

[62] J R Roth. Experimental observation of continuity-equation oscillations in slightly ionized deuterium, neon and helium gas, Plasma Phys. 11 (9) (1969) 763-778, doi: 10.1088/0032-1028/11/9/007.

[63] A V Nedospasov, V D Khait. Oscillations and instabilities of low-temperature plasmas (1979).

[64] J H Cahn. Electron velocity distribution function in high frequency alternating fields including electronic interactions, Phys. Rev. 75 (1949) 838-841, doi: 10.1103/PhysRev.75.838.

[65] M J Druyvesteyn, F M Penning. The mechanism of electrical discharges in gases of low pressure, Rev. Mod. Phys. 12 (1940) 87-174, doi: 10.1103/RevModPhys.12.87.

[66] G Bekefi. Principles of laser plasmas, J. Electrochem. Soc. 124 (12) (1976) 435C, doi: 10.1149/1.2133218.

[67] W L Nighan, W J Wiegand. Influence of negative-ion processes on steady-state properties and striations in molecular gas discharges, Phys. Rev. A 10 (1974) 922-945,

doi: 10.1103/PhysRevA.10.922.

[68] R A Haas. Plasma stability of electric discharges in molecular gases, Phys. Rev. A 8 (1973) 1017 - 1043, doi: 10.1103/PhysRevA.8.1017.

[69] V E Fortov. Encyclopedia of low - temperature plasma, High Temp. 46 (1) (2008) 1 - 2, doi: 10.1134/S0018151X0801001X.

[70] Smirnov, M Boris. Cluster plasma, Physics - Uspekhi 43 (5) (2000) 453, doi: 10.1070/PU2000v043n05ABEH000722.

[71] J Shang, S T Surzhikov. Plasma Dynamics for Aerospace Engineering, Plasma Dynamics for Aerospace Engineering, 2018.

[72] B M Smirnov. Cluster Plasma, Plasma Processes and Plasma Kinetics: 586 Worked Out Problems for Science and Technology, 2008.

[73] V R Soloviev, I V Selivonin, I A Moralev. Breakdown voltage for surface dielectric barrier discharge ignition in atmospheric air, Phys. Plasmas 24 (10) (2017) 103528, doi: 10.1063/1.5001136.

Reconfigurable Plasma Composite Absorber Coupled with Pixelated Frequency Selective Surface Generated by FD – CGAN

(2022 年 12 月)

Abstract Conventional plasma absorbers are challenging to obtain high electron density and sizeable spatial scale for effective absorption while meeting the applied requirements of low profile and low power consumption. Although the frequency selective surface (FSS) has proved to realize a lower profile of plasma absorber with some empirical patterns adequately, the issue of the FSS design matching the dispersion distribution of complicated plasmas is still in suspense. A reverse prediction method referenced as the forecast and design Conditional Generative Adversarial Network (FD – CGAN) is proposed to generate a pixelated FSS between double – layer plasma periodic arrays. The reflection attenuation characteristics examined by experiments show that the addition of the FSS makes the coupling absorption effect surpass that of either pixelated FSS or plasma solely.

本文发表于 IEEE Access。合作作者：Mengjie Yu，Haitao Wang，Xiangxiang Gao，Xi Ren，Zunyi Tian，Henggao Ding，Zhongyu Hou。

Measurements in reconfigurable working modes and array arrangements demonstrate that the proposed configuration maximizes absorption effectiveness in the same profile, accompanied by the simulation. An interfacial void model is proposed to assist the design of the composite absorbing structure, together with an equivalent circuit for the hybrid absorber including periodic patterns with stochastic distribution characteristics, which analyze the absorption effect of the composite structure. The study provides a new approach for various microwave applications, including multilayer radar-absorbing structures, plasma-based stealth technology, and reconfigurable filters.

Index Terms　Microwave absorption, Plasma periodic structures, Pixelated frequency selective surface, Generative adversarial network, Reconfigurable absorber.

1 Introduction

As a dispersive and lossy medium, the collisional plasma has a complex dielectric constant and can be used as an effective absorber of electromagnetic (EM) waves in a wide range of frequencies[1]. Besides exhibiting high attenuation over a large bandwidth, another advantage of a plasma absorber has also been emphasized: thanks to the sub-nanosecond collisional energy transition processes among particles in plasmas[2], the spatial refractivity of plasmas is tunable and fast-switchable, so its reconfigurability can be achieved by the electrical control[3]. Three particularities of plasma place technical demands on the preparation of microwave-active plasmas that are difficult to implement: the absorption capacity of a collisional plasma is determined by the oscillation and collision behavior of electronic dipoles[4], i. e., the high absorption capacity depends on the plasma with high densities in the certain wavelength range; the scale of volumetric plasma needs to be large enough compared to the wavelength of the incident EM wave[5]; and achieving high electron number density is unavoidably accompanied by the high power and heating scenarios[6].

It has been discovered that the inclusion of the frequency selective surface (FSS) can circumvent the above limitations to achieve high absorption with low power consumption and low profile. A large-area, lightweight, conformal plasma device which was electrically excited by FSSs was proposed to interact with X-band microwave energy[7]. An equivalent circuit model was demonstrated to analyze the wideband absorber with the combination of plasma and resistive FSS[8]. Payne et al. have carried out a series of studies on the low-profile plasma-based tunable absorber integrated with FSS[9-11]. In addition to forming the controlled wavefront distribution on a large scale, FSS can be coupled with plasmas to modulate the phase cancellation of EM waves. It seems to convert part of the bulk plasma into a thin, low-power, and highly controllable metallic periodic structure, which can be considered as a "quasi-plasma" with high density, yielding greater absorption effects while reducing the thickness of the plasma itself.

The conventional design process of FSS is empirical and usually consists of model design, parameter sweeping, and optimization, in which assembled patterns are mainly restricted to linear type, loop type, solid patch, or mixture, allowing for little flexibility in the optimizing procedure[12,13]. In addition, under the condition that the interlayer coupling of the composite absorbing structure cannot be neglected, synthesizing the satisfactory FSS with adaptive features of coupling with plasma becomes difficult[14]. Generative adversarial networks (GANs) of Deep learning (DL) method[15], which are characterized by training a pair of networks in competition with each other, have been proposed to innovate the design of an FSS to match the desired scattering properties[16-18]. Compared with other DL – based design approaches, GANs can partly solve the information loss of the convolution neural networks and generate new samples that are similar to those in the dataset but have a singularity, which indicates the advantage in generating unpredictable periodic structures[19]. Hence, the approach can be effective in matching the dispersion distribution of complicated plasmas during the design of a composite structure, which has no attempt to be found in the literature. Moreover, to increase the degrees of freedom of the structure, a pixelated pattern is used to exhibit great flexibility in the optimization procedure due to the use of binary inversion of pixel values under the certain frequency response requirement by splitting the unit cell up into a grid of sub pixels[20,21].

 The paper presents a reconfigurable microwave absorber composed of plasma arrays and the pixelated frequency selective surface for the first time. We propose an adaptive reverse design strategy with the conditional GAN model to optimize pixelated structures, which is enabled to generate optimal geometric patterns to reduce the return wave of the hybrid structure. It is found that the coupled structure has induced a wideband attenuation effect and reconfigurable - switching effect. The reflection characteristics of the absorber and pixelated FSS or plasma solely are explored using a free space measurement facility, accompanied by working modes and array arrangements. The reason for the intensive coupling effect is discussed through simulation analysis and the lumped circuit model. The method exhibits the potential to control the propagation of microwaves via a reconfigurable hybrid plasma system.

2 Experimental setup and method

2.1 Configurations of the proposed reconfigurable composite absorbing structure

Fig. 1 (a) provides the experimental diagram of the absorbing structure and the measuring system. The experiment is conducted in the anechoic chamber. The propagation of free space EM waves is realized by a pair of broadband horn antennas (LB - 10180 - NF, A - INFO Co., Ltd.) as the microwave source and receiver, which has frequency ranges of 1~18 GHz and the gain of 11 dB. The antennas are connected to an Agilent 8720ES vector network analyzer to measure the reflection characteristics. The distance from the symmetrical antennas to the absorber is about 110 cm, and the structure is fixed above the foam holder with the dielectric constant considered as 1. It is noticed that the electric field is polarized along the y - axis, which is transverse to the axis along plasma cylinders. The EM wave propagates in the negative direction of the z-axis, that is, normal incident on the target. The time-gating method is used to eliminate the influence of unnecessary scattering in free space.

The top view of the proposed composite absorbing structure under the working state is illustrated in Fig. 1 (b). In order to further reduce scattering noise, the absorbing wedges hide the electrical connection at the tube tip and leave a 20 cm × 20 cm testing window. As shown in Fig. 1 (c), the absorber is composed of a multi-layer composite structure, with the sequence from top to bottom of "top plasma layer - pixelated FSS layer - bottom plasma layer - metallic ground". The thickness of the metallic ground is much larger than the skin depth, reflecting all the incident electromagnetic waves, which can be considered as perfect electrical conductor (PEC). Each layer of plasma arrays consists of twelve individual discharge tubes with quartz enveloped ($\varepsilon = 3.8$), which are 15 mm and 13 mm in outer and inner diameter, respectively. The visible plasma length of discharge tubes filled with low-pressure argon and mercury vapor is 200 mm, and the distance between each tube is 2 mm. Each cylindrical plasma tube is driven individually by an AC ballast, which controls the plasma state by adjusting the plasma density and frequency. In accordance with the triangular-shaped current waveform of the discharge tube, the root mean square (RMS) current I_{RMS} is 154.73 mA at a frequency of 22.68 kHz.

It is well known that there are two characteristic parameters of plasmas, including the plasma density and the momentum transfer collision frequency of the plasma electrons. A gas discharge model is

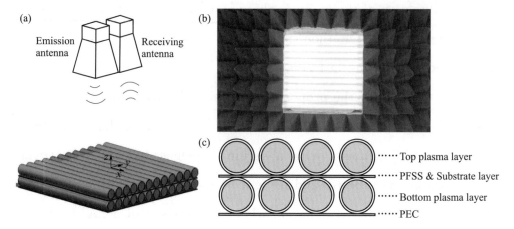

Fig. 1 The designed experimental system: (a) schematic of experimental apparatus and measuring system, (b) photograph of the test zone with absorbing wedges surrounding, and (c) side view of part of the proposed structure

used to estimate the resulting plasma density in the system[22]. The relationship between the measured I_{RMS} and the average electron density n_e is given as follows

$$n_e = \frac{I_{RMS}}{eA\mu_d} \tag{1}$$

where e is the charge of an electron, A is the inside cross-sectional area of each discharge tube, and μ_d is the electron mobility. μ_d is calculated through the Boltzmann equation using BOLSIG+[23], and the required time-averaged reduced electric field (E/N) is obtained from the discharge voltage and the total pressure of the mixed gas. Hence, the estimated electron density is roughly between 7.54×10^{16} m^{-3} and 1.23×10^{17} m^{-3}, which is in the range of the conventional mercury-based fluorescent lamp plasma[24].

As a lossy dispersive dielectric material, plasma has the dielectric constant described by the Drude model

$$\varepsilon_p = 1 - \frac{\omega_p^2}{\omega(\omega - jv)} = 1 - \frac{\omega_p^2}{\omega^2 + v^2} - j\frac{v}{\omega}\frac{\omega_p^2}{\omega^2 + v^2} \quad (2)$$

where ω_p is the plasma frequency, ω is the EM wave frequency, and v is the momentum transfer collision frequency of the plasma electrons, which is about the value of $v = 5.6 \times P(\text{torr}) \times 10^9 (\text{Hz})$ based on the low-pressure argon glow discharge technology[25]. In the above equation, ω_p is a function of the electron density

$$\omega_p = \sqrt{\frac{n_e e^2}{m_e \varepsilon_0}} \quad (3)$$

with m_e being the electron mass and ε_0 being the vacuum permittivity. Using the above relation, the propagation constant $k(\omega)$ for EM waves in a collisional plasma can be expressed as[26]

$$k(\omega) = \frac{\omega}{c}\sqrt{\varepsilon_p} \quad (4)$$

where c is the speed of light for vacuum.

2.2 FD-CGAN model and geometrical construction

Replacing the traditional design procedure of periodic patterns on FSSs, the FD-CGAN architecture, which combines the deep convolutional neural network (CNN) and the generative adversarial network (GAN), enables forward prediction of the S-parameters and designs geometrical patterns inversely under different frequency response requirements. The training and testing datasets of the model are constructed with the full wave simulator. Each square patch is represented by a binary bit, where "1" indicates the presence of PEC metal, and "0" indicates the absence. The coded patterns build a unique map from spatial terms to numerical values, simplifying the traditional modeling process. Fig. 2 provides an overview of FD－CGAN architecture, which is a semi-supervised learning model based on game theory. It comprises three different networks: the generator (G), the discriminator (D), and the predictor (P). The generative neural network G generates synthetic data represented as pixelwise images from the random noise and real spectra, and the discriminative neural network D estimates the probability that the sample came from the training data rather than G. If the D identifies a fake structure, the results are backpropagated to G, which the learnable parameters and growth rules will be adjusted to reduce the error. While if the D cannot distinguish errors of the data distribution between the fake and real structures, the information of matrices will be fed to the predictive neural network P. The P is a pretrained network for a given pattern as its input, and a fake reflection spectrum is generated as the output, which is learning to be close to the real response continuously[27]. Furthermore, extensive experiments demonstrate that the pixelated images generated by the proposed FD-CGAN model can obtain the preset electromagnetic responses.

The optimized pixelated FSS structure, including unit cell and screen geometry, is shown in Fig. 3. The fabricated FSS contains 20×20 elements on the substrate, and the unit cell area is split up into a 32×32 grids of pixels. The standard PCB process is done on the upper

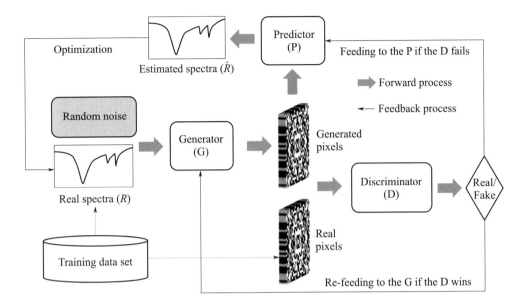

Fig. 2　The complete architecture of the proposed FD-CGAN model includes a generator (G), a discriminator (D), and a predictor (P). In the process of training G, the generated patterns vary depending on the feedback from D and P. The distribution of pixel points that meets the reflection response expectations is captured with iterations.

surface of a 1 mm FR-4 substrate with a permittivity of 4.4 and a loss tangent of 0.02. The other design parameters include the wire trace width $\omega = 0.315$ mm, the thickness of the copper clad $t_\omega = 30$ μm, and the periodic separation $s = 10.2$ mm.

3　Results and discussion

3.1　Reflection characteristics of plasma arrays or pixelated FSS solely and the coupled structure

The reflection spectrum of double-layer plasma arrays solely is

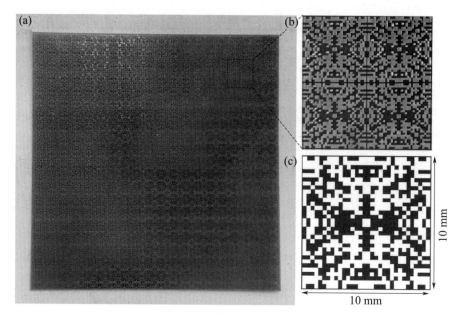

Fig. 3 (a) FSS screen structure with the unit cells of 20×20; (b) Detailed view of four units in FSS screen; (c) Unit cell geometry with the size of 10 mm by 10 mm, in which white areas indicate the metal

demonstrated in Fig. 4. Notice that the top plasma layer is closer to the microwave source than the bottom one. Quartz tubes that exist throughout the experiment serve as the structural framework, which induces a slight resonance of -4.8 dB located around 13.2 GHz. It can be attributed to the phase modification and scattering caused by the dielectric constant and shape of quartz tubes, respectively. The other three states have broadband attenuation across the test band. Comparing the single layer of the "ON" state, the resonance peak of -10.5 dB located around 8.9 GHz appears when the top plasma layer is excited. In contrast, the largest peak occurs at 13.2 GHz with the value of -10 dB as the bottom layer is excited. In the case of both discharge layers turned on, the intense resonance effect appears at 12.8 GHz with the value of −16.5 dB. The absorption peaks are the result of multiple reflections between double-layer plasma arrays, which are due to the cavity resonance effect and collisional absorption[28]. One can see that due to the cavity resonance effect of enclosed plasmas, the attenuation and layers are no longer a simple linear relationship. It can be inferred that there is a large density gradient at the boundary of plasma arrays, and the interface of plasmas and air is not completely satisfied with the impedance matching condition, leading to an attenuation of the incident wave.

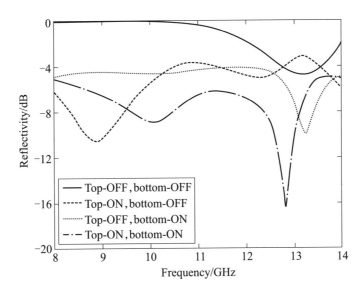

Fig. 4 Reflection spectrum comparisons for various states of plasma arrays with double layers in experimental measurement. The "OFF" state means the plasma layer is not working, while the "ON" state is the opposite

The reflectivity is calculated over a frequency range from 8 to

14 GHz, and use minimum/maximum tolerable values to define the solution's fitness. Fig. 5 (a) presents the theoretically calculated and experimentally measured reflection spectrum of the pixelated FSS with constraints. Despite the existence of value differences, the experimental result shows a congruent changing trend with the simulated one, both exhibiting the resonance at around 11.7 GHz. The experimental curve displays the attenuation peak of -10.7 dB, which is a reduction of $+46.0\%$ from the simulated one in the absolute value. Deviations in resonant frequencies and attenuation are due to the difference between the finite measurement and the infinite simulation.

Fig. 5 (b) depicts the surface current distribution on the pixelated unit cell at the resonant frequency of 11.7 GHz. It is observed that the magnetic field distribution of the cell where the resonances around $f = 11.7$ GHz originates from the surface current flow inside the cell rather than on the boundaries bordering other cells. In other frequency ranges, the induced surface current becomes remarkably weak. One can see that the intensity of the surface current distribution and the fringing field that give rise to the capacitive effect are low at the resonant frequency of 11.7 GHz, which is an advantage from a power-handling point of view[29].

With the consideration of the adaptive features of coupling with plasma, a composite absorbing structure composed of plasma arrays and the pixelated FSS is designed. The reflection coefficient magnitude of the coupled structure for normal incidence in the working frequency range of 8~14 GHz is shown in Fig. 6. At the OFF state, namely, the

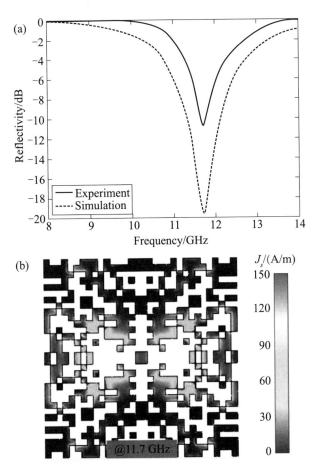

Fig. 5 (a) Experimental measured and simulated reflection spectrum of the designed pixelated FSS alone; (b) The surface current distribution on the unit cell at the resonance

situation of pixelated FSS combined with quartz tubes, a little absorption point of −6.6 dB appears around the central frequency of 12.2 GHz, which is the synthetic result of the modulation of EM wave by FSS and the structure reflection of the quartz tube. When the FSS and double-layer plasma arrays coexist, the reflection spectrum is far beneath that of the former. The wideband response occurs during the whole working frequency bands (below -6 dB), showing resonant frequencies of 9.6 GHz and 12.3 GHz with the corresponding absorption peak values of -38.5 dB and -12.4 dB, respectively. More concretely, the -10 dB frequency bandwidth for the primary resonance reaches 1.4 GHz, and the absorption band below -10 dB is marked in the figure.

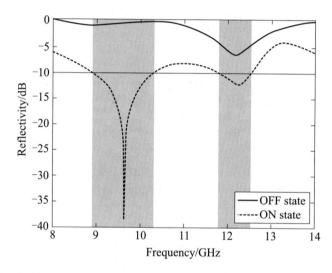

Fig. 6 Reflection response of the proposed composite structure for two states in experimental measurement. The "OFF" state means plasma layers are not working, while the "ON" state is the opposite. The parts marked in red are absorption bands below -10 dB

To illustrate the joint attenuation effect of pixelated FSS and plasmas, additional comparisons of absorption are also given in Fig. 7. First, the effect of FSS and plasmas alone on the incident wave will cause two resonant peaks with 1.1 GHz apart. It is observed that the pixelated FSS does not show an excellent absorptive response below

10 GHz, while double-layer plasma arrays lead to an attenuation of more than −5 dB over the entire operating frequency. Second, the central frequency and the resonance effect of FSS change with and without quartz tubes. It can be attributed to the fact that the outside layer with quartz modifies the phase of EM waves. The exterior materials can modify the phase of the incident wave so as to move the peak into the desired frequency range[28]. However, the cylindrical shape inevitably aggravates the scattering of EM waves. Third, one can see that the composite absorbing structure demonstrates remarkable resonance and bandwidth improvement. Apparently, double-layer plasmas on the grounded dielectric have an acceptable absorption value reaching −16.5 dB, and by adding the pixelated FSS, the attenuation is extended by almost 128%. In addition, the resonance points of the coupled structure are not made in the region where they overlap each other. The frequency shift is due to the effect of combination and mutual coupling between different parts of the absorber. As the presence of multiple reflections, the incident wave interacts with each layer of the model, and it is enhanced by the total reflection of PEC. The combined attenuation effect of this structure is better than the effect of either plasma arrays or pixelated FSS solely.

3.2 Reconfigurability characteristics of plasma composite absorbing structure coupled with pixelated FSS

Under the reduced power consumption of the proposed absorption structure, six working modes exhibiting reconfigurability characteristics are employed, as shown in Fig. 8. In general, the absorbing ability of the

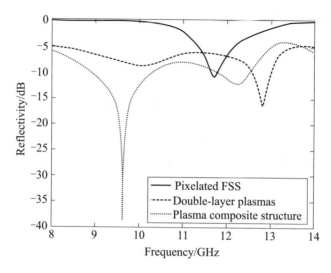

Fig. 7 Comparisons of reflection response of the proposed coupled structure with pixelated FSS (PFSS) or plasma solely in experimental measurement

discharge on the interval side is better than that on the same side. Nevertheless, the coupled structure consisting of double-layer plasma arrays and pixelated FSS still exhibits unattainable performance at bandwidth and attenuation. According to the absorption theory of plasmas, the wave energy transfers to charged particles in the plasma, and subsequently to neutral particles (atoms and molecules) by elastic and inelastic collisions. It is thought that the hindering effect of plasmas on EM wave propagation decreases as the number of discharge tubes in working conditions decreases[30]. As the normal incident wave reaches the top layer, part of the wave is absorbed by plasmas, and the rest of the energy can be transmitted from the non-working discharge tubes since the dielectric constant of the gas inside the tube is extremely close to that of air when it is not working. In the case of the same side discharge, nearly half of the incident energy is not exposed to plasmas and is only modulated by the FSS, which causes less effect on the attenuation of EM waves than the interval discharge. The comparison of the single-layer plasma with the "ON" state is as follows. When the plasma layer is closer to the wave source, there is a broadband absorption of −6 dB exhibiting in most of the working frequency bands, accompanied by the absence of attenuation peaks. In the case of plasmas further away from the wave source, two resonance points with central frequencies of about 10.7 and 12.6 GHz exist, displaying the most significant attenuation value of −14.3 dB. It can be concluded that the proposed absorbing structure reconfigures in six modes as plasma layers switching with the same low power consumption: a broadband absorption at low frequencies, a broadband absorption at high frequencies, a broadband absorption without peaks, and an absorption with two peaks.

3.3 The absorption analysis with changing arrangements

In order to analyze the absorption mechanism of this "sandwich-type" structure, two other structures with the same profile are added for discussion seen in Fig. 9. In case 1, the pixelated FSS and plasma region modulate the incident wave successively, resulting in a resonance with a peak of −8.6 dB around 10.8 GHz. Conversely, when the incident wave first passes through thick plasma layers as shown in case 2, the reflection spectrum displays a wideband attenuation of about −5 dB over the whole frequency band, even with the presence of FSS.

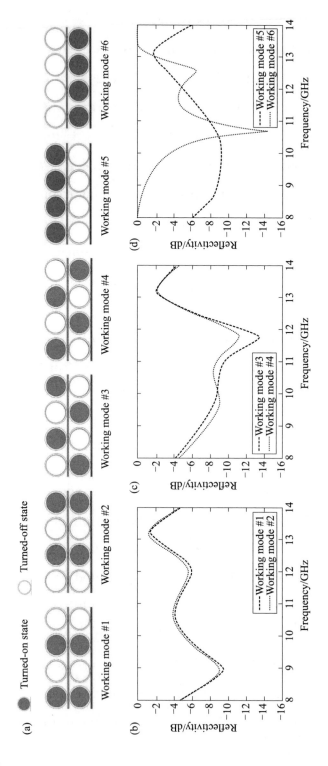

Fig.8 (a) Schematic of different working modes; (b) ~ (d) Reconfigurable reflection response of the proposed composite absorbing structure with six working modes in experimental measurement.

It can be seen that the front structures both react poorly in terms of broadband and attenuation, while the proposed structure maximizes the effectiveness of absorption over a wide frequency range in a relatively low profile.

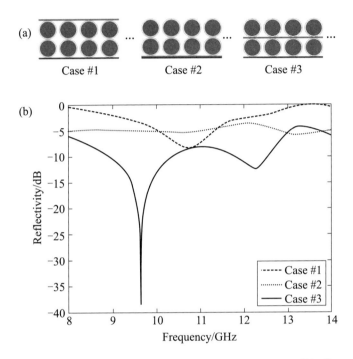

Fig. 9 (a) Schematic of different array arrangements; (b) Comparisons of reflection response of the composite absorbing structure in three cases in experimental measurement

To account for the absorption mechanism, the simulated average electric field distribution of each absorber at the cross-section in the y-z plane or on the FSS plane is shown in Figs. 10 (a) to (f). The simulation is conducted with the frequency domain solver. The dielectric constant of plasma is defined by the Drude model, which the plasma frequency is assumed to be 6.8 GHz as well as the collision frequency is 7.4 GHz based on the calculation in Sec. 2.1. Compared with the electric field distribution of the sandwich-type structure at the y-z section, the decay of average superimposed electric fields of the other two structures are significantly weak in the free space, even though both are accompanied by the coupling effect, corresponding to the reflection feature in Fig. 9. In Fig. 10 (e), a strong electric field is generated at the contact region of the FSS and two separate plasma layers, indicating a large energy dissipation, which results in a decay of the superimposed field strength above the structure. Moreover, there are strong local field enhancements in the center of the pixelated cell in all three structures, while the average induced electric field of the proposed structure is maximum in the whole plane. The pixelated metal arrays in case 1 are induced with the electric field concentrated in the center due to the contact with the incident wave initially, with a large field gradient to the surrounding. However, the coverage region of the high-intensity field in case 2 is rarely resulting from absorption and scattering by thick plasma arrays. Due to the existence of the multilayer structure, part of the waves bounces back and forth in plasmas. An excessively thick plasma layer will cause the lower EM wave energy throughout to reach the FSS, reducing the modulation effect. On the other hand, the incident wave cannot achieve a deep penetration depth in plasma after being modulated by FSS. It can be seen that the strong coupling effect requires proper placement matching. The result can also explain the propagating features in Fig. 6

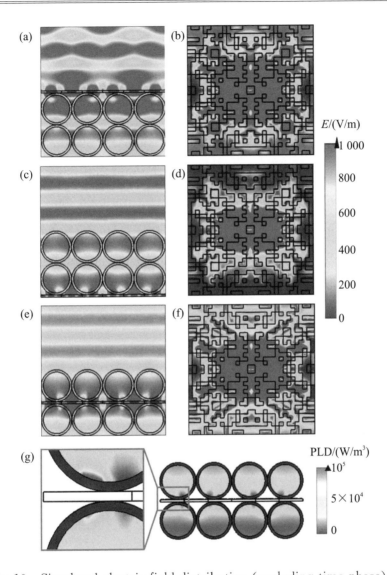

Fig. 10　Simulated electric field distribution (excluding time phase) of three absorbers and the power loss density of the proposed structure observed at respective resonance points. (a), (c), and (e) E-field distribution of three structures in the $y-z$ plane, respectively; (b), (d), and (f) E-field distribution of three structures on the FSS plane, respectively; (g) power loss density of the proposed composite absorbing structure in the $y-z$ plane and the partially enlarged illustration

and Fig. 8 (d), in which the absorption effect with double excited layer seems to be a superposition of monolayer excitation. The power loss density distribution of the proposed composite absorbing structure in the $y-z$ plane is depicted in Fig. 10 (g), indicating the sum of the electric and magnetic power dissipated inside the plasma. It is noticed that the projection of the $y-z$ plane on the unit cell passes horizontally through the center of the surface. The inset not only demonstrates that the power dissipation is concentrated at the contact region between the plasma and FSS, but also emphasizes the absorption effect in the center of the FSS structure, which is exactly the result we expect. The local maximum achieves 1.54×10^5 W/m^3.

An interfacial void model for electromagnetic structures is proposed based on the field distribution. Unlike conventional interfaces of matter, the contact surface between electromagnetic layers is discontinuous and unitary, and for practical consideration, layers containing plasmas can only be used in a discrete manner. It leads to the existence of an interfacial void with little absorption contribution between the plasma layer and FSS layer, i. e. , the air and the glass in this case, which modulates the incident wave through the curvature of plasma columns.

Fig. 11 depicts the definition of the interfacial void by the incidence of EM waves and a simplified one-dimensional model, which the equivalent dielectric constant is denoted by ε_{eff}. As dissipative media, plasmas and the FSS layer sandwich the interfacial void zone, which is dominated by the dielectric loss. Periodic changes in topology lead to regular variations in ε_{eff}, and most dielectric materials have few losses so that the attenuation caused by the interfacial void under the plane wave condition is related to the wave number and the equivalent loss tangent [31]

$$\alpha \sim k\,(\tan\delta)_{eff} \tag{5}$$

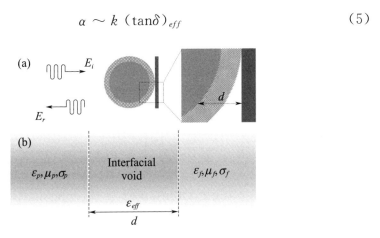

Fig. 11 Schematic diagram of (a) the two-dimension interfacial void and (b) the simplified one-dimension interfacial void

Combining three matching layouts under the same profile, the following analysis can be derived. Firstly, the larger the contact area there is, the more wave energy is dissipated. The sandwich-type structure exists with two times the interfacial void area than the other two cases. In addition, periodic changes in the interfacial void affect the interfacial effects of the composite structure. The morphology of the plasma layer controls the scale of the interfacial void and thus influences interfacial capacitance. Given that the interfacial void cannot be obliterated, combining the particularity of the interfacial void to rationalize the absorber's topology can effectively control the absorption effect.

3.4 Power consumption estimation

It is known that plasmas are reconfigurable due to the physical parameters and input power, and some special applications, such as flying vehicle surfaces, require high attenuation with power and profile limitations. Thus, it is necessary to evaluate power conservation in this design.

Power required to maintain the electron density over the volume V is given by[4]

$$P = \frac{n_e E_i V}{\tau} = k_r n_e^2 E_i V \qquad (6)$$

where E_i is the ionization potential, τ is the plasma life time, and k_r is the recombination rate.

It is observed that the power is positively correlated with the volume assuming the uniform distribution and the same physical parameters of each enclosed plasma. The reflection characteristics for plasma periodic structures with different numbers of layers are given in Fig. 12, in which the pixelated FSS is absent. As the number of layers increases, the absorption center frequencies move towards the lower-frequency end and the absorption peak gradually decreases, with multiple peaks appearing from the fourth layer. The average reflection tends to decrease and then increase, achieving the same level as the proposed coupled structure at the fifth layer in the simulation. In other words, ignoring the thickness of the pixelated FSS, it gives a rough estimation that the periodic plasma arrays without coupling require 2.5 times the power than the hybrid design, including the height

$$\frac{P_{\text{plasma}}}{P_{\text{hybrid}}} = \frac{H_{\text{plasma}}}{H_{\text{hybrid}}} = \frac{5}{2} \qquad (7)$$

with P_{plasma} and H_{plasma} being the Power consumption and the height of the simulated plasma periodic structures, P_{hybrid} and H_{hybrid} being the Power consumption and the height of the proposed composite absorbing structure with double-layer plasma arrays sandwiched the pixelated FSS.

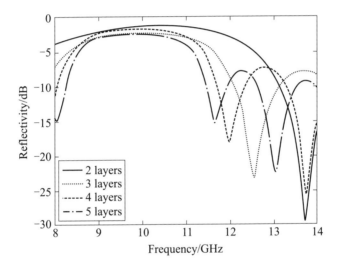

Fig. 12　Simulated reflection (S11) for plasma periodic structures with different layers

3.5　Equivalent circuit modeling

To provide an excellent physical insight into the design properties of the structure, a lumped circuit model is adopted based on the transmission line theory in Fig. 13. It is assumed that the FSS layer embedded between two plasma layers has a similar configuration in both layers, which the coupling effect is experienced by the mutual capacitance with the resonant eddy currents in plasmas nearby ($C_{\text{couple 1}}$ and $C_{\text{couple 2}}$). For each layer of the plasma tube array, there is an intrinsic impedance in the axial direction of the tube due to the plasma conductivity, i.e., a series of resistive $R_{\text{top(bot)}}$ and inductive $L_{\text{top(bot)}}$, and the mutual inductance in the radial direction is represented by a parallel capacitive $C_{\text{top(bot)}}$. An equivalent circuit model of the periodic metallic structure in a unit cell consists of a series of inductive L_{cell} and capacitive C_{cell}, ignoring the effect of surface resistivity and the loss of medium. Each unit also has capacitive interactions C_{chond} with other unit structures. There is another capacitive interaction C_{board} that needs to be considered, which is the interaction between the absorber and the total reflective metal plate[6]. In addition, Z_0 is the wave impedance in the free space and is equal to 377 Ω.

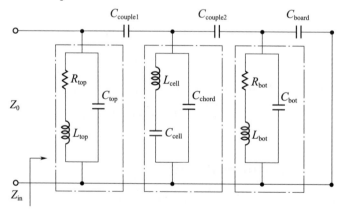

Fig. 13　The equivalent circuit model of the proposed multilayer absorbing structure based on transmission line approach. The blue dashed boxes represent plasma layers, while the orange one represents the FSS layer

The lumped parameters of the equivalent circuit model are determined so that an acceptable agreement is observed between the results of the equivalent circuit model and those of the experimental results, as shown in Table 1. The equivalent resistance of glow discharge plasma can be approximated by the resistivity model[32]

$$\rho \approx \frac{\pi e^2 \, m^{1/2}}{(4\pi\varepsilon_0)^2 \, (KT_e)^{3/2}} \ln\Lambda \qquad (8)$$

where

$$\Lambda = \lambda_D / r_0 = 12\pi n_e \lambda_D^3 \qquad (9)$$

with Λ representing the maximum impact parameter, which depends on the electron number density and Debye length. In this experiment, $\ln\Lambda$ is given to be 10, and the electron temperature T_e is determined to be 0.34 ~ 1 eV through the conventional microwave diagnostic method. Thus, the resistance of each plasma tube ranges between 0.3 Ω and 3.1 Ω.

Table 1 The main equivalent elements and the corresponding numerical values used in the model

Parameters	Values
R_{top}/Ω	0.46
L_{top}/nH	6.71
C_{top}/fF	5.11
R_{bot}/Ω	0.79
L_{bot}/nH	0.20
C_{bot}/fF	153.20
$C_{couple1}/fF$	33.06

续表

Parameters	Values
$C_{\text{couple 2}}/\text{pF}$	2.65
$C_{\text{board}}/\text{pF}$	0.78
$L_{\text{cell}}/\text{nH}$	10.53
$C_{\text{cell}}/\text{fF}$	901.08
$C_{\text{chord}}/\text{fF}$	0.39

Three deductions can be summarized by comparing the circuit parameters of the top and bottom plasma layers. First, plasma at the top layer has a more negligible impedance affected by the coupling effect with the periodic metallic structure. It is known that the electron number density in the tube increases directly, leading to an increase in conductivity and a decrease in impedance. Through the parameter sensitivity analysis, microwave absorption is enhanced by the reduction of the resistance value of plasmas. Second, the inductive impedance of the top layer is greater as the electric field increases. Through parameter sensitivity analysis, it can be seen that as the inductance decreases, the absorption center frequency will shift to the high-frequency end, and otherwise, it shifts to the low-frequency end. Finally, the top layer has a stronger interaction between tubes, which is exhibited by the reduction of the capacitance. The parameter sensitivity analysis shows that as the capacitance decreases, the absorption center frequency will also shift to the high-frequency end; otherwise, it shifts to the low-frequency end. Corresponding to the electric field analysis in Fig. 10, the inductive impedance of the top layer is greater, and the capacitive impedance is smaller, which results in stronger resonant eddy currents. The control of frequency shift and bandwidth results from multi-parameter synthesis[6].

Furthermore, in conjunction with the interfacial void model presented in the previous section, the two coupling capacitances show the interfacial effect of interfacial voids. The coupling capacitance of the upper layer is smaller, showing the tighter interaction between the plasma and the FSS. If the curvature radius of the plasma tube increases, and thus the scale of the interfacial void increases, leading to a decrease in the interfacial capacitance. Fig. 14 documents changing processes of S-parameters for several sets of typical values, in which the solid line plots the initial case. In general, as C_{couple} decreases, the absorption center frequency will shift to the high-frequency end, and otherwise, it shifts to the low-frequency end. Thereinto, $C_{couple1}$ shows a more sensitive response to the EM wave; however, $C_{couple2}$ has little effect on microwave absorption until below 0.1. The upper plasma layer makes the effect of the interfacial void more significant due to being closer to the wave source, but the lower plasma layer is not negligible either. It demonstrates that the shape of the designed layer also influences absorption characteristics.

At normal incidence, the grating lobes wavelength equals the FSS periodicity, leading to a 30 GHz propagating Floquet harmonic in the proposed pixelated FSS[33]. However, when the FSS is embedded with dielectric media, the first high-order phenomenon is represented by the onset of trapped dielectric modes (or trapped surface waves) that

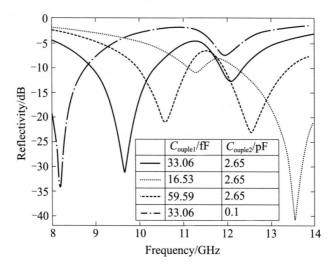

Fig. 14　S-parameters for the circuit parameter adjustment of coupling capacitances

occur well below the propagation of the first grating lobe. Since the "quasi-plasma" high-density metallic periodic structure has a random distribution with tiny discrete elements, the frequency-dependent impedance network with a multi-scale characteristic structure is discussed below.

In order to describe the response behavior of the pixelated FSS with continuously adjustable parameters, several additional elements are necessary for the circuit model, which is characterized as frequency-dependent additional terms. In addition to the intrinsic lumped elements (L_{cell}, C_{cell} and C_{chord}), the capacitive impedance of the FSS coupled to plasma layers on both sides ($C_{couple1}$ and $C_{couple2}$) should also have frequency-dependent characteristics. Taking L_{cell} as an example, the improved circuit model is expressed as

$$L_{cell}(\omega) = L_{cell0} + P(\omega)\hat{L}_{cell} \qquad (10)$$

where
$$P(\omega)\hat{L}_{cell} < L_{cell0} \qquad (11)$$

with L_{cell0} referring to the value in Table 1. The frequency-dependent excitation factor $P(\omega)$ is assumed to be the product of the weight p and the frequency for simplicity. The perturbation term $\hat{L}_{cell}$ is assumed to be one-tenth of the base value, which is acceptable for discussion with a definite level.

Four statuses have been calculated based on the improved lumped circuit model with different weights, and the results are shown in Fig. 15. The status without frequency independence is plotted by the solid line. It is observed that as the weight increases, the first absorption center frequency shifts toward the low-frequency end and the absorption enhances, while the second absorption center frequency has a slightly reduced shift and the absorption decreases. As the frequency dependence increases, the frequency response characteristics of the operating band gradually disappear. The approach is a matching procedure mainly aiming at understanding physical mechanisms in the dynamic frequency response of the coupled structure.

4 Conclusion

In this study, a novel design of a reconfigurable composite absorbing structure is implemented to control the propagation of EM waves in 8～14 GHz bands. The FD-CGAN model based on conditional GANs is proposed to generate optimal geometric structures

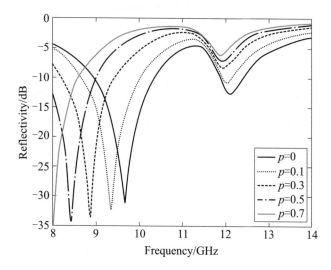

Fig. 15　S-parameters for frequency-dependent circuit parameter adjustment.

according to the adaptability of the coupled structure. The incident wave power is reduced significantly in a wide frequency bandwidth via a proper design of the absorber with enclosed plasmas and pixelated FSS, accompanied by the dynamically adjustable attenuation level. Measurements indicate that the addition of metallic periodic structures makes the coupling absorption effect surpass that of either pixelated FSS or plasma solely, and the reconfigurability is validated in six working modes. The reflection responses of the three absorbers with various arrangements demonstrate that the "sandwich-type" structure maximizes the effectiveness of absorption over a wide frequency range in a relatively low profile, accompanied by the electric field and the power loss density distribution. An interfacial void model is proposed to assist in the design of the composite absorbing structure. The power consumption estimation shows that the periodic plasma arrays without coupling require 2.5 times the power to achieve the same absorption as the hybrid design. The coupled structure is described based on the transmission line method, which characteristic lumped elements have been analyzed using experimental data fitting, and the frequency-dependent impedance network with a multi-scale characteristic structure is discussed. As the plasma generation technology becomes mature, the method in this paper is promising to be introduced into the application such as plasma-based stealth technology. In addition, combining DL with the coupled absorber structure design is of practical significance to overcome the limitation of the traditional rule-based strategies with trial-and-error.

Appendix oblique incident performance

In addition to the propagation of normally incident EM waves in the proposed reconfigurable absorber, we have also investigated the oblique incident performance at various angles in both transverse electric (TE) and transverse magnetic (TM) polarizations. The measured results are depicted in Fig. 16, which shows the stable dual resonance peaks up to 30° with center frequencies slightly shifted. While at an incident angle of 45°, especially for the TM polarization, the absorption response produces large frequency shifts, which may be due to the cavity resonance effect of the array structure at large angles.

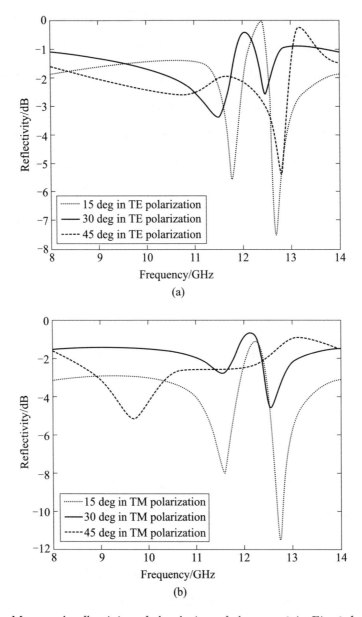

Fig. 16 Measured reflectivity of the design of the case 3 in Fig. 9 for various oblique angles of incident wave in (a) TE polarization and (b) TM polarization

References

[1] R J Vidmar. On the use of atmospheric pressure plasmas as electromagnetic reflectors and absorbers. IEEE Trans. Plasma Sci., vol. 18, no. 4. pp. 733–741, Aug. 1990.

[2] T Jiang and Z Hou. A reconfigurable frequency selective structure using plasma arrays. AIP Adv., vol. 10, no. 6, Jun. 2020, Art. no. 065102.

[3] R Kumar and D Bora. A reconfigurable plasma antenna. J. Appl. Phys., vol. 107, no. 5, p. 3272, 2010.

[4] H Singh, S Antony, and R M Jha. Plasma – Based Radar Cross Section Reduction. Singapore: Springer, 2016, pp. 1–46.

[5] K R Stalder, R J Vidmar, and D J Eckstrom. Observations of strong microwave absorption in collisional plasmas with gradual density gradients. J. Appl. Phys., vol. 72, no. 11, pp. 5089–5094, 1992.

[6] Y Hu, M Fang, Y Qian, M Liu, H Shen, and Z Hou. Electrode cooling and microwave absorption by phase change fluid discharge plasma. IEEE Trans. Microw. Theory Techn., vol. 70, no. 7, pp. 3472–3485, Jul. 2022.

[7] L W Cross, M J Almalkawi, and V K Devabhaktuni. Development of large – area switchable plasma device for X – band applications. IEEE Trans. Plasma Sci., vol. 41, no. 4, pp. 948–954, Sep. 2013.

[8] M Z Joozdani and M K Amirhosseini. Wideband absorber with combination of plasma and resistive frequency selective surface. IEEE Trans. Plasma Sci., vol. 44, no. 12, pp. 3254–3261, Dec. 2016.

[9] K Payne, E F Peters, J Brunett, D K Wedding, C A Wedding, and J H Choi. Second – order plasma enabled tunable low – profile frequency selective surface based on coupling inter – layer. in Proc. 46th Eur. Microw. Conf. (EuMC), Oct. 2016, pp. 309–312.

[10] K Payne, J K Lee, K Xu, and J H Choi. Low – profile plasma – based tunable absorber. in Proc. IEEE Int. Symp. Antennas Propag. USNCIURSI Nat. Radio Sci. Meeting, Jul.

2018, pp. 2065-2066.

[11] K Payne, K Xu, J H Choi, and J K Lee. Plasma-enabled adaptive absorber for high-power microwave applications. IEEE Trans. Plasma Sci., vol. 46, no. 4. pp. 934-942, Apr. 2018.

[12] B A Munk, Freq. Selective Surfaces: Theory Design. Hoboken, NJ, USA: Wiley, 2005.

[13] D G Holtby, K L Ford, and B Chambers. Genetic algorithm optimisation of dual polarised pyramidal absorbers loaded with a binary FSS. in Proc. Loughborough Antennas Propag. Conf., Nov. 2009, pp. 3878-3880.

[14] P Naseri and S V Hum. A generative machine learning-based approach for inverse design of multilayer metasurfaces. IEEE Trans. Antennas Propag., vol. 69, no, 9, pp. 5725-5739, Sep, 2021.

[15] I Goodfellow, J Pouget-Abadie, M Mirza, B Xu, D Warde-Farley, S Ozair, A Courville, and Y Bengio. Generative adversarial nets. in Proc. 27th Int. Conf. Neural Inf. Process. Syst., 2014, pp. 2672-2680.

[16] Z Liu, D Zhu, S Rodrigues, K Lee, and W Cai. Generative model for the inverse design of metasurfaces. Nano Lett., vol. 10, no. 18, p. 65706576, May 2018.

[17] Z Zhang, D Han, L Zhang, X Wang, and X Chen. Adaptively reverse design of terahertz metamaterial for electromagnetically induced trans-parency with generative adversarial network. J. Appl. Phys., vol. 130, no. 3, Jul. 2021, Art. no. 033101.

[18] J A Hodge, K V Mishra, and A I Zaghloul. RF metasurface array design using deep convolutional generative adversarial networks. in Proc. IEEE Int. Symp. Phased Array Syst. Technol. (PAST), Oct. 2019, pp. 1-6.

[19] Y Mao, Q He, and X Zhao. Designing complex architectured materials with generative adversarial networks. Sci. Adv, vol. 6, no, 17, Apr. 2020. Art. no. eaaz4169.

[20] D J Kern and D H Werner. A genetic algorithm approach to the design of ultra-thin electromagnetic bandgap absorbers. Microw. Opt. Technol. Lett., vol. 38, no. 1, pp. 61-64, Jul. 2003.

[21] S Genovesi, R Mittra, A Monorchio, and G Manara. Particle swarm optimization for the design of frequency selective surfaces. IEEE Antennas Wireless Propag. Lett., vol. 5,

pp. 277–279, 2006.

[22] B Wang, J A Rodriguez, O Miller, and M A Cappelli. Reconfigurable plasma–dielectric hybrid photonic crystal as a platform for electromagnetic wave manipulation and computing. Phys. Plasmas, vol. 28, no. 4, Apr. 2021, Art. no. 043502.

[23] G J M Hagelaar and L C Pitchford. Solving the Boltzmann equation to obtain electron transport coefficients and rate coefficients for fluid models. Plasma Sources Sci. Technol., vol. 14, no. 4, pp. 722–733, Oct. 2005.

[24] M K Howlader, Y Yang, and J R Roth. Time–resolved measurements of electron number density and collision frequency for a fluorescent lamp plasma using microwave diagnostics. IEEE Trans. Plasma Sci., vol. 33, no. 3, pp. 1093–1099, Jun. 2005.

[25] A Ghayekhloo, A Abdolali, and S H Mohseni Armaki. Observation of radar cross–section reduction using low–pressure plasma–arrayed coating structure. IEEE Trans. Antennas Propag., vol. 65, no. 6, pp. 3058–3064. Jun. 2017.

[26] B Chaudhury and S Chaturvedi. Three–dimensional computation of reduction in Radar cross section using plasma shielding. IEEE Trans. Plasma Sci., vol. 33, no. 6, pp. 2027–2034, Dec. 2005.

[27] X Ren, C Liu, and M Zeng. S11 parameter calculation of frequency selective surface based on deep learning. J. Phys., Conf. Ser., vol. 1865. no. 4, Apr. 2021, Art. no. 042022.

[28] C X Yuan, Z X Zhou, J W Zhang, X L Xiang, Y Feng, and H G Sun. Properties of propagation of electromagnetic wave in a multilayer radar–absorbing structure with plasma–and radar–absorbing material. IEEE Trans. Plasma Sci., vol. 39, no. 9, pp. 1768–1775, Sep. 2011.

[29] F Costa, A Monorchio, and G Manara. Efficient analysis of frequency–selective surfaces by a simple equivalent–circuit model. IEEE Antennas Propag. Mag., vol. 54, no. 4, pp. 35–48, Sep. 2012.

[30] Y Liang, Z Liu, L Lin, J Peng, R Liu, and Q Lin. Transmission characteristics of electromagnetic waves in 2D tunable plasma photonic crystals. Appl. Opt., vol. 60, no. 9, p. 2510, 2021.

[31] D M Pozar. Microwave Engineering. Hoboken, NJ, USA: Wiley, 2006.

[32] F F Chen. Introduction to Plasma Physics and Controlled Fusion, vol. 1. 2nd ed. New York, NY, USA: Plenum Press, 1984, ch. 5, pp. 178–183.

[33] F Costa, A Monorchio, and G Manara. An overview of equivalent circuit modeling techniques of frequency selective surfaces and metasurfaces. Appl. Comput. Electromagn. Soc. J., vol. 29, no. 12, pp. 960–976. Dec. 2014.